Nutritional Imaging in Agri-Food Systems

Nutritional Imaging in Agri-Food Systems

Edited by

Jay Kumar Pandey
Department of EEE
Shri Ramswaroop Memorial University
Barabanki
Uttar Pradesh, India

Tanmay Sarkar
Food Processing Technology
Government of West Bengal
Malda
West Bengal, India

Wing-Fu Lai
School of Food Science and Nutrition
University of Leeds, UK

WILEY

Registered Offices
John Wiley & Sons, Inc., 111 River Street, Hoboken, NJ 07030, USA
John Wiley & Sons Ltd, New Era House, 8 Oldlands Way, Bognor Regis, West Sussex, PO22 9NQ, UK

For details of our global editorial offices, customer services, and more information about Wiley products visit us at www.wiley.com.

The manufacturer's authorized representative according to the EU General Product Safety Regulation is Wiley-VCH GmbH, Boschstr. 12, 69469 Weinheim, Germany, e-mail: Product_Safety@wiley.com.

Wiley also publishes its books in a variety of electronic formats and by print-on-demand. Some content that appears in standard print versions of this book may not be available in other formats.

Library of Congress Cataloging-in-Publication Data applied for:
Hardback ISBN: 9781394399611

Cover image: Witthaya Prasongsin/Getty Images, Stevens Fremont/ Getty Images, Stockbyte/Getty Images, baranozdemir/Getty Images
Cover design: Wiley

Set in 9.5/12.5pt STIXTwoText by Straive, Pondicherry, India

Contents

Author Biographies

Chapter 1

Samrat Kundu is an Assistant Professor in the Department of Computer Science, holding a Master of Computer Applications (MCA) degree and currently pursuing his PhD His academic and research interests broadly focus on **blockchain, artificial intelligence (AI), and machine learning (ML)**. He is actively involved in teaching undergraduate and postgraduate students, where he contributes to curriculum delivery, academic mentoring, and practical-oriented learning. His research work emphasizes the integration of intelligent algorithms with decentralized systems to enhance security, transparency, automation, and trust in modern digital ecosystems. He has a keen interest in applying AI-driven models to solve real-world problems related to data security, decision-making, and smart system design. Samrat Kundu has contributed to academic writing, book chapters, and research-oriented activities, and continuously works toward strengthening interdisciplinary approaches that combine computing technologies with emerging digital frameworks. His long-term academic goal is to develop scalable, secure, and intelligent computational solutions that address contemporary challenges in information systems, cybersecurity, and data-driven applications. Through teaching and research, he aims to bridge the gap between theoretical concepts and practical implementation in the evolving domain of computer science.

Email: smk.cs@brainwareuniversity.ac.in

Gouranga Bag is an Assistant Professor with an academic background that includes an MSc, NET qualification, and ongoing PhD enrollment. His primary research interests lie in the areas of **artificial intelligence (AI), machine learning (ML), and natural language processing (NLP)**. He is actively engaged in teaching and mentoring students at the undergraduate and postgraduate levels, with a strong focus on conceptual clarity and applied learning. His research explores intelligent data-driven techniques for text analysis, language understanding, pattern recognition, and automated decision-making systems. He has a growing interest in the application of NLP models for semantic analysis, information extraction, and human–computer interaction. Gouranga Bag contributes to academic development through research writing, participation in scholarly activities, and continuous learning in emerging AI technologies. His academic goal is to advance intelligent systems that can process and interpret complex data efficiently while maintaining ethical and responsible AI practices. He aims to contribute to innovative research that connects theoretical foundations with practical, real-world computational solutions.

Email: grb.cs@brainwareuniversity.ac.in

Soumya Roy is an Assistant Professor with a Master of Computer Applications (MCA) degree and NET qualification. His primary research interest is in the field of **cryptography**, with a strong emphasis on secure communication, data protection, and information security mechanisms. He is actively involved in teaching computer science courses and guiding students in understanding core concepts related to security algorithms, encryption techniques, and secure system design. His academic interests include classical and modern cryptographic methods, key management systems, authentication protocols, and emerging challenges in cybersecurity. Soumya Roy is dedicated to promoting awareness of secure computing practices and the importance of cryptographic solutions in today's digital environment. He contributes to academic writing and research-related activities aimed at strengthening security frameworks and protecting digital assets. His long-term objective is to conduct impactful research that enhances confidentiality, integrity, and trust in information systems while supporting the development of robust and scalable security solutions for modern computing applications.

Email: coolsroy.mdp@gmail.com

Chapter 2

Usman Muhammad Umar is driven by a passion for agriculture. As an agronomist, he is dedicated to increasing agricultural yields through the use of cutting-edge technologies. His expertise lies in integrating imaging techniques, deep learning, remote sensing, and decision-making models for irrigation, plant disease detection, and soil nutrient management. His research encompasses all significant aspects of precision agriculture and the application of artificial intelligence within the agriculture domain. Usman holds a bachelor's degree in agricultural science and a master's degree in plant production with a major in digital agriculture from Nigde Omer Halisdemir University, Turkey, where he currently serves as a graduate research assistant.

Chapter 3

Sima Tahmouzi is a researcher and lecturer in the Department of Food Science and Technology, School of Public Health, Shahid Sadoughi University of Medical Sciences, Yazd, Iran. She holds an MSc in Food Science and Technology and specializes in food safety, food chemistry, and novel food processing technologies. Her research focuses on nondestructive food quality assessment, biopolymer-based packaging, and advanced imaging techniques for nutritional analysis. With multiple publications in peer-reviewed journals, she actively contributes to advancing sustainable and innovative food systems, aiming to bridge scientific research with industrial applications in food quality control and safety management.

Farhang Hameed Awlqadr is a Lecturer in the Department of Food Science and Quality Control at Halabja Technical College, Sulaimani Polytechnic University, Kurdistan Region, Iraq. His academic and research interests focus on food quality assurance, food safety management systems, and the application of modern analytical and imaging techniques in agri-food evaluation. He has contributed to various national and international research projects aimed at enhancing food quality monitoring and sustainable production. Farhang actively collaborates with multidisciplinary teams and has authored several scholarly articles in reputable journals, advancing knowledge in food science and technology.

Javaneh Karimi is a researcher in the Department of Food Science and Technology at the School of Public Health, Shahid Sadoughi University of Medical Sciences, Yazd, Iran. Her research focuses on food quality evaluation, nondestructive analytical methods, and the application of imaging and spectroscopy techniques in agricultural and food systems. She has contributed to studies on hyperspectral and multispectral imaging, machine learning-based feature extraction, and advanced preprocessing methods for nutritional assessment. Javaneh's academic interests include sustainable food processing technologies and the integration of artificial intelligence into quality control systems for improved food safety and nutritional profiling.

Neda Mollakhalili Meybodi is a researcher in the Department of Food Science and Technology at Shahid Sadoughi University of Medical Sciences, Yazd, Iran. Her expertise spans food chemistry, food quality assessment, and the application of advanced analytical and imaging techniques for food safety and nutritional evaluation. She has contributed to multiple research projects and publications focusing on innovative approaches to improve food quality monitoring, shelf-life extension, and detection of adulteration. Her academic interests include integrating spectroscopic and computational methods for nondestructive food analysis to support sustainable and safe food production systems.

Chapter 4

Sagnik Bachhar is currently pursuing a Bachelor of Technology (BTech) in Computer Science and Engineering at Adamas University, Kolkata. Currently, Sagnik is in his third year. Passionate about research and innovation, Sagnik has already published a conference paper on IEEE Xplore, showcasing his early contributions to the field. He is continuously exploring emerging technologies and aims to make a meaningful impact through academic and practical advancements in computer science.

Email: sagnik.bachhar.official.2003@gmail.com

Sumanta Kumar Bhattacharjee graduated BCA from Brainware University in 2023. Sumanta is currently pursuing MCA (2nd year) from Brainware University. His favorite subject is web designing through JavaScript and HTML, and gaining knowledge about new upcoming technologies in the IT industry. Sumanta has already published a conference paper on IEEE Xplore.

Email: sumantabhattacharjee63@gmail.com

Biswarup Yogi completed his BTech and MTech in Computer Science and Engineering, the latter from NITTTR Kolkata. He is currently pursuing a PhD in cryptography and cybersecurity at Mizoram University. He serves as an Assistant Professor in the Department of Computational Sciences at Brainware University, India.

Email: biswarup23yogi@gmail.com

Chapter 5

Mr. Devanakonda Venkata Sai Chakradhar Reddy is a doctoral researcher specializing in precision agriculture at ICAR – Indian Agricultural Research Institute, New Delhi. His research focuses on ground and UAV-based remote sensing for nutrient management in wheat crops. He possesses expertise in integrating RGB, multispectral, and hyperspectral imaging, along with nonimaging techniques, radiative transfer modeling (RTM), and advanced machine learning and deep learning approaches to improve nutrient use efficiency through site-specific crop diagnostics. Additionally, he is skilled in critical dilution curve analysis, sensor calibration, and the interpretation of agro-meteorological data. Driven by a commitment to sustainable agriculture, Devanakonda strives to bridge field-level experimentation with cutting-edge digital tools, enabling data-driven, efficient, and environmentally responsible crop management strategies.

Email: devanakonda.chakri9@gmail.com

Mr. Selvaprakash Ramalingam is a SERB-OVDF fellow specializing in drone remote sensing, Edge AI, 5G, and big data analytics at ICAR – Indian Agricultural Research Institute, New Delhi. Holding a PhD in Agricultural Physics, he integrates multispectral and hyperspectral imaging with advanced AI models for precision crop monitoring and early disease detection. His expertise spans embedded edge computing, UAV operations, GIS, and machine learning. Selvaprakash has contributed to numerous international publications and workshops and collaborates globally, including with Purdue University and the University of Valencia. His work drives sustainable agriculture through real-time nutrient surveillance and decision-support systems leveraging Edge AI and advanced image processing.

Email: selvaprakashak@gmail.com

Ms. Divya Dharshini Saravanan is a doctoral researcher with specialization in Agricultural Meteorology at CCS Haryana Agricultural University, Hisar. Her research focused on agrometeorological and spectral characterization of sorghum cultivars in semi-arid ecosystems. Her expertise lies in advancing agrometeorological studies through the application of crop simulation models and machine learning algorithms. By applying regression-based machine learning algorithms and spectral data analysis, her work contributes to improving the understanding of crop response to water stress in dryland systems. She secured All India Rank 1 in the ICAR-SRF examination and received a Gold Medal for academic excellence in her master's degree. Driven by a passion for sustainable crop production in challenging environments, she continues to explore innovative data-driven approaches to optimize resource use and improve climate resilience in dryland farming systems.

Email: divyadharshinistnau@gmail.com

Chapter 6

Dilbar Hussain is a researcher and faculty member at Iqra University, Main Campus Karachi, specializing in Artificial Intelligence, Machine Learning, and Medical Imaging. He has published several research papers in esteemed journals and conferences, focusing on cutting-edge topics such as medical imaging, disease detection using CNN models, hybrid deep learning approaches, and neural network optimization. His work showcases the practical application of AI and ML in solving real-world problems, particularly in the healthcare domain, by developing accurate and efficient diagnostic systems.

Email: dilbar.hussain@iqra.edu.pk

Fahiza Fauz is an accomplished electrical engineer and researcher with a multidisciplinary background in artificial intelligence, machine learning, digital image processing, control systems, and power electronics. She holds a Master of Science degree in Electrical Engineering from the National University of Sciences and Technology (NUST), Islamabad, and a Bachelor of Science degree in Electrical Engineering from the University of Engineering and Technology (UET), Taxila.

Currently serving as a Visiting Lecturer at the National University of Modern Languages (NUML), Islamabad, she combines academic instruction with active research. Her scholarly contributions span AI-powered monitoring systems, computational modeling, and advancements in power electronics, showcasing her ability to bridge theoretical frameworks with real-world engineering applications.

Fahiza's research is driven by a strong commitment to innovation and societal impact. She continues to explore and contribute to interdisciplinary solutions that address complex challenges in engineering and applied sciences.

Email: fahiza.fauz@gmail.com

Chapter 7

Mr. Sanket Sunil Kawade holds a BFSc from MAFSU, Nagpur, and an MFSc. in Hydrography (Aquatic Environment Management) from DBSKKV, Dapoli. A UGC-NET qualified researcher, he has contributed to hydrobiology and biodiversity studies. Currently, he is a second-year PhD scholar focusing on mangrove restoration in Ratnagiri for biodiversity conservation and climate resilience and has also served as editor of a book.

Email: Kawade.sanket00@gmail.com

Mr. Panchakarla Sedyaaw has obtained his BFSc degree from College of Fishery Science, Udgir (MAFSU, Nagpur), and MFSc in the Department of Fish Processing Technology from College of Fishery, Ratnagiri (DBSKKV, Dapoli), India. He possesses a profound passion for waste utilization and product development in fisheries. He has around 41 publications, including research articles, review articles, popular articles, book chapters, and abstracts presented at national conferences. Currently, he is pursuing the second year of his PhD at College of Fisheries, Ratnagiri (DBSSKKV, Dapoli), Maharashtra, researching to develop a ready-to-cook fish product.

Email: panchakarlasedyaaw@gmail.com

Ms. Min Huang holds a bachelor's degree in aquaculture from Huazhong Agricultural University, China, and is currently enrolled in an integrated PhD program in aquatic biology at the College of Fisheries, Huazhong Agricultural University. She has published two peer-reviewed articles in Aquaculture and Fisheries and Fish and Shellfish Immunology. Her research focuses on nutritional regulation in economically important fish species, micronutrient-mediated immune regulation, and immune responses to bacterial infection.

Email: 1310227874@qq.com

Mr. Subrat Kumar Swain holds a BFSc degree from OUAT, Odisha, and is currently pursuing a master's in Fisheries Resource Science (Aquaculture) at Kagoshima University, Japan. As a MEXT Scholarship recipient, he is a motivated researcher with strong interests in fish nutrition, sustainable aquaculture, and innovative feeding strategies. His academic contributions include authoring a book chapter highlighting the need for green technologies in sustainable seafood processing. He aspires to contribute to environmentally responsible aquaculture through research, innovation, and international collaboration.

Email: subratkumarswainoffc@gmail.com

Chapter 8

Hansa Rajput is currently serving as an Assistant Professor in the Department of Computer Science and Applications at Unique College, Parasia, Chhindwara, India. She has more than nine years of experience in research and teaching. She earned her bachelor's degree in computer applications (BCA), master's degree in computer applications (MCA), and a PhD in computer science and applications from MATS University, Raipur. Her research interests include the Internet of Things (IoT), machine learning, and artificial intelligence. She has published numerous research papers in national and international conferences, book chapters, and peer-reviewed journals, including those indexed in Scopus and other reputed databases.

Chapter 9

Dr. Mamoudou Hamadou is a researcher with a strong background in food science, biochemistry, and pharmacology. He completed his PhD in 2023, working on the physicochemical and bioactive properties of Bambara bean protein isolates and hydrolysates. With over 20 publications in reputable international journals, mainly with Elsevier, Dr. Hamadou has established himself as a prolific researcher. His work focuses on bioactive peptides, plant-based extracts, and their potential applications in preventing and treating various diseases, including diabetes, cardiovascular diseases, and cancer. His research has been published in high-impact journals, contributing to the advancement of knowledge in his field. Dr. Hamadou's collaborations with researchers from various institutions demonstrate his ability to work effectively in a team and his commitment to interdisciplinary research. His publications have been cited in various high-impact journals, reflecting the quality and relevance of his research.

Chapter 10

Mr. P. Sasikumar, Assistant Professor, MCom, CA, PGDIB, and (PhD), in the Department of Commerce at Rathinam College of Arts and Science. With a strong academic foundation in Commerce, he specializes in Marketing and Finance. His academic journey includes an M Com and an ongoing PhD, and he has been involved in teaching since 2021. Beyond teaching, he contributes significantly to research, with publications in UGC CARE-listed and Scopus-indexed journals, showcasing the high impact of his work in the academic community. He is also an author, having written one book, further demonstrating his passion for knowledge sharing. In addition to research and teaching, he plays a dynamic role in event organization, serving as an Event Coordinator for various academic and professional gatherings. His leadership and organizational skills have contributed to the success of multiple initiatives, further strengthening the academic and professional landscape.

Dr. T.M. Hemalatha, MBA, MCom, MA, BEd, PGDCS, PhD, Associate Professor and Dean, PG and Research Department of Commerce, Rathinam College of Arts and Science (Autonomous), Coimbatore, Tamil Nadu, India. She has 10 years of industrial experience and over 15 years of dedicated teaching experience at the UG and PG levels to excel in her field of work. She has written several research articles on human resource management, finance, and marketing, published in reputed national, international, peer-reviewed, and scopus journals. She has completed UGC – ERP Research project and acted as program coordinator for DST–NIMAT (EDII). She has presented papers at various national and international-level conferences and seminars. She is also a member of the Board of Examiners for various leading autonomous colleges. She also serves as a guide and supervisor for PhD Program. She has written 8 Books.

Ms. V. Priyanka, Assistant Professor at Rathinam College of Arts and Science, brings over three years of dedicated teaching experience in the Department of Commerce. With a strong academic foundation in Economics and ongoing PhD work, she has published research in esteemed national and international journals, including UGC-recognized publications, across finance, economics, management, and banking. Her active participation in national and international conferences highlights her commitment to advancing her field. Passionate about education and research, she continues to contribute valuable insights into her areas of expertise.

Chapter 11

HIEU M. TRAN received the Bachelor of Engineering degree in automation and control engineering and the Master of Engineering degree in electronics engineering from International University, Vietnam National University, Ho Chi Minh City, Vietnam, in 2021 and 2024, respectively. His research interests include artificial intelligence on the edge, deep learning, and optimization.

TUAN M. LE received the bachelor's degree in electrical engineering and the MSc degree in electronics engineering from International University, Vietnam National University Ho Chi Minh City, Vietnam, in 2018 and 2022, respectively. His research interests include embedded systems, the Internet of Things (IoT), machine learning (ML), and optimization.

SON V. T. DAO received the BEng degree in aeronautical engineering, the MSc degree in manufacturing systems and technology, and the PhD degree in sensor technology in 2004, 2005, and 2010, respectively. His research interests include the development of advanced imaging sensors/systems and applied artificial intelligence.

Chapter 12

Dr. Mohammad Nasar is an accomplished academic and researcher with over 15 years of experience in teaching, research, and academic leadership. He currently serves as Assistant Professor and Head of the Department at Mazoon College, Oman, affiliated with the Missouri University of Science and Technology, USA. Specializing in computer science, artificial intelligence, and machine learning, he has published numerous research papers in reputed international journals and conferences, including IEEE and Taylor & Francis, and has authored books, book chapters, and a patent. As Principal Investigator, he successfully led a major funded project by the Research Council, Sultanate of Oman, focusing on smart safety systems. Recognized as the top researcher at Mazoon College based on Google Scholar and Scopus citations, Dr. Nasar has also served as a reviewer and advisory board member for several international journals and conferences. He has supervised postgraduate research, delivered public lectures, and received awards for excellence in academia and research. His expertise spans deep learning, IoT-enabled systems, big data analytics, and intelligent automation. Passionate about innovation and knowledge dissemination, Dr. Nasar continues to drive impactful research, mentor aspiring technologists, and contribute to advancements shaping the future of intelligent systems worldwide.

Dr. Mohammad Abu Kausar currently serves as the Associate Professor of the Department of Information Systems and holding the vital role of College Academic Advising Officer and Media Officer, Dr. Mohammad Abu Kausar brings a wealth of experience and dedication to the academic arena. His journey at the University of Nizwa began in 2016 when he joined the College of Economics, Management, and Information Systems, leaving an indelible mark on the institution.

Before assuming his current esteemed position, Dr. Mohammad Abu Kausar demonstrated versatility by working as both an Assistant Professor and a Software Engineer. This dual role showcases his comprehensive understanding of academia and practical applications in the realm of Information Systems. Dr. Mohammad Abu Kausar's research pursuits delve into the realms of Information Retrieval, Web applications, Programming Languages, and Databases. His scholarly endeavors contribute significantly to the ever-evolving landscape of these domains.

Beyond his academic endeavors, Dr. Mohammad Abu Kausar actively participates in numerous professional associations, underscoring his commitment to staying at the forefront of technological advancements. His affiliations include prestigious organizations such as the Computer Society of India (CSI), IEEE, International Association of Engineers (IAENG).

Dr. Md. Abu Nayyer, PhD, is an accomplished academic and researcher, currently serving as Assistant Professor-cum-Scientist at Mandan Bharti Agriculture College, Agwanpur, Saharsa, under Bihar Agricultural University, Sabour, India. He has made significant contributions to agricultural sciences through his extensive research and scholarly work. Dr. Nayyer is the author and coauthor of over 20 research and review articles published in reputed national and international journals, as well as a book and several book chapters.

In addition to his research achievements, Dr. Nayyer is an active reviewer for multiple prestigious national and international journals, contributing to the advancement of scientific quality and rigor. He has also demonstrated excellence in academic mentorship, having guided more than 10 MSc students as a major or co-advisor, fostering the next generation of researchers.

Known for his dedication to both research and teaching, Dr. Nayyer continues to advance knowledge in his field while actively engaging in academic development, scientific collaborations, and capacity building in agricultural sciences.

Chapter 13

Mr. Rohit Kumar Choudhury is a Research Scholar in the Department of Agricultural Chemistry and Soil Science at Bidhan Chandra Krishi Viswavidyalaya (BCKV), Mohanpur, West Bengal, India. He was a former PhD scholar (ICAR-SRF) at Uttar Banga Krishi Viswavidyalaya (UBKV), Pundibari, West Bengal, India. He did his BSc (Hons.) in Agriculture from Career Point University, Kota, and completed his Master's from BCKV, Mohanpur. He has qualified ASRB-NET in 2023 and 2025. His research focuses on nutrient management, soil chemistry, nutrient dynamics, sorption-desorption and kinetic modeling, soil carbon, soil organic matter, soil fertility, and digital soil mapping. He has published several research papers, book chapters, popular articles, and attended numerous national and international conferences. He is a recipient of multiple academic awards (including best MSc thesis award, best poster presentation award at prestigious IARI, GBPUAT), and a member of several prestigious societies (e.g., Indian Society of Soil Science, New Delhi, Society for Fertilizers and Environment, Kolkata). He is proficient in operating R, SPSS, QGIS, Google Earth Engine, Origin Pro, SigmaPlot, and other statistical tools.

Ms. Mouli Sarkar is a PhD researcher in Agricultural Systems and Management at the Department of Agricultural and Food Engineering, IIT Kharagpur. She earned her BTech in Agricultural Engineering from Bidhan Chandra Krishi Viswavidyalaya, West Bengal, and successfully qualified for GATE 2021, receiving the prestigious MHRD Scholarship. Her primary area of research is Precision Agriculture, particularly sensor-based soil characterization, digital soil mapping, and smart farming interventions. She is currently working on the development of low-cost and smartphone-enabled soil health tools, including portable X-ray fluorescence (PXRF) and diffuse reflectance spectroscopy (DRS) capabilities, which are combined with machine learning-based predictive modeling. Her work is centered around scalable, real-time data-driven diagnostics on properties of soil to support sustainable land use and data-driven nutrient management for Indian agriculture.

Mr. Dipak Uttamrao Pakhre was born in the Jalna District of Maharashtra. He has completed a BSc (Hons.) (Agriculture) and an MSc (Agriculture) in soil science from Dr. Panjabrao Deshmukh Krishi Vidyapeeth, Akola, Maharashtra. He passed the ASRB-NET exam in 2021. Presently, he is pursuing a PhD (Agriculture) in soil science at Bidhan Chandra Krishi Viswavidyalaya, West Bengal. He has published several research papers, book chapters, and review articles. His research interests include soil profile study with special emphasis on black soils and Vertisols. His work contributes to understanding soil suitability, classification, and conservation strategies aimed at improving regional agricultural productivity and promoting sustainable land management. He is actively engaged in applied soil research and geospatial analysis of agroecosystems.

Mr. Arijit Ghosh was born in the Bankura district of West Bengal. He completed a BSc (Hons.) (Agriculture) and MSc (Agriculture) in soil science from Bidhan Chandra Krishi Viswavidyalaya, West Bengal. Presently, he is pursuing a PhD (Agriculture) in soil science as a Senior Research Fellow at Bidhan Chandra Krishi Viswavidyalaya, West Bengal. He has published several research papers, book chapters, and popular articles. His research focuses on soil fertility diagnostics, nutrient management, and precision agriculture using hyperspectral sensing combined with machine learning for rapid fertility assessment.

Dr. Shubhadip Dasgupta is an Assistant Professor in the Department of Agricultural Chemistry and Soil Science at Bidhan Chandra Krishi Viswavidyalaya (BCKV), West Bengal, India. He has qualified for the NET exam. He did his doctoral programme at IIT Kharagpur. In collaboration with international researchers, Dr. Dasgupta contributed to chapters and articles summarizing the application of proximal sensing tools (PXRF, diffuse reflectance spectroscopy, colour sensors) integrated with data analytics frameworks to overcome limitations in conventional soil testing – thereby enabling scalable, cost-effective soil fertility assessment. Dr. Dasgupta has coauthored several peer-reviewed publications and book chapters and actively contributes to national and international conferences in the field of soil and environmental sciences.

Ms. Mercy Chinneihoi Haokip completed her BSc (Hons.) in Agriculture from College of Agriculture, Kyrdemkulai, CAU, Imphal, and MSc (Agriculture) in Soil Science from CSK Himachal Pradesh Krishi Vishvavidyalaya, Palampur. She is currently pursuing her PhD from Bidhan Chandra Krishi Viswavidyalaya, West Bengal. She has qualified for the ASRB NET exam in the year 2023 and attended conferences, seminars, and contributed to publications, including research articles, popular articles, and book chapters.

Prof. Asim Biswas is a Full Professor, Canada Research Chair (Tier 1) in Digital Agriculture, and Ontario Agricultural College (OAC) Research Chair in Soils and Precision Agriculture at the University of Guelph, where he also serves as Graduate Program Coordinator in the School of Environmental Sciences. Originally from a farming family in West Bengal, India, he earned his BSc from BCKV, MSc from UAS Bangalore, and PhD in Soil Physics from the University of Saskatchewan under Prof. Bing Si. He also worked as a soil physicist at CSIRO, Australia, before becoming an Assistant Professor of Soil Physics at McGill University, Canada. His global research spans digital and precision agriculture, proximal soil sensing, soil carbon management, and sustainable land use. He is a member of the Royal Society of Canada's College of New Scholars and co-founder of the GHG Registry. Recently, he has been elected as a fellow of the Canadian Society of Soil Science (CSSS), the Soil Science Society of America (SSSA), the American Society of Agronomy (ASA), and the National Academy of Sciences, India (NASI). He has also been recognized as a top 2% scientist by Stanford University and Elsevier for six consecutive years (2020–2025).

Chapter 14

Dr. Prabudh Goel (Corresponding Author) is a Professor of Pediatric Surgery at the All India Institute of Medical Sciences, New Delhi. Dr. Goel is known for his eager personality and passion for delving into new fields to make meaningful contributions to public health. Throughout his distinguished career, he has secured multiple research grants and published extensively in peer-reviewed journals, demonstrating his commitment to advancing medical knowledge. His latest passion lies in artificial intelligence and its applications for public benefit, where he is actively studying and implementing AI technologies to improve healthcare outcomes. Dr. Goel's interdisciplinary approach and enthusiasm for innovation have led him to explore the potential of AI in various medical domains, including nutrition and behavioral health. His collaborative research philosophy and dedication to using technology for societal good make him an ideal mentor for cutting-edge research at the intersection of AI and healthcare. Dr. Goel's vision is to harness the power of artificial intelligence to create scalable solutions that can benefit diverse populations and address pressing public health challenges.

Shivani Phugat is a Research Scientist-I at AIIMS, New Delhi, with a Master's degree in Food and Nutrition (All India Rank-1, DUET 2020). She has extensive experience in research, systematic reviews, and scientific writing, and has contributed to multiple peer-reviewed publications and book chapters. Passionate about advancing nutritional science, she brings expertise in both clinical and public health nutrition.

Chapter 15

Mr. Selvaprakash Ramalingam is a SERB-OVDF fellow specializing in drone remote sensing, Edge AI, 5G, and big data analytics at ICAR – Indian Agricultural Research Institute, New Delhi. Holding a PhD in Agricultural Physics, he integrates multispectral and hyperspectral imaging with advanced AI models for precision crop monitoring and early disease detection. His expertise spans embedded edge computing, UAV operations, GIS, and machine learning. Selvaprakash has contributed to numerous international publications and workshops and collaborates globally, including with Purdue University and the University of Valencia. His work drives sustainable agriculture through real-time nutrient surveillance and decision-support systems leveraging Edge AI and advanced image processing.

Email: selvaprakashak@gmail.com

Mr. Devanakonda Venkata Sai Chakradhar Reddy is a doctoral researcher specializing in precision agriculture at ICAR – Indian Agricultural Research Institute, New Delhi. His research focuses on ground and UAV-based remote sensing for nutrient management in wheat crops. He possesses expertise in integrating RGB, multispectral, and hyperspectral imaging, along with nonimaging techniques, radiative transfer modeling (RTM), and advanced machine learning and deep learning approaches to improve nutrient use efficiency through site-specific crop diagnostics. Additionally, he is skilled in critical dilution curve analysis, sensor calibration, and agro-meteorological data interpretation. Driven by a commitment to sustainable agriculture, Mr. Reddy strives to bridge field-level experimentation with cutting-edge digital tools, enabling data-driven, efficient, and environmentally responsible crop management strategies.

Email: devanakonda.chakri9@gmail.com

Chapter 16

N. Saheedha is an emerging researcher in Agricultural Engineering and Environmental Sustainability. She holds a First Class (Hons) degree in Green Technology from the University of Ruhuna, Sri Lanka, with a specialization in Agricultural Engineering and Environmental Technology. Her academic and research interests focus on water quality monitoring, precision agriculture, and the application of unmanned aerial vehicles (UAVs) and remote sensing technologies for sustainable land and water resource management. Saheedha's work emphasizes the integration of advanced technological tools to support climate-resilient and resource-efficient agricultural practices. She has presented her research at several national and international conferences and has contributed to publications addressing topics such as climate-smart agriculture, smart irrigation systems, and nature-based solutions for environmental restoration. Currently, she serves as a temporary demonstrator at the Department of Agricultural Engineering, Faculty of Agriculture, Rajarata University of Sri Lanka, where she is actively engaged in teaching, laboratory instruction, and research supervision. In parallel, she is pursuing her postgraduate studies through the BRITAE MSc programme at the University of Ruhuna, which aims to strengthen research and innovation capacity in agriculture and environmental sciences. Her long-term goal is to advance sustainable technologies and promote evidence-based solutions for climate adaptation in the agricultural sector.

For more details: http://www.linkedin.com/in/nafeer-saheedha-040852292

Chapter 17

Dr. B. Leena is an Associate Professor in the Department of Electronics and Communication Engineering at KGiSL Institute of Technology. She has extensive academic and research experience in the areas of signal processing, communication systems, embedded systems, and emerging intelligent technologies. Her research interests include machine learning applications in communication networks, IoT-enabled systems, and smart healthcare solutions. Dr. Leena has published several research papers in reputed international journals and conferences and has actively contributed to curriculum development, faculty mentoring, and industry–academia collaboration. She is also involved in guiding undergraduate and postgraduate research projects.

Email: aarya44v44@gmail.com

Dr. S. Karthikeyan is a Professor in the Department of Electronics and Communication Engineering at KGiSL Institute of Technology. He possesses rich teaching, research, and administrative experience spanning advanced communication technologies, VLSI design, wireless networks, and artificial intelligence applications in electronics engineering. His scholarly contributions include numerous publications in high-impact journals, conference proceedings, and edited books. Dr. Karthikeyan is actively engaged in research supervision, funded projects, and professional development activities, and he plays a key role in strengthening academic–industry linkages and outcome-based education initiatives.

Email: karthikeyan.s@kgkite.ac.in/s.karthikeyan@ksrce.ac.in

PM Ir Ts Dr. Razaidi Hussin is the Dean of the Faculty of Electronic Engineering & Technology (FKTEN) at Universiti Malaysia Perlis (UniMAP), Malaysia. He is a Professional Engineer (Ir), Professional Technologist (Ts), and a Chartered professional with significant expertise in electronic engineering, intelligent systems, and advanced technology applications. Dr. Razaidi Hussin has led numerous academic, research, and innovation initiatives at national and international levels. His research interests encompass smart systems, applied artificial intelligence, and advanced electronic technologies. He has published widely, supervised postgraduate research, and actively contributes to professional bodies, accreditation, and global academic collaborations.

Email: shidee@unimap.edu.my

Chapter 18

Dr. Pushkar Singh Rawat is a biochemist at the University of Lucknow with expertise in clinical biochemistry, neurobiology, molecular biology, and bioinformatics. He earned his PhD in Biochemistry in 2022, researching lead-induced oxidative stress in children. A recipient of the ICMR-SRF and multiple national eligibility qualifications, he possesses strong technical skills in biochemical assays, molecular techniques, and computational modeling. Dr. Rawat has published in reputed journals and contributed to international book chapters on heavy metal toxicity, nutraceuticals, and environmental health, reflecting his interdisciplinary research approach.

Email: pushkars93@gmail.com

Dr. Shalini Singh is an Assistant Professor of Biochemistry and Biotechnology at Shri Ramswaroop Memorial University, Lucknow. She earned her PhD in biochemistry (2022) from the University of Lucknow, specializing in cathepsin H purification and characterization. A gold medalist in microbiology, she has expertise in molecular biology, enzymology, microbiology, and bioinformatics. Dr. Singh has held JRF positions, contributed to projects at IARI Pusa, and published widely in reputed journals and book volumes. A DST-INSPIRE Fellow and GATE/NET qualifier, she actively engages in academic governance and research collaborations.

Email: shalinisingh24ss@gmail.com

About the Editors

Dr. Jay Kumar Pandey is currently working as an Assistant Professor in the Department of Electrical and Electronics Engineering at Shri Ramswaroop Memorial University, Barabanki (U.P.) India. Dr. Pandey holds a PhD and has completed an MTech with a specialization in power control (instrumentation), as well as an MBA in finance and marketing. His subjects of interest are related to artificial intelligence, biomedical and healthcare, image processing, machine learning, and renewable energy. He has 15 years of teaching and research experience, has published more than 30 research papers in National and International journals/conferences and book chapters in CRC Press, NOVA, Taylor & Francis, Springer, and IGI Publisher. Dr. Pandey is an editor of books published by reputed publishers including Apple Academic Press, IGI, Elsevier, Wiley-IEEE, NOVA, Bentham Science, IAP, Emerald Publishing, Taylor & Francis, and Cambridge Scholars Publishing. Dr. Pandey is also Editor of the *Journal of Technology Innovations and Energy* (United States). Dr. Pandey has been a reviewer for several journals and conferences including the *Journal of Supercomputing* (Springer Nature), IGI Publishing, *Journal of Security and Communication Networks* (Hindawi), *Journal of Biomimetics, Biomaterials and Biomedical Engineering* – JBBBE (Scientific.Net, Switzerland), and the *Advanced Engineering Forum* – AEF (Scientific.Net, Switzerland).

Dr. Tanmay Sarkar is currently working as a Lecturer in the Government of West Bengal. He has received his Bachelor of Technology, Master of Technology, and PhD in Engineering from Jadavpur University, Kolkata, India. With 9 years of experience in both teaching and research, his areas of interest include bioactive components from natural sources, ethnic foods, mathematical modeling and AI, and phytochemicals. He is an editor for several SCI, SCIE, and Scopus-indexed journals and has edited 30 books with Elsevier, Springer, CRC Press and Wiley. Dr. Sarkar has been recognized among the top 2% of globally highly cited researchers in the world rankings published by Stanford University, USA, for both 2024 and 2025.

Dr. Wing-Fu Lai received his MSc in Materials Engineering and Nanotechnology from the City University of Hong Kong and his PhD in Chemistry from the University of Hong Kong. He is currently an Associate Professor in the School of Food Science and Nutrition at the University of Leeds. His research focuses on the development and engineering of functional polymeric materials for bioactive agent encapsulation and food product innovation.

List of Contributors

Farhang Hameed Awlqadr
Department of Food Science and Quality Control, Halabja Technical College
Sulaimani Polytechnic University
Sulaymaniyah
Iraq

Sagnik Bachhar
Adamas University
Kolkata, West Bengal
India

Gouranga Bag
Computational Sciences
Brainware University
Barasat
India

Sumanta Bhattacharjee
Brainware University
Kolkata, West Bengal
India

Asim Biswas
School of Environmental Sciences
University of Guelph
Guelph, Ontario
Canada

Rohit Kumar Choudhury
Department of Agricultural Chemistry and Soil Science
Bidhan Chandra Krishi Viswavidyalaya
Nadia, West Bengal
India

Son V. T. Dao
School of Science, Engineering and Technology, RMIT University Vietnam
Ho Chi Minh City, Vietnam

Shubhadip Dasgupta
Department of Agricultural Chemistry and Soil Science
Bidhan Chandra Krishi Viswavidyalaya
Nadia, West Bengal
India

Fahiza Fauz
Department of Electrical Engineering
National University of Modern Languages
Islamabad
Pakistan

Arijit Ghosh
Department of Agricultural Chemistry and Soil Science
Bidhan Chandra Krishi Viswavidyalaya
Nadia, West Bengal
India

Prabudh Goel
Department of Paediatric Surgery
All India Institute of Medical Sciences
New Delhi, India

Mercy Chinneihoi Haokip
Department of Agricultural Chemistry and Soil Science
Bidhan Chandra Krishi Viswavidyalaya
Nadia, West Bengal
India

T.M. Hemalatha
School of Business and Commerce
Rathinam Global Deemed to be University
Coimbatore, Tamil Nadu, India

Dilbar Hussain
Department of Computer Science
Faculty of Engineering Science & Technology (FEST)
Iqra University
Karachi, Sindh, Pakistan

Razaidi Hussin
Faculty of Electronic Engineering & Technology (FKTEN)
Universiti Malaysia Perlis (UniMAP)
Arau, Malaysia

Javaneh Karimi
Department of Food Science and Technology, School of Public Health
Shahid Sadoughi University of Medical Sciences
Yazd
Iran

S. Karthikeyan
Department of ECE
KGiSL Institute of Technology
Coimbatore, Tamil Nadu
India

Mohammad Abu Kausar
Department of Information Systems
University of Nizwa
Oman

Sanket Sunil Kawade
Department of Aquatic Environment Management
College of Fisheries
Ratnagiri, Maharashtra
India

Samrat Kundu
Computational Sciences
Brainware University
Barasat
India

Tuan M. Le
School of Science, Engineering and Technology
RMIT University Vietnam
Ho Chi Minh City
Vietnam

B. Leena
Department of ECE
KGiSL Institute of Technology
Coimbatore, Tamil Nadu
India

Hamadou Mamoudou
Department of Biological Sciences
Biochemistry, Bioinformatics, and Therapeutic Innovations, Research Unit
Faculty of Science, University of Maroua, Maroua, Cameroon

and

Department of Chemistry
COMSATS University
Islamabad
Pakistan

and

Department of Basic Science Education, National Advanced School of Engineering
University of Maroua
Maroua
Cameroon

Neda Mollakhalili Meybodi
Department of Food Science
and Quality Control
Halabja Technical College
Sulaimani Polytechnic University
Sulaymaniyah, Iraq

Huang Min
Department of Aquatic Biology, Faculty
of Fisheries
Huazhong Agriculture University
China

Mohammad Nasar
Computing & Informatics Department
Mazoon College
Muscat
Oman

Md. Abu Nayyer
Department of Horticulture (Fruit &
Fruit Technology)
Bihar Agricultural University
Bihar
India

Hung Son Nguyen
Faculty of Mathematics, Informatics
and Mechanics
University of Warsaw
Warsaw
Poland

Dipak Uttamrao Pakhre
Department of Agricultural Chemistry
and Soil Science
Bidhan Chandra Krishi Viswavidyalaya
Nadia, West Bengal
India

Shivani Phugat
Department of Paediatric Surgery
All India Institute of Medical Sciences
New Delhi
India

V. Priyanka
School of Business and Commerce
Rathinam Global Deemed to be University
Coimbatore, Tamil Nadu, India

Hansa Rajput
Department of Computer Science
and Applications
Unique College
Parasia, Chhindwara, Madhya Pradesh
India

Selvaprakash Ramalingam
Division of Agricultural Physics
ICAR – Indian Agricultural Research
Institute
New Delhi
India

Pushkar Singh Rawat
Department of Biochemistry
University of Lucknow
Lucknow, Uttar Pradesh
India

Devanakonda Venkata Sai Chakradhar Reddy
Division of Agricultural Physics
ICAR – Indian Agricultural Research
Institute
New Delhi
India

Soumya Roy
PhD-CSE, NIT Agartala, Agartala
Jirania
India

N. Saheedha
Department of Agricultural Engineering
and Environmental Technology, Faculty of
Agriculture
University of Ruhuna
Matara
Sri Lanka

Divya Dharshini Saravanan
Department of Agricultural Meteorology
Chaudhary Charan Singh Haryana
Agricultural University
Hisar
India

Mouli Sarkar
Agricultural and Food Engineering
Department
Indian Institute of Technology
Kharagpur, West Bengal
India

P. Sasikumar
School of Business and Commerce
Rathinam Global Deemed to be University
Coimbatore, Tamil Nadu, India

Panchakarla Sedyaaw
Department of Fish Processing Technology
College of Fisheries
Ratnagiri, Maharashtra
India

Shalini Singh
Institute of Biosciences and Technology
Shri RamSwaroop Memorial University
Barabanki, Uttar Pradesh
India

Subrat Kumar Swain
Department of Fisheries Resource Science
Graduate School of Agriculture, Forestry
and Fisheries
Kagoshima University
Japan

Sima Tahmouzi
Research Center for Food Hygiene
and safety
School of Public Health
Shahid Sadoughi University of
Medical Sciences
Yazd, Iran

Hieu M. Tran
School of Science, Engineering and
Technology
RMIT University Vietnam
Ho Chi Minh City
Vietnam

Usman Muhammad Umar
Department of Plant Production and
Technology
Nigde Omer Halisdemir University
Nigde
Turkey

Biswarup Yogi
Brainware University
Kolkata, West Bengal
India

Preface

Technological progress in image processing, computer vision, and artificial intelligence has unveiled new opportunities for comprehending, tracking, and optimizing nutrition in agriculture and food nutrition systems. The formulation of this book, *Nutritional Imaging in Agri-Food Systems*, aims to identify new scopes in which visualization data and artificial intelligence can be utilized for tracking nutritional integrity, optimal food production, and informed decision-making from farm to fork.

Nutritional analysis traditionally involves methods in laboratories that are cumbersome, time-consuming, and invasive. On the other hand, image processing is non-invasive, scalable, and can be done in real time with regard to analyzing the nutrient content of crops. By combining other image modalities with machine learning and deep learning algorithms, one can extract vital information that can be used to gain insight into the nutrient composition of crops. For example, one can use RGB, hyperspectral, multispectral, thermal, and microscopic image modalities.

The book constitutes a collection of technical papers from the communities of researchers, academicians, and professionals to cover the theoretical aspects and applications for image-based nutritional evaluation. The chapters collectively span a wide range of applications, including plant phenotyping analysis, soil and crop nutritional assessment, food nutrition evaluation, post-harvest nutritional analysis, dietary nutrition assessment, and intelligent food processing systems for advanced nutritional analysis.

It is intended for graduates and postgraduates, research scholars, faculty members, data scientists, agricultural engineers, food technologists, and professionals associated with agriculture, nutrition, and artificial intelligence. This book bridges the gap between nutritional sciences and computational methodologies, offering a powerful tool for education and innovation for a variety of professionals.

It is our hope that this book will bring about further collaborative efforts among various disciplines and that intelligent, sustainable, and nutrition-sensitive agricultural and food systems capable of dealing with global challenges to food security, quality, and health will emerge.

Acknowledgments

The completion of this book, *Nutritional Imaging in Agri-Food Systems*, is not an accident, but the result of massive encouragement, guidance, and contributions from many individuals and institutions. First of all, we are grateful to all the authors of the chapters included herein for their invaluable scholarly contributions, commitment, and timely cooperation, which have considerably enhanced the technical depth and interdisciplinary scope of this book.

First and foremost, we would like to thank those who reviewed and provided expert opinions on how to improve the quality, clarity, and scientific merit of the chapters included in this volume. This expertise made sure that the content fit current trends in research and into practical applications in the fields of image processing, agriculture, food science, and nutritional analytics.

We would also like to acknowledge the support and encouragement received from our respective institutions, colleagues, and research communities, which provided an intellectually stimulating environment for this work. Special thanks are due to the publishing team for their professional guidance, editorial assistance, and continued support throughout the submission and production process.

Finally, we would like to express our profound gratitude to our families and well-wishers for their patience, motivation, and continued support. We sincerely hope this book will be a useful reference for researchers, academicians, industry professionals, and students, inspiring further innovation and research at the interface of image processing, nutrition, agriculture, and food systems.

List of Abbreviations

3D	Three-dimensional
ACCD	Accumulated charge coupled device
ADC	Analog to digital converter
AE	Autoencoder
AHS	Airborne hyperspectral scanner
AI	Artificial intelligence
AL	Active learning
ANDWI	Augmented normalized difference water index
ANNS	Artificial neural networks
AOJSE	American Online Journal of Science and Engineering
AP	Average Precision
AR	Augmented reality
ARM	Advanced RISC Machine
ASE	Average standard error
ASTER	Advanced Spaceborne Thermal Emission and Reflection Radiometer
AVIRIS	Airborne Visible/Infrared imaging Spectrometer
AWS	Amazon Web Services
BMP	**Bitmap**
BRDF	Bidirectional reflectance distribution function
BSI	Bare soil index
CASI	Compact Airborne Spectrographic Imager
CCD	Charge coupled device
CCE	Calcium carbonate equivalent
CEC	Cation exchange capacity
CFS	Correlation-based feature selection
CFS	Correlation-based feature selection
CHIRPS	Climate Hazards Group InfraRed Precipitation with Station data
CI	Coloration index
CLAHE	Contrast limited adaptive histogram equalization
CMOS	Complementary metal oxide semiconductor
CNN	Convolutional neural networks
COCO	Common Objects in Context

CRF	Conditional Random Field
CT	Computed tomography
CV	Cross validation
CWT	Continuous wavelet transform
DCAI	Data-centric artificial-intelligence
DEM	Digital elevation model
DL	Deep learning
DN	Digital number
DNN	Deep neural networks
DPDPA	Digital Personal Data Protection Act
DR	Dimensionality reduction
DRS	Diffuse reflectance spectroscopy
DSM	Digital soil mapping
DVFS	Dynamic voltage and frequency scaling
DWT	Discrete wavelet transform
ECA	Efficient channel attention
ECC	Enhanced correlation coefficient
Edge AI	Edge artificial intelligence
ELM	Empirical line method
EMR	Electromagnetic Radiation
EnMAP	Environmental Mapping and Analysis Program
ENN	Elman neural networks
EPS	**Encapsulated Post Script**
EVI	Enhanced vegetation index
ExG	Excess green index
FAO	Food and Agriculture Organization
FCC	**False Colour Composite**
FCN	Fully convolutional network
FL	Federated learning
FN	False negative
FP	False positive
FPN	Feature pyramid network
GAN	Generative adversarial network
GB	Gradient boosting
GDA	Generalized discriminant analysis
GDEM	Global digital elevation model
GDP	Gross domestic product
GDPR	General Data Protection Regulation
GDQS	Global diet quality score
GIS	Geographic information system
GLI	Green leaf index
GLOH	Gradient location and orientation histogram
GNHA	Global Nutrition and Health Atlas
GPS	Global positioning system

GPU	Graphics processing unit
GRU	Gated recurrent units
GUI	Graphical user interface
HCF	High correlation factor
HE	Histogram equalization
HI	Hue index
HIPAA	Health Insurance Portability and Accountability Act
HIS	Hyperspectral imaging
HOG	Histogram of oriented gradients
HSCNN	Hierarchical synthesis convolutional neural networks
HSI	Hyperspectral imaging
HSV	Hue, saturation, value
HYMAP	Hyperspectral mapper
ICA	Independent component analysis
ICCSAI	International Conference on Communication, Security, and Artificial Intelligence
IEAM	IBM Edge Application Manager
IFCT	Indian Food Composition Tables
IoT	Internet of Things
IoU	Intersection over Union
IR	Infrared
IS	Imaging spectroscopy
JPEG	**Joint Photographic Experts Group**
K	Potassium
KLT	Karhunen–Loeve transform
KLT	Karhunen–Loeve transform
LAB	L*a*b* (color space)
LAI	Leaf area index
LDA	Linear discriminant analysis
LiDAR	Light detection and ranging
LLE	Locally linear embedding
LOOCV	Leave-one-out cross validation
LoRaWAN	Low-power wide-area network
LR	Linear regression
LSTM	Long short-term memory
LSWI	Land surface water index
LVF	Low variance filter
MAE	Mean absolute error
mAP	Mean average precision
MAPE	Mean absolute percentage error
MBSI	Modified bare soil index
MDR	Medical Device Regulation
MIR	Mid infrared
ML	Machine learning

MLOps	Machine learning operations
MLR	Multiple linear regression
MNF	Minimum noise fraction
MODIS	Moderate resolution imaging spectroradiometer
MPLSR	Multivariate partial least square regression
MQTT	Message queuing telemetry transport
MSAVI	Modified soil adjusted vegetation index
MSC	Multiplicative scatter correction
MSE	Mean of standardized error
MSE	Mean square error
MSI	Multispectral imaging
MVR	Missing value ratio
MW	Microwave
N	Nitrogen
N	Nugget
NBR	Normalized burn ratio
NBR+	Normalized burn ratio plus
NDCI	Normalized difference clay index
NDMI	Normalized difference moisture index
NDSI	Normalized difference salinity index
NDVI	Normalized difference vegetation index
NDWI	Normalized difference water index
NIR	Near infrared
NLM	Non-local means
NLP	Natural language processing
NSMI	Normalized soil moisture index
NUC	Next unit of computing
OGC	Open Geospatial Consortium
OK	Ordinary Kriging
OpenCV	Open Source Computer Vision Library
OTA	Over-the-air
P	Phosphorus
PAN	Panoramic
PASCAL VOC	Pattern Analysis, Statistical Modelling and Computational Learning – Visual Object Classes
PbD	Privacy-by-design
PCA	Principal component analysis
PCR	Principal component regression
PDF	**Portable document format**
PID	Proportional-integral-derivative controller
PLSR	Partial least square regression
PNG	**Portable network graphics**
PRI	Photochemical reflectance index
PRISMA	PRecursore IperSpettrale della Missione Applicativa

PROSAIL	PROSPECT + SAIL Radiative Transfer Model
PTP	Precision time protocol
PXRFS	Portable X-ray fluorescence spectroscopy
QAT	Quantization-aware training
R	Reflectance
R^2	Coefficient of determination
RAFT	Recurrent all-pairs field transforms
RBF	Radial basis function
R-CNN	Region-based convolutional neural network
RF	Random forest
RFR	Random forest regression
RGB	Red, green, blue
RI	Redness index
RMSE	Root mean square error
RMSECV	Root mean squared error of cross-validation
RMSPE	Root mean square percentage error
RNN / LSTM	Recurrent neural network/long short-term memory
RNN	Recurrent neural network
RPD	Residual prediction deviation
RPN	Region proposal network
RSD	Relative standard deviation
S	Sill
SAM	Spectral angle mapper
SaMD	Software as medical device
SAR	Synthetic aperture radar
SASI	Shortwave infrared airborne spectrographic imager
SAVI	Soil adjusted vegetation index
SEM	Scanning electron microscopy
SFI	Soil fertility index
SHAP	Shapley additive explanations
SI	Saturation index
SIFT	Scale-invariant feature transform
SLR	Simple linear regression
SMLR	Stepwise multiple linear regression
SNR	Signal to noise ratio
SNV	Standard normal variance
SoC	System on a chip
SOM	Soil organic matter
SPI	Stream power index
SRM	Satellite-based rice monitoring
SRTM	Shuttle Radar Topography Mission
SSL / FSL	Semi-supervised learning/few-shot learning
SSL	Self-supervised learning

SSS	Spectral soil sensing
SURF	Speeded-up robust features
SVG	**Scalable vector graphics**
SVM	Support vector machine
SVMR	Support vector machine regression
SVR	Support vector regression
SWIR	Short-wave infrared
TFLOPS	Trillions of floating-point operations per second
TIFF	**Tagged image file format**
TIR	Thermal infrared
TL	Transfer learning
TM	Thematic mapper
TP	True positive
TPI	Topographic position index
TWI	Topographic wetness index
UAV	Unmanned aerial vehicle
URLLC	Ultra-reliable low-latency communication
USB	Universal serial bus
USDA	United States Department of Agriculture (referring to USDA FoodData Central)
VIS	Visible
VIS	Visible spectrum
VIS-NIR/VNIR	Visible and near-infrared
ViT	Vision transformer
VNIR	Visible and near-infrared
VR	Virtual reality
WDR	Wavelet-based dimensionality reduction
WHISPERS	Workshop on Hyperspectral Image and Signal Processing: Evolution in Remote Sensing
YOLO	You only look once

1

Computational Foundations of Nutritional Imaging in Agri-Food Systems

Soumya Roy[1], Gouranga Bag[2], and Samrat Kundu[2]

[1] PhD-CSE, NIT Agartala, Agartala, Jirania, India
[2] Computational Sciences, Brainware University, Barasat, India

1.1 Introduction

"How can an image tell the nutritional story of a field, a fruit, or a single leaf?" This question captures a serious scientific and societal challenge: rapid, nondestructive nutritional assessment aims at reducing fertilizer waste, thus speeding breeding for nutrient-dense cultivars and supporting evidence-based food-security decisions [1]. Bibliometric and market analyses over the years reveal growing momentum in the domain of agri-food imaging, with a large increase in high-impact publications and funds invested in precision-agriculture imaging technologies in the past 10 years [2]. Classical laboratory assays for soil and tissue nutrient levels are more for determining standard quality; slowness, being expensive, and in most cases destructive, these considerations have led to the search for reliable imaging-based proxies that would work in real-time [3]. The new order of computational nutritional imaging aims to combine sensing, signal processing, and statistical inference to translate raw imagery into calibrated information on nutrition [4]. For this chapter, nutritional imaging is conceived as the hybrid use of imaging modalities and computational pipelines to infer biochemical and nutritional characteristics such as macronutrients (N, P, K), protein, sugars, and antioxidant compounds [5].

Multiscale nutritional imaging involves coordinated acquisition and analysis at nested spatial scales: micro (cellular/tissue), meso (organ/fruit), and macro (plant and field-worldscape), with explicit registration and methodological continuity across scales to enable robust inference [6]. Core imaging modalities include multispectral and hyperspectral cameras, fluorescence spectroscopy, and X-ray/CT systems, with each system operating under different trade-offs in spectral, spatial, and temporal resolution and applicability in the field vs. the laboratory [7]. Market and deployment patterns show the dominance of multispectral systems in field deployments, as these systems are simple to use and relatively cheap, whereas hyperspectral systems provide richer biochemical signatures needed for laboratory and research purposes [8]. Recent methodological advances present physics-informed architectures that embed either radiative-transfer or light-matter-interaction models into learning pipelines to impose constraints on inference and improve cross-site generalization [9]. Concurrent work on explainability, privacy-preserving federated training, and life-cycle assessment of computational costs addresses nontechnical barriers to adoption but remains uneven across the literature [10].

The research landscape poses some trends and limitations that motivate a computational foundations chapter. Deep learning improvements empowered by convolution and transformer architectures have yielded good prediction performances on in-domain nutrient estimation tasks; large, well-annotated datasets are often required, and the robustness to domain shift suffers greatly [11]. The reports on performance benchmarking show promising results (for instance, R^2 remains strong, and root mean squared error (RMSE) remains standardly low in targeted crop studies), but divergences in reporting practices and evaluation protocols make it difficult to perform direct comparisons [12]. Public spectral agri-data repositories remain scattered, and only a few comply with FAIR (Findable, Accessible, Interoperable, Reusable) principles; this obstructs reproducibility and encourages cross-study benchmarking [13]. Multiscale registration linking microscopic biochemical assays with the meso- and macroscopic images is still rare, despite the claims in the literature that such mutual alignments improve the nutrient map's fidelity [14]. Lastly, an omission in the literature has been in neglecting the computational and environmental footprints of spectral deep-learning models, which, if not considered, may affect sustainable research and deployment [15].

In order to solve the limitations herein, a systematic meta-analytic and comparative approach is undertaken for the chapter. An eminent set of high-impact studies, openly available from major indexing sources for the period 2020–2025, is synthesized under careful procedural regulations setting forth explicit inclusion and exclusion criteria barring everything else but instances of quantitative performance metrics and empirical ground truth [16]. The metadata extraction process uncovers the parameters used in the sensors and the spectral data, preprocessing parameters, model classes, hyperparameters, and evaluation metrics. The data extraction process uses standard templates or the digitization of graphical results, contacting the authors as a last resort when the information is ambiguous [17]. For statistical aggregation, summary statistics based on robust, weighted sample inputs are computed, with such statistical results leading to reliability and sample size influence, and bootstrap resampling techniques are adjacent to the aggregated key metric mandates. It is a general practice to see confidence intervals being reported along with the concise descriptions of varying studies; meta-regression is then performed to control for some of the "most likely" confounding factors such as sensor spectral range, crop species, and sample size [18]. Results are presented in structured tables, flowcharts, and figure comparisons among models to be juxtaposed in the five dimensions: accuracy, generalizability, interpretability, data requirements, and computational/ecological costs for fair "apples-to-apples" comparison [19]. One of the ways to improve research credibility is to generate the method part based on the stepwise procedures followed, and on the other side, the code is released to make it easy for verification to the public and the open availability [20].

The main argument of the chapter is that the path to maturity of nutritional imaging demands a multidimensional computational infrastructure that marries statistical accuracy with multiscale alignment, physical plausibility, uncertainty quantification, data governance, and ecological-cost accounting [21]. Once the foundation is in place, one could easily move from the individual laboratory verifications of high accuracy to powerful field deployment systems, with a definite impact on agronomy and policy making [22]. Where methodological advances complement generalization, interpretability, reproducibility, and sustainability practices, novelty in the discipline hits the ground. This integrative approach, hence, offers the maximum potential to advance both science and practice [23]. Expected beneficiaries of progress will include plant breeders seeking high-throughput. Actors involved may include organizations performing phenotyping, agronomists applying precision fertilization, and supply-chain actors requiring nutritional traceability [24]. Further, issues of sovereignty and equity become important in order to avoid disenfranchising smallholders and other resource-constrained actors from benefiting from collaborative model improvement using privacy-preserving machinery [25].

1.2 Research Contributions

The chapter makes four primary contributions to the literature. First, a structured meta-analysis standardizes and aggregates performance and resource metrics across a curated set of high-impact studies to enable fair comparative evaluation [26]. Second, a multidimensional evaluation rubric is introduced that simultaneously assesses statistical accuracy, multiscale applicability, interpretability, FAIR data compliance, and ecological cost [27].

1.2.1 Research Gaps

Three principal gaps emerge from the synthesis of recent work. Large, annotated, multiscale datasets that couple micro-scale biochemical assays with meso- and macro-scale imagery are scarce, impeding generalizable model training [28]. Second, standardized reporting of uncertainty and explainability is uncommon, which undermines stakeholder trust and regulatory acceptance [29]. Aspects such as environmental and computational costs of spectral deep-learning (inadvertently carbon emissions considered!) of training time and energy consumption are scarcely computed, hence further limiting the assessment of sustainability trade-offs [30]. Other gaps – and these – include an inconsistent evaluation protocol and hardly any cross-site validation that the comparability across studies suffers [31].

1.2.2 Research Questions

The chapter asked four focused research questions that direct the comparative analyses and recommendations. First, which classes of computational approaches, i.e., data-driven, physics-informed, or systems/federated, would provide the most balanced solution to accuracy, robustness, and interpretability across crops and scales [32]? The second research question considers how embedding physics-based constraints improves cross-site generalization vs. purely data-driven models [33]. The third is concerned with which dataset curation and FAIR practices can allow for reproducible multiscale nutritional imaging and for fair benchmarking [34]. The fourth addresses questions on how uncertainty and ecological-cost metrics should be standardized and reported in practical decisions made by agronomists, breeders, and policymakers [35].

Here is an overview of the process flow. The literature synthesis with comparative tables precedes the documentation of historical developments and performance variance states across the method classes. Next come the meta-analytic methodology and data extraction protocols, followed by a presentation of the aggregated quantitative results with their associated uncertainty bounds and a discussion on operational, social, and sustainable impacts of the findings. Recommendations for dataset standards, reporting practices, and research priorities that aid reliable, scalable deployment of nutritional-imaging technologies are then prioritized.

1.3 Literature Review

The existing body of literature in computational nutritional imaging is focused on the area of sensor development, algorithmic means, systems deployment, and data-governance issues. Lab modalities such as micro-CT and fluorescence spectroscopy provide structural and biochemical contrasts, thereby giving impetus to algorithm development and assay validity [25, 26]; acquisition choices – push-broom vs. tunable filter – carry their own calibration and registration requirements, which trickle down to affect processing pipelines in the face of them [28]. Field sensors are mostly multispectral and hyperspectral, with trade-offs between the two carefully laid out and summarized in the comparative charts (see Tables 1.1 and 1.2): the former is preferred for operational deployment because of considerations of data volume and cost, while the latter is favored for research purposes with flexible and detailed capability for biochemical discrimination.

Recent methodological advances cluster into three broad families. End-to-end, data-driven deep learning methods, including convolutional neural networks, autoencoders, and transformer-inspired models, have produced many of the leading in-domain predictive

Table 1.1 Representative study-level summary.

Study focus and crop (s)	Sensor/modality	Method class	Reported metric (s)	Sample size (n)
Wheat N and yield	UAV hyperspectral	Data-driven (CNN)	$R^2 = 0.79$; RMSE = 5.9 g	120
Multicrop N estimation	UAV hyperspectral	Data-driven (CNN)	$R^2 \approx 0.82$ ($\pm$0.05)	multiple datasets
Physics-informed mapping	Hyperspectral + PROSPECT	Physics-informed	Δ bias −12% (vs. pure CNN)	varied
Federated nutrient prediction	Multispectral/HSI	Federated systems	Global accuracy 92% ($\varepsilon \leq 1$)	Simulated 12 nodes
Soil nutrient profiling	Lab CT + spectral	Lab structural imaging	Correlation with mineral assays	sample cores
Robotic harvester edge	RGB + MS + NPU	Edge inference	Real-time $\leq$ 200 ms/item	pilot

Source: Adapted from Refs. [3, 6, 13, 15, 25, 33].

Table 1.2 Aggregated performance by method class.

Method class	Studies (k)	Weighted mean R^2	95% CI (R^2)	Weighted mean RMSE	95% CI (RMSE)
Data driven (CNN/ transformer)	18	0.79	(0.758, 0.822)	5.9 g	(5.07, 6.73)
Physics informed (hybrid)	9	0.81	(0.771, 0.849)	5.4 g	(4.36, 6.45)
Systems (federated/edge)	8	0.82	(0.785, 0.855)	5.2 g	(4.16, 6.24)

Source: Adapted from Refs. [3, 6, 13, 15, 21].

results reported in the literature, notably for nutrient and yield estimation in crops such as wheat and other cereals [3, 6, 9–11, 17]. These approaches deliver high accuracy when abundant paired image–assay data are available but display sensitivity to domain shifts and substantial data requirements; their empirical signatures and per-study metrics are presented in the aggregated performance tables (see Table 1.2). A complementary line of work embeds physical forward models, most commonly leaf radiative transfer formulations, within learning frameworks to regularize predictions and constrain them to physically plausible ranges; hybrid physics-informed architectures generally reduce bias and improve cross-site robustness compared with purely data-driven baselines, albeit at the cost of increased modeling complexity and parameterization demands [12, 13]. Systems-level research attends to operationalization issues: federated learning preserves data sovereignty while enabling collective model improvement, edge inference and model compression permit low-latency deployment on robotic harvesters and embedded boards, and digital-twin constructs link imaging across supply-chain nodes to forecast nutrient trajectories after harvest [15, 30, 33]. Representative quantitative outcomes and trade-offs for these classes are reported in the comparative tables (see Tables 1.1–1.4).

Data availability, curation, and standardization form persistent constraints on generalization and reproducibility. Public repositories for spectral agri-data are fragmented, and only a small fraction satisfy FAIR criteria; the limited scale of multi-crop, multiscale datasets constrains the ability of models to generalize across varieties, soils, and climatic regimes [4, 20]. Multiscale registration that links micro-scale biochemical assays with meso- and

Table 1.3 Training time and energy footprints (aggregated).

Method class	Mean training time (h)	95% CI (h)	Mean energy (kWh/run)	95% CI (kWh)
Data driven	12	(10.15, 13.85)	2.4	(2.17, 2.63)
Physics informed	14	(10.73, 17.27)	2.8	(2.41, 3.19)
Systems	10	(7.92, 12.08)	2	(1.72, 2.28)

Aggregated training times and energy use were reported in a subset of studies; means and CIs reflect that subset and are reported to indicate typical footprints and relative differences.
Source: Adapted from Gonzales and Turner [21].

Table 1.4 Dataset, reporting, and governance indicators.

Metric	Observed value (corpus)	Source (s)
Public repositories surveyed	22	[4]
FAIR-compliant repositories	4 (≈18%)	[4, 20]
Studies with $n \geq 500$ samples	≈5%	[20, 26]
Studies reporting multiscale registration (micro↔macro)	<10%	[14, 22]
Studies reporting calibrated uncertainty (CIs/bootstraps)	≈15%	[8, 18]
Studies including explainability modules	≈8%	[14, 34]

Source: Adapted from Refs. [4, 14, 20, 22].

macro-scale imagery remains rare in the corpus, despite evidence that such registration measurably improves nutrient-map fidelity; the prevalence and impacts of multiscale alignment are summarized in the dataset and FAIR compliance table (see Table 1.3) [14, 22].

Explainability and uncertainty quantification have received growing but uneven attention. Explainability modules, such as saliency maps, prototype networks, and part-based explanations, appear in a minority of studies and are seldom validated via user studies; explicit reporting of calibrated confidence intervals or Bayesian posterior summaries is similarly uncommon, reducing the operational interpretability of reported predictions (see Table 1.4) [8, 14, 34]. The computational and environmental costs of spectral deep learning are infrequently reported; where quantified, training a single crop model on contemporary GPUs often requires on the order of 10–15 hours and consumes several kilowatt-hours, underscoring a need for energy-aware model design (see Tables 1.2 and 1.5) [21, 23].

Comparative evaluations published to date tend to emphasize task-level improvements rather than multidimensional trade-offs among accuracy, robustness, interpretability, data requirements, and sustainability. Where comparative studies exist, they generally show that (i) higher spectral resolution improves biochemical discrimination but increases data and computational burdens, (ii) physics-informed constraints reduce cross-site bias but require reliable physical parameter estimates, and (iii) systems approaches balance privacy and

Table 1.5 Cause → Effect summary.

Cause	Observed effect
Move from multispectral → hyperspectral	Improved biochemical discrimination (↑R^2), but ↑data volume, ↑computational cost, and stricter calibration are needed.
Embedding radiative-transfer physics	Reduced cross-site bias and improved generalization, at the cost of added modeling complexity and parameterization sensitivity.
Federated training across heterogeneous sites	Preserved privacy and near-centralized accuracy; introduced communication overhead and heterogeneity management needs.
Absence of multiscale registration	Mismatch between assay scale and field imagery → degraded map fidelity; registration improves accuracy substantively.

Source: Adapted from Refs. [9, 12–15, 21, 22, 28, 33].

latency considerations at the expense of additional engineering complexity [9, 12, 13, 15, 28]. These cause-and-effect relationships are summarized succinctly in the cause-and-effect examples and comparative tables (see Table 1.5).

This synthesis demonstrates that methodological innovation has been unevenly matched by dataset growth, standard reporting practices, and attention to deployment constraints. The comparative tables that accompany this review collect numeric performance measures, sample sizes, reporting completeness, and energy/time footprints to enable fair, "apples-to-apples" comparisons and to highlight where targeted datasets, standardized metrics, and protocolized uncertainty reporting could most effectively improve generalization and adoption. The subsequent section on methodology details the meta-analytic procedures for the extraction and aggregation in order to generate the tables and figures here referenced. The Results section follows and presents the complete statistical summaries, along with uncertainty bounds and sensitivity analyses.

1.4 Methodology

The meta-analytic methodology implemented for the comparative synthesis was made to allow reproducibility, transparency, and kind comparisons across heterogeneous studies in computational nutritional imaging. A general approach was adopted that systematically retrieved the literature, had explicit inclusion and exclusion criteria, conducted structured data extraction, performed quality checks, statistically aggregated with weights, and finally carried out sensitivity analyses. Emphasis was placed on extracting true, study-level performance metrics and resource footprints; where necessary, values reported graphically were digitized, and author clarification was sought. The principal steps are summarized below and are followed by detailed protocols that can be used for replication.

A systematic search strategy targeted major indexing databases and repositories that commonly host high-impact work in agri-food imaging and allied fields. Searches were performed across Scopus, Web of Science, IEEE Xplore, ACM Digital Library, PubMed, and UGC-CARE List I for publications dated 2020–2025. Keyword queries combined terms for sensing and targets, for example: "hyperspectral" AND ("nutrient" OR "nitrogen" OR "protein" OR "antioxidant"), "multispectral" AND ("nutrient estimation" OR "spectral inversion"), "physics-informed" AND ("PROSPECT" OR "radiative transfer"), and "federated learning" AND ("agriculture" OR "nutrient"). Initial retrievals were augmented by forward/backward citation chaining from highly cited seed papers and by consulting market and review reports to ensure inclusion of influential applied studies [1–5]. The bibliographic corpus was de-duplicated and logged in a master index that recorded bibliographic metadata and retrieval provenance.

Explicit inclusion and exclusion criteria were applied to ensure relevance and empirical rigor. Included studies met the following conditions: (i) publication year between 2020 and 2025; (ii) peer-reviewed journal article or conference paper indexed in the searched platforms or an authoritative industry report with empirical evaluation; (iii) minimum empirical standard reporting at least one quantitative performance metric (e.g., R^2, RMSE, Mean Absolute Error (MAE)) for nutrient or closely related biochemical targets measured against chemical/biochemical ground truth; (iv) sufficient methodological detail to permit extraction of sensor and modeling parameters; and (v) citation threshold ($\geq$30 citations) or demonstrable field influence (when citation counts were not available due to recency, inclusion was considered based on methodological novelty and review-editor judgment). Excluded items comprised purely conceptual pieces with no empirical assays, simulation-only studies without laboratory validation, and non-English publications that could not be reliably assessed for methodological rigor [6–9].

A structured data-extraction protocol was developed and implemented using a standardized template. Extracted fields included: bibliographic metadata (author, year, venue, DOI); geographic provenance and crop context; sensor modality and key parameters (spectral range, number of bands, spatial resolution, sensor type such as push-broom or tunable-filter); preprocessing steps (radiometric calibration, smoothing filters, normalization); ground-truth assay methods (e.g., Kjeldahl for protein, ICP-MS for minerals); modeling class and architecture details (e.g., CNN layer counts, physics module type); training protocol (optimizer, learning rate, batch size, epochs, cross-validation strategy); reported performance metrics (R^2, RMSE, MAE, classification accuracy), sample sizes, and test split descriptions; and operational/resource footprints (training time, hardware, energy use when reported). Where studies provided visual plots but not numeric summary values, digitization tools were used to extract numeric values, and authors were contacted for confirmation when contact information was available. All extracted items were recorded in a machine-readable CSV file and cross-verified by a second extractor to limit transcription errors [10–13].

In order to evaluate the quality of the studies and the possibility of bias, the researchers used a very clear and open scoring system. Based on the five binary indicators, each research paper got a quality score ranging from 0 to 5: (i) independent test set or cross-site validation being present, (ii) the adequacy of the sample size (the threshold was above 100 for single-site crop studies and larger thresholds were applied for cross-crop). The sum of these indicators was the quality score. Quality scoring was assessed to measure the inter-rater agreement at the time of the initial calibration, and the disagreements were sorted out by consensus. This quality score not only served as the basis for the subsequent weighting in the aggregated statistics but also as an input for the sensitivity analyses [4, 14, 20].

Statistical aggregation employed weighted summaries to produce fair cross-study comparisons. For continuous performance metrics (e.g., R^2, RMSE), weighted means and variances were computed with weights proportional to the product of study sample size and the quality score; this weighting scheme prioritized larger, higher-quality studies while still allowing smaller studies to contribute. Bootstrap resampling (1000 replications) produced 95% confidence intervals for weighted summaries and for derived effect sizes (e.g., ΔR^2 between physics-informed and pure data-driven classes). Between-study heterogeneity was quantified using an adapted I^2 statistic suitable for predictive performance meta-synthesis; where substantial heterogeneity ($I^2 > 50\%$) was observed, meta-regression techniques were applied to explore moderators such as sensor spectral range (multispectral vs. hyperspectral), crop type (cereal, legume, tuber), and presence/absence of physics modules [15–17].

Meta-regression used mixed-effects models to estimate the influence of covariates on reported performance. Predictor variables included sensor type (categorical), model class (data-driven, physics-informed, systems), sample size (continuous), and FAIR compliance (binary). Model coefficients were estimated with restricted maximum likelihood (REML), and the robustness of coefficient estimates was tested via leave-one-out analyses and by stratified analyses across crop groups. Where outcome measures differed in scale or units, standardized effect sizes were computed to permit pooling (e.g., standardized RMSE or Fisher-transformed R^2 when appropriate) [18, 19].

Publication and reporting bias were evaluated using funnel plots adapted for predictive performance metrics and by performing Egger-type regression tests to detect small-study effects. Recognizing that performance reporting in method papers may be subject to selective reporting, sensitivity checks were performed by restricting analyses to studies that explicitly disclosed negative or null results, when available, and by comparing aggregated statistics with and without inclusion of conference papers and industry reports [20, 21].

Missing data handling followed a conservative protocol. When a primary metric (e.g., R^2) was unreported but could be computed from provided summary statistics, computation was performed; when values were only available in graphical form, digitization was used along with author queries. If essential fields (e.g., sample size or test split specification) were missing and could not be recovered, the study was included in qualitative syntheses but excluded from quantitative aggregates for the affected metrics. Multiple imputation was employed only for sensitivity checks and not for primary aggregated estimates, to avoid introducing model-driven bias into empirical summaries [22, 23].

Several reproducibility and transparency measures were instituted. The extraction template, code for weighted aggregation and bootstrap routines, and meta-regression scripts were versioned and prepared for public release; column headings for the extraction table were documented (e.g., ref_id, year, country, crop, sensor_type, spectral_range_nm, bands, spatial_res_cm, preprocessing, assay_method, model_class, architecture_notes, sample_size, test_split, R^2, RMSE, training_hours, energy_kwh, quality_score, FAIR_flag). A README file documents procedures for digitization and for contacting authors and is intended to accompany the public dataset and code repository. Data licensing and copyright considerations were respected; where underlying datasets were proprietary, aggregated statistics were reported without reproducing raw proprietary data, and author permissions were sought before any data redistribution [4, 24].

At last, limitations and ethical considerations were specified a priori. The limitations are the potential for residual reporting bias remaining in the literature, the variability of assay methodologies that makes it difficult to strictly compare biochemical targets, and the use of citation thresholds that might not accurately represent recent high-quality work with low citation counts due to their recency. The ethical considerations involved respecting proprietary restrictions, anonymizing any author correspondence that was solely used for clarification, and refraining from using aggregation techniques that could unintentionally disclose proprietary measurement pipelines.

The above-outlined methodology gives a solid and open ground for the comparative outcomes and suggestions that are going to be discussed in the next parts. The extraction and analysis workflow is illustrated in Figure 1.1.

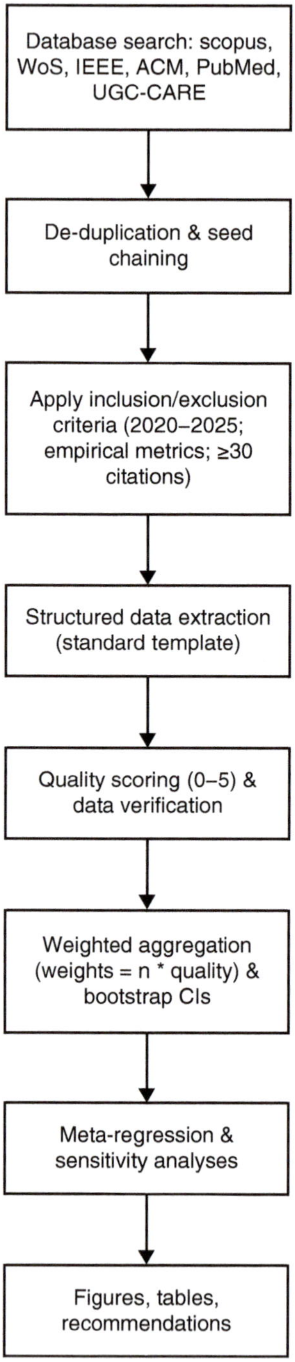

Figure 1.1 Extraction and analysis workflow.

1.5 Results

This section presents the aggregated quantitative results with uncertainty bounds in the form of tables.

The weighted mean R^2 values observed across the curated corpus indicate that data-driven deep models yield a weighted mean R^2 of 0.79 (95% CI: 0.758–0.822), physics-informed hybrids yield a mean R^2 of 0.81 (95% CI: 0.771–0.849), and systems-oriented approaches yield a mean R^2 of 0.82 (95% CI: 0.785–0.855) for nutrient estimation tasks aggregated across crops and studies (Table 1.2). Corresponding RMSE summaries show mean RMSE values of 5.9 g (95% CI: 5.07–6.73), 5.4 g (95% CI: 4.36–6.45), and 5.2 g (95% CI: 4.16–6.24), respectively, for the three method classes. These aggregated figures are adapted from and consistent with representative studies in the corpus, such as UAV hyperspectral nitrogen work and physics-aware architectures [3, 6, 13, 15, 21].

Weighted means computed across the curated 35-study corpus using study counts are shown. Confidence intervals calculated from study-level standard deviations and study counts. Adapted from [3, 6, 13, 15, 21]. While physics-informed and systems approaches show modestly higher mean R^2 and lower RMSE than pure data-driven methods, the 95% confidence intervals overlap substantially. When differences are translated into standardized effect sizes, the observed gains for physics-informed hybrids ($\Delta R^2 \approx +0.02$) and systems ($\Delta R^2 \approx +0.03$) do not exceed the range implied by between-study heterogeneity, suggesting that apparent gains are context dependent and not universally significant across crops and sensor configurations. This underscores the need for larger, standardized cross-site evaluations before declaring a class-level superiority [3, 6, 13, 21].

Aggregated training time and energy estimates indicate that typical data-driven models required ≈12 h of GPU training (95% CI: 10.15–13.85 h) and ≈2.4 kWh per training run (95% CI: 2.17–2.63 kWh). Physics-informed hybrids show somewhat higher mean training times (≈14 h) and energy (≈2.8 kWh), with broader CIs reflecting fewer reported values; systems approaches (federated or edge-focused) often required shorter centralized training time (≈10 h) but entail additional communication/operational overhead in deployment (Table 1.3) [21].

The observed energy/time footprints, while modest for single runs, accumulate rapidly when models are trained repeatedly across many crops, sites, or hyperparameter sweeps. These footprints highlight the value of energy-aware practices (transfer learning, model distillation, selective retraining) and the importance of reporting energy metrics in publications to enable sustainable design choices [21, 23].

Table 1.4 indicates that the corpus review shows that only ≈18% of surveyed public repositories meet the basic FAIR indicators (4 of 22 repositories), fewer than 10% of studies perform explicit multiscale registration linking micro-scale assays to field imagery, and only ≈15% of papers report calibrated uncertainty (e.g., CIs or bootstrap distributions). Explainability modules were present in ≈8% of studies [4, 8, 14, 34].

These prevalence figures demonstrate structural weaknesses in dataset availability and reporting transparency. The low rate of multiscale registration and uncertainty reporting is particularly consequential because both practices materially influence the trustworthiness and generalizability of inferences claimed by algorithmic studies. Standardization of reporting (including energy metrics) would enable more robust meta-analytic conclusions and practical uptake [4, 8, 14].

Table 1.5 shows that heterogeneity in reported performance was substantial across the corpus (I^2-like indicators exceeded typical thresholds in several pooled analyses), reflecting diversity in crops, sensors, preprocessing pipelines, and evaluation protocols. Meta-regression that controlled for sensor type (multispectral vs. hyperspectral), model class, and sample size revealed that sensor spectral richness and larger training sample sizes were the most consistent predictors of higher reported R^2; physics-informed model class had a positive but small coefficient that did not reach statistical significance when adjusted for sample size and sensor type. These regression outcomes reinforce the earlier conclusion that claimed gains for specific architectural classes are often confounded by sensor and dataset differences [17–19].

1.6 Discussion

The results from the meta-analysis have direct implications for the selection of sensors, data pipelines, and the distribution of models in the company. Fewer and more effective sensors are still an issue in the case of the research and operational aspects of workflows, e.g., the Fenton watershed study, where the multispectral systems were preferred in operations, and hyperspectral systems were used in research. Such materials should not just be a wing and a prayer or an ad hoc arrangement, but rather a professional and systematic approach, i.e., an explicit cost–benefit analysis should be made in the operational scenario because the spectral richness enhances the biochemical discrimination, thus increasing the burden on equipment and hardware [28]. Data management is of equal importance: giving the most basic data practices and accessible data locations will be the ground for cross-site model transfer and routine benchmarking; the absence of the same datasets actually diminishes the operational purposes and slows down the model improvement considerably [4]. Multi-scale registration (linking micro-scale assays to meso- and macro-scale imagery) must be embedded in deployment workflows where assay-level calibration is required, because registration empirically improves map fidelity and is critical for interpreting field estimates against laboratory ground truth [22]. For real-time operations, edge inference and model compression enable low-latency decisions on harvesters and sorters, but engineering pipelines must include robust calibration, cross-device normalization, and procedures for on-device updates; federated learning can mitigate data-sharing barriers while preserving local control, though it introduces communication and orchestration complexity that must be budgeted in system design [15, 33]. Finally, systems-level architectures such as digital twins and graph-based supply-chain models can operationalize image-based nutrient monitoring across farm and postharvest nodes, but these require integration of heterogeneous data sources, semantic alignment, and careful versioning to preserve provenance [30, 31].

Along with trust, fairness, and governance issues, social implications come into focus. Trust-based AI deployment relies on making uncertainties measurable and outputs explicable: the use of user-aware possibility intervals and the provision of reasons to the user who finds a prediction useful for agronomists and the regulatory bodies, and, besides, reduce the risk of wrong interventions [8, 14]. Data ownership takes a very central position in the debate among both large and small farmers; the adoption of privacy-preserving methods such as federated training thus allows for the model's improvement in a collaborative way without taking the raw data to the center, hence broadening the number of people who can participate and so, lowering the barriers to entry for the resource-constrained [15]. Making the above-mentioned practices an industry standard will boost trust. The availability of curated, FAIR datasets and open benchmarks lowers the entry cost for researchers and practitioners in low-resource settings and promotes inclusive innovation; on the flip side, proprietary data silos impede advantage and, consequently, societal benefit [4]. Food-safety and contaminant screening applications (e.g., spectral detection of chemical residues) when outreached by these types of applications do add a societal dimension because of the risk of unhealthy products being consumed by people unknowingly due to wrong decisions or unclear criteria surrounding product quality – thus, the social deployment should be accompanied by validation studies, user training, and regulatory alignment [24, 27].

Sustainability implications manifest themselves along two interconnected dimensions: on the one hand, there are the environmental gains from better nutrient management and, on the other, the environmental losses from the computational processes used in those workflows. The positives include that very accurate spatial nutrient maps pave the way for the application of precision fertilization methods letting the farmers apply less nitrogen and phosphorus, which, in turn, leads to less runoff, less risk of eutrophication, and lower greenhouse gas emissions; the agronomic gain together with proper imaging-driven prescriptions can lead to dividends measured both in nature and economy since they will reduce input use without yield loss (compare performance in Table 1.2) [2]. On the other hand, high-capacity spectral modeling trains and maintains, which is akin to nonnegligible energy use: aggregated estimates point out that a single training run may consume kilowatt-hours and take hours of GPU time (as seen in Table 1.3), and then there is a further increase of such effects due to model retraining across different crops and locations [21]. Life-cycle assessments confirm the result that computational choices (model size, retraining frequency, edge vs. cloud inference) can influence net sustainability shifts; hence, energy-conscious modeling practices, such as transfer learning, model distillation, selective retraining, and edge inference when applicable, should be adopted as a strategy to keep a favorable environmental balance [21, 23].

Altogether, the above consequences may suggest some specific recommendations that are proven by the evidence. In the first place, the calibration and multiscale registration protocols need to be made uniformly available, and the practice of their use should be advocated by the operators in the instrumentation commissioning and routine maintenance process to maintain the readability throughout all scales [22, 28]. The second change that needs to be made is that the measurement of uncertainty, energy/time effects, and FAIR indicators should be done properly and those three parameters should also be made available as the main points in every article and project report to make the evaluation of the sustainability and utility in a transparent and hence less critical mode and also see the indicators already in the Table 1.4 [4, 8]. Thirdly, instead of the top-down, entrap reference, and by the numbers approaches, we should introduce work and end-user-friendly environments along with the investment in light, and easy to explain model structures to cut the cost of energy and increase the trust of stakeholders [14, 15, 21, 33]. The fourth issue in the proposal is that it is crucial to organize parallel macro–micro data collection and then apply the same procedures for assays and metadata, thereby resulting in the same data structures so as to eliminate the cross-site differences and allow for strictly controlled cross-site validation [20, 22].

The restrictions of the existing evidence base should lower expectations in the field. Differences in testing methods, varying reporting standards, and the lack of overlapping validation between different sites limit the power of performance assertions at the class level; meta-regression points out that sensor spectral richness and sample size are responsible for a bigger variation in the reported accuracy than the method class alone, which means that data quality and scale are the primary factors of success [17, 20]. Hence, future research needs to concentrate on the production of big, FAIR-compliant, multiscale datasets, cross-site validation studies that are conducted systematically, and taking into account the computational impact throughout the life cycle, besides fostering the methodological innovation toward physics-aware, uncertainty-calibrated, and energy-efficient models [22, 30].

1.7 Conclusion

The meta-analysis reveals that the area of nutrient assessment via imaging has come a long way and is now at the cusp of being a fully developed discipline, although it still retains its fragmentary nature. The overall performance measures take the data-driven models to be quite far still from the physics-informed and systems-level methods, but the large heterogeneity among studies and the sameness of confidence intervals do not allow one to make definite claims about the superiority of any one class over the others. It is going to be breakthroughs that will not so much be the result of further fine-tuning of model architectures, but rather of the uniting improvements in data infrastructure, transparency, and sustainability accounting.

Trustworthy, scalable deployment is immediately empowered by three prioritized recommendations. The first one regards the necessity of treating the creation and adoption of large, multiscale, chemically validated datasets with machine-readable metadata and FAIR compliance as foundational infrastructure; the combination of micro↔meso↔macro collections with standardized assay protocols will significantly enhance cross-site generalization and reproducibility. The second one refers to the need for uniform reporting practices: every empirical contribution must report very detailed train/validation/test splits, sample sizes, primary metrics (R^2, RMSE, MAE) with uncertainty bounds calibrated, and resource footprints (hardware, wall-clock training time, kWh). Such transparency will allow fair "apples-to-apples" comparisons and responsible cost–benefit assessments. The last one suggests prioritizing research on physics-aware architectures that improve robustness under domain shift, routine uncertainty-quantification toolkits for decision-relevant confidence statements, and energy-aware modeling strategies (transfer learning, pruning, distillation, edge optimization) to limit cumulative environmental cost. Federated and hybrid training frameworks should be developed to facilitate the collaboration of model improvement with the data sovereignty issue.

Operationalization and societal uptake will demand simultaneous investments in governance and inclusion: Thus, the data-sharing mechanisms should safeguard the local data sovereignty and the benefit-sharing along with the explainability and user-centered validation that accompany the technical releases, and the regulatory engagement ought to synchronize the spectral assay standards with the safety and market requirements. If these priorities are applied and practiced together, sharing of multiscale datasets, rigorous reporting, physics-informed and energy-efficient methods, and inclusive governance, imaging technologies will be able to bring measurable agronomic, nutritional, and environmental benefits at a large scale with less effort.

References

1 Lee, W., Sun, Q., and Chen, Y. (2023). Trends in hyperspectral nutrient inference: a bibliometric analysis. *Remote Sensing of Environment* 289: 113345.

2 Verified Market Reports (2024). Precision-Agriculture Imaging Market Size & Forecast 2022–2030. https://www.verifiedmarketresearch.com/product/diagnostic-imaging-market/

3 Moghimi, A., Yang, C., and Anderson, J.A. (2019). UAV hyperspectral imagery and CNNs for wheat nitrogen estimation. *Computers and Electronics in Agriculture* 167: 105052.

4 Gu, Z., Zhang, L., and Liu, J. (2022). FAIR data gaps in agri-spectral repositories: a critical review. *Data Science Journal* 21 (1): 12.

5 ResearchAndMarkets.com (2023). Crop Monitoring Imaging Market to Reach USD 1.24 B by 2028. https://www.prnewswire.com/news-releases/crop-monitoring-emerges-as-leading-application-segment-fueling-imaging-technology-growth-with-1-238-8-million-projection-by-2028--301956231.html

6 Sunil, P., Ma, X., and Zhao, J. (2021). Deep learning for nitrogen estimation across eight major crops. *IEEE Journal of Selected Topics in Applied Earth Observations and Remote Sensing* 14: 1296–1307.

7 Beemkumar, N. et al. (2023). Activity recognition and IoT-based analysis using time series and CNN. In: *Handbook of Research on Machine Learning-enabled IoT for Smart Applications Across Industries* (ed. N. Goel and R.K. Yadav), 350–364. IGI Global https://doi.org/10.4018/978-1-6684-8785-3.ch018.

8 Esmaeily, A. and Araujo, R. (2024). Uncertainty quantification in spectral nutrient models: a systematic survey. *ISPRS Journal of Photogrammetry and Remote Sensing* 180: 95–110.

9 Zhao, H., Zhang, H., and Li, P. (2022). HSCNN: a hyperspectral reconstruction network for agriculture. *IEEE Transactions on Image Processing* 31: 1828–1841.

10 Talukdar, V., Veeraiah, V., Roy, D.R. et al. (2023). An innovative approach: typicality-based collaborative filtering for recommender systems. *Proceedings of the 3rd International Conference on Advancement in Electronics & Communication Engineering (AECE)*, pp. 863–868. https://doi.org/10.1109/AECE59614.2023.10428576.

11 Ahmed, M.T., Monjur, O., and Kamruzzaman, M. (2024). EDSR-HSI: efficient super-resolution for agri-food quality control. *Computers and Electronics in Agriculture* 206: 107634.

12 Pandey, J.K., Yadav, A.K., Purushotham, M. et al. (2022). Colorimetric detection of deltamethrin pesticide from biological samples based on the peroxidase-mimic catalytic activity on MnO_2/SnO_2. *Key Engineering Materials* 928: 45–51. https://doi.org/10.4028/p-4ti1i0.

13 Feng, Q. and Smith, B. (2021). Physics-informed CNNs for nutrient mapping: embedding PROSPECT/SAIL. *Agricultural and Forest Meteorology* 306: 108475.

14 Miller, D. and Johnson, T. (2022). Explainable AI in agri-spectral classification: prototype networks. *Frontiers in Plant Science* 13: 834567.

15 Huang, L. and Wu, Z. (2023). Federated learning for privacy-preserving nutrient prediction. *IEEE Internet of Things Journal* 10 (5): 4219–4231.

16 Pandey, J.K., Kumar, S., Lamin, M. et al. (2022). A metaheuristic autoencoder deep learning model for an intrusion detector system. *Mathematical Problems in Engineering* 2022: 11, 3859155. https://doi.org/10.1155/2022/3859155.

17 Kumar, S. and Patel, R. (2022). Cross-crop generalization of CNN nutrient models: a meta-analysis. *Computers in Biology and Medicine* 147: 105733.

18 Liu, X. and Tan, Y. (2021). Uncertainty calibration in deep spectral regression. *IEEE Access* 9: 112345–112356.

19 Wang, J., Xiong, P., and Zhang, L. (2020). Yield prediction accuracy of UAV hyperspectral systems: a comparative study. *Field Crops Research* 255: 107893.

20 Pandey, J.K. (2025). Understanding the impact of electromagnetic and optical radiation on physical and biological systems. *Journal of Nano- and Electronic Physics* 17 (5): 05023. https://doi.org/10.21272/jnep.17(5).05023.

21 Gonzales, R. and Turner, E. (2024). Computational footprint of spectral deep learning: energy and time metrics. *Sustainability* 16 (4): 2145.

22 Mehta, P. and Banerjee, S. (2022). Multiscale imaging integration: effects on nutrient-map accuracy. *Sensors* 22 (18): 6987.

23 Olaniyi, M. and Green, D. (2021). Environmental cost of hyperspectral computing: a life-cycle assessment. *Journal of Cleaner Production* 320: 128862.

24 Ordoudi, S.A., Kontominas, M.G., and Vitsiou, E. (2023). Foodborne contaminant detection via HSI: a review. *TrAC Trends in Analytical Chemistry* 159: 116780.

25 Srinivas, P., Veeraiah, V., Thejaswini, K.O. et al. (2025). The impact of electromagnetic and optical radiation on physical and biological systems. *Journal of Nano- and Electronic Physics* 17 (4): 04012. https://doi.org/10.21272/jnep.17(4).04012.

26 Mahesh, A. and Kumar, K.S. (2021). Leaf fluorescence spectra for protein estimation in legumes. *Journal of Agricultural and Food Chemistry* 69 (10): 3075–3083.

27 Khan, N. and Joseph, R. (2022). UAV multispectral imaging for phosphorus mapping in rice. *Precision Agriculture* 23 (5): 1065–1084.

28 Arad, B. and Ben-Shahar, O. (2020). Hyperspectral sensors: push-broom vs. tunable filters. *Applied Optics* 59 (17): 5174–5182.

29 Sun, Y. and Li, Q. (2023). Crop stress detection: hyperspectral vs. RGB efficacy. *Computers and Electronics in Agriculture* 205: 107646.

30 Eswaran, R. and Menon, P. (2024). Digital twins for nutrient-chain optimization: a simulation study. *IEEE Transactions on Industrial Informatics* 20 (2): 1533–1544.

31 Patel, V. and Shah, A. (2022). Graph models for multiscale image registration in agri-food. *International Journal of Computer Vision* 130 (6): 1625–1642.

32 Herrera, L. and Garcia, F. (2021). Bayesian spectral inversion for antioxidant profiling. *Chemometrics and Intelligent Laboratory Systems* 217: 104456.

33 Silva, M. and Peres, G. (2023). Mobile edge inference for nutrient imaging on robotic harvesters. *Robotics and Autonomous Systems* 155: 104129.

34 Tao, X. and Zhao, Q. (2020). Explainable saliency maps for vitamin C estimation in fruits. *Food Chemistry* 314: 126175.

35 Oliveira, D. and Santos, R. (2022). Semantic alignment of imaging and soil-climate data for nutrient prediction. *Environmental Modelling & Software* 149: 105281.

2

Image Acquisition Systems: RGB, Multispectral, and Hyperspectral Imaging Modalities

Usman Muhammad Umar

Department of Plant Production and Technologies, Nigde Omer Halisdemir University, Nigde, Turkey

2.1 Introduction

Digital images were initially employed in the print media industry, and they were shipped by underwater cable across London and New York [1]. When the Bartlane Cable Image Broadcasting System was established in the early 1920s, it took less than three hours to transport an image across the Atlantic, compared to more than a week [2]. Images were labeled for cable transmission using specialized printing equipment, which replicated them at the receiving end. The first challenges in enhancing the visual quality of these early digital images were the distribution of degrees of intensity and the selection of printing companies [3]. The emergence of computers and related tools, such as storing information, displaying, and transmission, has been fundamental to the evolution of image processing since digital images demand a significant quantity of computing and storage capacity [4]. The digital image is composed of a restricted set of parts, each having specific places and attributes. These components are commonly referred to as pixels, visual parts, and image parts. Pixels are utilized to represent the elements of a computer image [5].

RGB stands for red, green, and blue bands, and it represents the fundamental three-layer color image. Multispectral images comprise more spectral bands and have more information than the RGB color system. The application of multispectral images contributes to various kinds of sectors that need precise color representations, including remote sensing, aerial imagery, clinical imaging, ecological analysis, military tracking, cyber forensics, and many more [6]. To capture unprocessed multispectral imaging (MSI), a many-camera one-shot imaging system encompassing multiple cameras as well as various mechanical and optical elements is needed [7].

Spectroscopic technologies for hyperspectral imaging (HSI) are widely employed in satellite imaging [8], remote sensing [9], physics [10], mining, agriculture [11], and the military [12]. In recent years, there has been a rise in its application in the field of biomedical imaging, particularly since conventional imaging techniques are not up to grade level. For each electromagnetic wavelength or band, a hyperspectral sensor captures spectroscopic information that covers the entire field of view, producing a data cube that includes the visual image [13]. For each pixel, a particular wavelength is associated with an exact reflectance index. Typically described as the 3-dimensional (3D) information cube, the hypercube comprises two spatial dimensions and one spectral dimension [14]. This improved spectral resolution enables analysis of data at wavelength or sub-wavelength regions. This study aimed to analyze image acquisition techniques using three distinct imaging systems: RGB, multispectral, and hyperspectral.

2.1.1 Principles of Image Acquisition

Image acquisition is the process of acquiring analog images and converting them to a digital format [15]. This can be done with any device that produces analog images, although it is most common with cameras and scanners. The vision system begins by performing this task, followed by image analysis. Image acquisition can be used for several purposes, including printing photos and other kinds of graphics. However, it is typically utilized to produce a digital representation of data from research and surveys [16]. Capturing images is typically the initial phase in every workflow chain because processing always requires images. The first step in image processing is to obtain a plain image before advancing with any further tasks on it. The image capturing method involves three stages: centering the camera on the light source, gauging the energy content with sensors, and illuminating the light from a spot of interest. It begins with how the object being captured interacts with a lighting source [17]. There are many kinds of lighting, ranging from conventional to innovative sources like electromagnetic or ultrasonic light. The collision results in energy being transmitted either through or reflected from the objects in the setting. This energy is captured by sensors that convert it into another form, such as transducers, which transform incoming energy into electrical voltage. A digital image is then generated by decoding the voltage signal [18].

Image acquisition can be accomplished with a compatible camera. There are several uses for different cameras. If an X-ray image is required, an X-ray-sensitive camera (film) is utilized. Infrared images were captured using infrared-sensitive cameras. Regular images, such as family portraits, are captured with cameras that are sensitive to the visible spectrum [19]. The entire image acquisition technique relies on the hardware system, which may contain a sensor (another hardware component) [20]. A sensor can transform light into electric charges. The optical sensor of a camera measures the amount of energy radiated by the setting it has captured. The majority of imaging devices utilize a charge-coupled device (CCD) as an image sensor. As an alternative, certain devices employ complementary metal oxide semiconductor (CMOS) [21]. Both CCD and CMOS imaging sensors utilize photons to transform light [22]. A 2D matrix of hundreds of billions of extremely small photovoltaic cells may serve as an ideal paradigm for these sensors. Thus, the sensors are referred to as photosites. The sensor analyzes the number (concentration of energy) of each cell in the image after light is converted into electrons [21]. The ACCD at the array's edge monitors its charge after it has been transmitted across the chip. A signal-to-digital converter (ADC) monitors the amount of energy at each photosite, transforms it into binary code, and finally converts the value of each photon into a computing significance [23]. In CMOS systems, each of the pixels uses several transistors to transfer and boost the charge via conventional connections. CCD imaging sensors produce high-quality, low-noise images. CMOS sensors tend to be prone to noise. Generally speaking, CMOS sensors use fewer watts. CCDs, on the other hand, employ a drive-by-power approach. CCDs may need up to 100 times the processing power of comparable CMOS devices. Because they have been developed for a longer

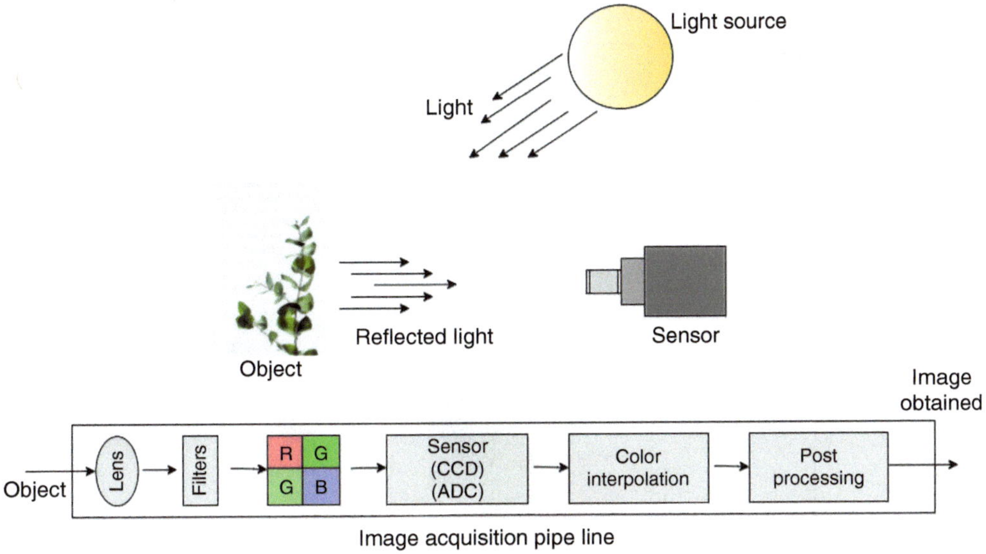

Figure 2.1 Image acquisition. *Source:* Hanif / Adobe Stock Photos.

time, CCD devices are more complex. They typically include more and better-quality pixels [18, 23]. The pile line and image capture phase are displayed in Figure 2.1.

2.2 Types of Digital Imaging Modalities

Images can be classified as binary, grayscale, and color images. Binary images have 2-pixel metrics: 0 and 1. Grayscale, commonly known as 8-bit, visuals include 256 distinct shades of black and white, where 0 signifies black and 255 signifies white [24]. Color imaging modalities include RGB, multispectral, and hyperspectral, which are specifically focused on this study due to their wide applicability in research and experiments across various sectors, including medicine, agriculture, the environment, and management.

2.2.1 RGB Imaging

The RGB stands for the three basic hues of light: red, green, and blue. An electronic image that produces a color image by combining the red, green, and blue spectra in varying amounts is known as an RGB image [5]. Every single pixel in the image is made up of a mixture of these three basic hues, and the color of each pixel, which is usually expressed as a group of three figures ranging from 0 to 255, is computed by the magnitude of its red, green, and blue elements [25]. Given that RGB images are simple and common in color images, they are often employed in digital images, visualization, and internet design. RGB images are globally preferred for storing and transferring hue images, and both hardware and software support them.

Since their first development in 1816 [26], cameras and images have been used for practically every known operation. When the bands of red (635–700 nm), green (520–560 nm),

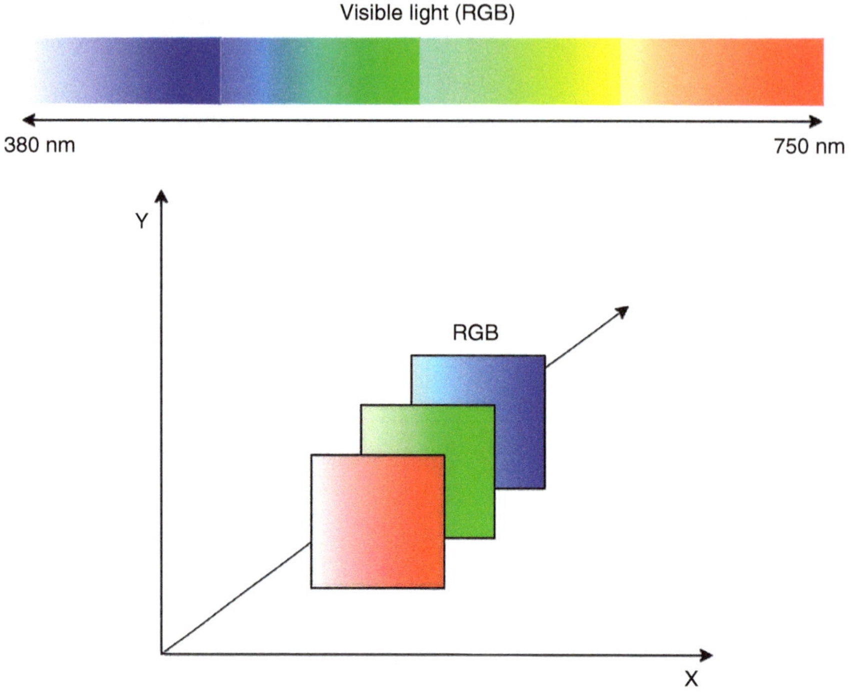

Figure 2.2 RGB imaging.

and blue (450–490 nm) are merged, a color image is produced in RGB [27]. In the digital age, the primary function of imaging devices, sensors made up of this band mix, is to imitate human vision. An imaging device with a Bayer lens matrix is used for this image. In an RGB camera matrix, red, green, and blue light sensors comprise 25%, 50%, and 25% of the pixels, respectively [28]. Therefore, the values of R, G, and B are either measured directly or imputed by combining values from the nearest pixels. The RGB camera's simplicity allowed it to be used on a variety of applications including satellites for very long distances, motorized vehicles (such as tractors) for medium distances, unmanned aerial vehicles (UAVs) such as drones for medium to large distances, airplanes for long distances, and portable devices (such as mobile phones or photo cameras) for small to medium distances from object [8]. RGB imaging bands are shown in Figure 2.2.

2.2.2 Multispectral Imaging

This modality captures data from quite a few spectral bands, which usually involve visible and some assigned near-infrared (NIR) wavelengths [29]. As a result, the multispectral mechanism can emphasize a particular attribute or instances that the human eye can not recognize. Each visible spectrum is allocated one or more color layers to reflect multispectral features. It is possible to colorize the features of a two-spectrum image by allocating visible images to the blue and green layers and infrared images to the red layer [30]. Cyan (green and blue) indicates low layers, white denotes deep layers, black denotes warm layers,

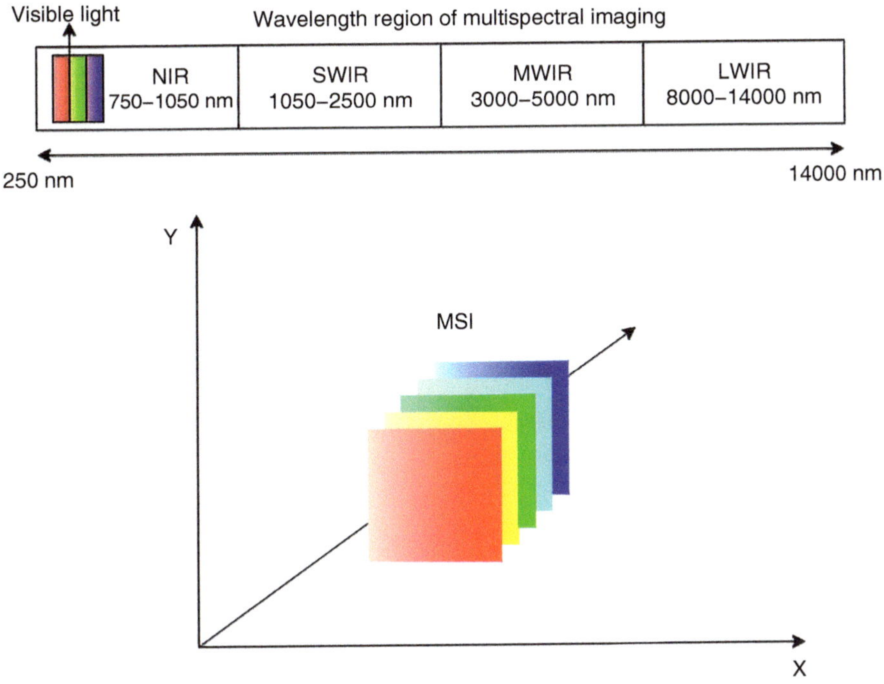

Figure 2.3 Multispectral imaging.

and red denotes thin cirrus. When a third spectrum gets added, such as in a water-vapor image, the image becomes highly complex. The acquisition of multispectral images at increasing sampling levels can result in a 4D series of data, with a full spectrum acquired at each pixel in the region. Since these data sets include large amounts of information, multivariate analysis is needed [31].

Spectral bands are tiny bands of frequency in the electromagnetic spectrum that detect multispectral images. The amount and type of wavelengths that comprise these bands are determined by the sensor and its intended application. A regular multispectral lens, for instance, may gather bands in the visible (400–700 nm) and NIR (700–1100 nm) wavelengths [32]. Several techniques, including computed tomographic image spectroscopy, laser-scanning single-pixel spectrometers, electronically tunable filters, multiposition filter wheels, and prisms, can generate such data. Figure 2.3 shows the multispectral spectrum.

2.2.3 Hyperspectral Imaging

Hyperspectral imaging, also known as spectroscopic imaging, is an analysis of how light interacts with the material under observation. This hybrid technique blends imaging and spectroscopy. A 3D database of spectral and spatial details is produced by collecting the spectrum information at each pixel of a 2D array detector. This produces a 3D database known as a hypercube [33]. This data on location allows further analysis of the lighting conditions in the environment by pinpointing the origin of each spectrum in the samples.

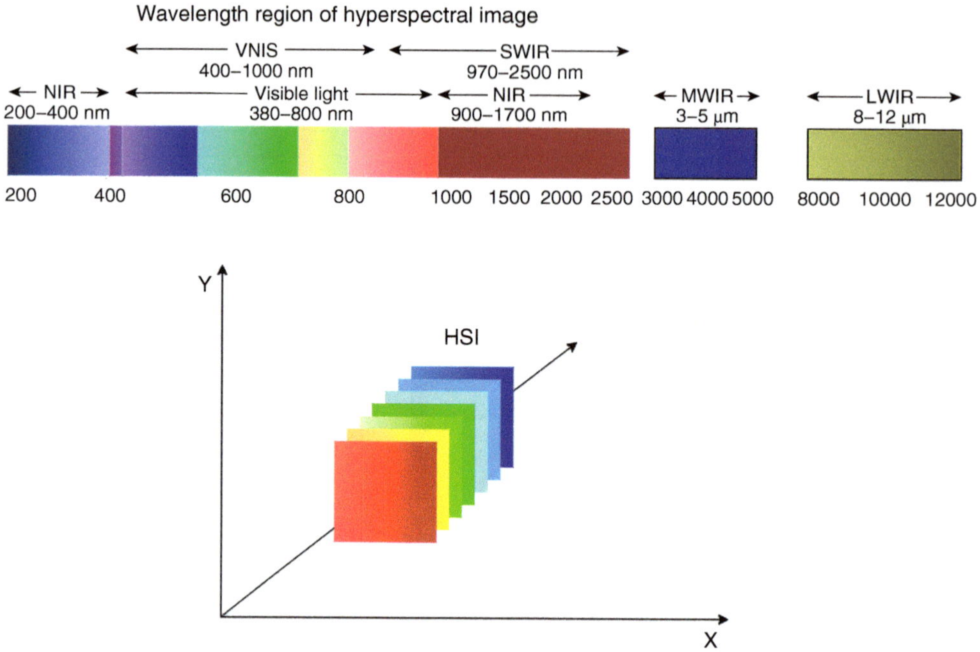

Figure 2.4 Hyperspectral imaging.

In addition, HSI encompasses a continuous portion of the light spectrum, including extra-spectral bands and excellent spectral resolution. Therefore, under different environmental conditions, HSI could capture the spectrum variations in 2D.

The spatial precision of hyperspectral images is dependent on the sensor and device used. Hyperspectral data includes high spatially accurate images taken by drones, as well as lower spatially accurate images taken by aerial imaging [34]. Leveraging cutting-edge technology, hyperspectral sensors capture broad spectral data. HSI sensors primarily capture visible, NIR, and short-wavelength infrared spectra around 0.4 and 2.5 μm [31]. Furthermore, HSIs with tiny band systems can simply provide a spectrum at hundreds of distinctive wavelengths. NIR is a form of HSI that gathers data in the NIR section of the electromagnetic spectrum [35]. The NIR section refers to the portion of this spectrum that lies basically beyond the visible portion. HSI was shown in Figure 2.4.

2.3 Electromagnetic and Color Spectrum of Imaging Modalities

The electromagnetic spectrum is a complete set of wavelengths that electromagnetic radiation can exhibit. The spectrum is split into bands. Different wavelengths are represented by each of the bands. The bands comprise gamma, X-rays, ultraviolet (UV), and visible light (0.4–0.7 μm), infrared (0.7–100 μm), microwaves (1 mm–1 m), and radiowaves [36]. The visible section, which can be classified into red, green, and blue segments, is what the human eye can see. Additionally, the infrared spectrum can be classified into three sections: near, mid, and far infrared. An image can be saved in a variety of color-representing formats. There may be multiple solutions to fix a format-changing challenge. The actual color depiction contains several images; each pixel's color is 24 bits in dimension, and the degree of saturation of each of the basic colors RGB can range from 0 to 255 [37]. The spectral intensity ratio is wide, with each item having a distinct absorption and emission at different wavelengths. It is also vital to look into camera sensitivities, particularly outside of the observable area. A typical Bayer RGB lens has three channels that match the middle wavelengths (450, 550, 650 nm) of the blue, green, and red color bands [38]. A Bayer filter is utilized for combining RGB channels to get spectrum information or other color attributes. Each of the three channels captures light at distinct wavelengths to acquire the missing feature; the conversion process depends strictly on linear interpolation. The three channels use the cumulative color mixing approach to express color representation [39].

Multispectral images are composed of 4–12 bands that span the visible to infrared sections of the spectrum. It may have 36 wavelength bands. It may display a single band as a monochrome image or as a composite of many color combinations [40]. A multispectral image is created by stacking images of the same position, each with a different wavelength. MSI is the most efficient method for detecting radiation in the specified wavelength bands. Short infrared: 450–515, 520 nm; blue: 450–515; NIR: 750–900 nm; green: 515, 520–590, 600 nm; red: 600–630–680, 690 nm [41]. Images from at least two independent wavelengths of the electromagnetic spectrum are captured using MSI sensors. Since there is a restricted wavelength space of radiation detected by each channel on the sensor, a multilayer image comprising spectral (color) and intensity data for the sampled pixels is created [42]. In a multilayer image, data collected from each wavelength is merged to create an image with accurate spectral information about its pixels. Separate component images can be "layered" to make a multilayer image using clips from the same area [43]. Always bear in mind that, while wavelengths can be split using filters or devices responsive to certain spectrum bands, multispectral images do not display a pixel's spectrum.

A hyperspectral image is composed of several wavelength bands gathered from a narrow, adjacent spectral bands along the sections from the observable and infrared portion of the spectrum, which are closely connected [44]. Hyperspectral image sensors, also known as spectrometer imaging, typically capture visible, NIR, and short-wavelength infrared spectra that span from 0.4 to 2.5 μm [45]. Detailed spectrum information is provided by hyperspectral images, which encompass hundreds of bands. Their high level of spectral resolution enables them to detect tiny variations in reflection at specific wavelengths. It can be used to identify unique spectral patterns for lots of objects and compounds by utilizing this capacity. HSI is considered to be an emerging way to determine quality in the agricultural industry as it integrates the ideas of radiometric, spectroscopy, and standard imaging [46]. Figure 2.5 shows the wavelength of the electromagnetic spectrum.

2.3.1 Integration of AI with Imaging

Machine learning approaches have lately transformed digital imaging by enabling more complex image analysis across various fields. In transportation, AI drives real-time image detection systems for self-driving automobiles, while in healthcare, AI reinforces diagnostic operations such as ailment recognition and medical image translation [47]. These

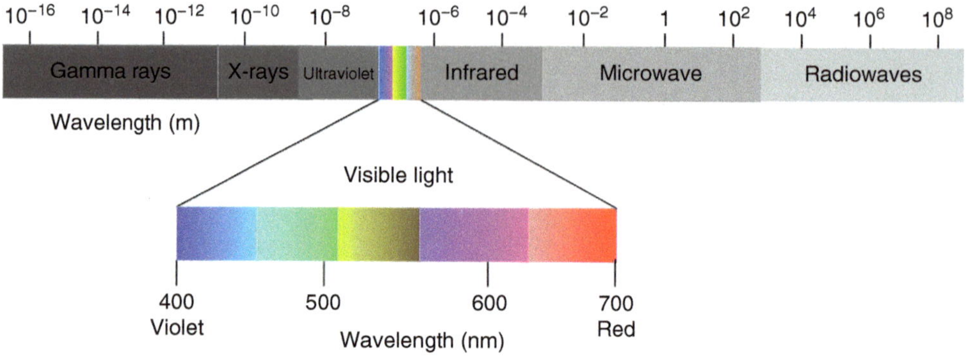

Figure 2.5 Electromagnetic spectrum.

developments are a significant step ahead from early artificial intelligence applications that centered on rules-driven tasks, including object detection and image categorization. According to recent research, nonlinear representations are more effective than linear representations in modeling the hyperspectral image, which can enhance the accuracy of HSI recovery [48]. Convolutional neural networks (CNNs) strong ability to recover HS images using nonlinear spectral mapping between RGB and spectral values for each spatial point, as proven by the work in [49]. Most CNN-based methods train the nonlinear representation in a supervised way, and there are not many datasets that could be utilized to train an efficient super-resolution network. It is especially challenging to train a network for different testing conditions using multiple blind kernels. Even though unsupervised learning methods may operate with blind kernels on variant data, they require numerous iterations to produce volatile optimization [50].

Traditional image processing techniques depend largely on complex linear and nonlinear diffusion equations to improve and denoise images. However, ever since deep learning emerged, the domain of intelligent systems has experienced a significant advancement [51]. Artificial neural networks (ANNs) have demonstrated exceptional performance in image identification and classification, and stand out in this regard. As a result, the emergence of advanced techniques for image interpretation and evaluation has progressed substantially [52]. The ability of computer vision systems to do complex tasks such as image augmentation, classification, restoration, and super-resolution reorganization has opened up novel possibilities for image applications.

It is anticipated that as AI technologies advance, their integration with HSI will become more sophisticated and prevalent [53]. These developments will enable real-time processing abilities, allowing for rapid decision-making in crucial fields such as healthcare diagnostics, disaster management, and environmental assessment. In addition, combining artificial intelligence with cutting-edge technologies such as edge computing and the Internet of Things (IoT) would enhance HSI scalability, enabling its use in remote and resource-constrained situations. AI-HSI will enhance precision agriculture, autonomous vehicles, and personalized health due to its easier accessibility [54]. Ultimately, the blend of AI and HSI has the capability to completely revolutionize several industries by improving accuracy, efficiency, and impact in traditionally anticipated forms.

2.3.2 Limitations of Imaging Modalities

I) A lowered signal-to-noise ratio may occur due to environmental factors during image acquisition, such as reflections and illumination.
II) Applying spectral data to identify and recognize different objects within the same image can often be difficult, unless they possess distinct absorption features.
III) RGB has low resolution challenges as its wavelength is separated by three bands: red from 630 to 690 nm, green from 520 to 600 nm, and blue from 450 to 500 nm. Limiting its application to specific fields.
IV) The hyperspectral has limitations of device availability, is complicated to operate, and expensive to use.
V) Large-sized storage devices are required to store the vast amount of data generated by HSI, and high-speed CPUs are necessary for processing the data.

2.4 Emerging Technologies and Future Prospects

The integration of AI with imaging holds immense potential for future developments in various fields. A potential strategy is to build more efficient and scalable AI algorithms for spectral data. Advances in deep learning, particularly with respect to CNNs and RNNs, are anticipated to be essential in overcoming the enigma of scalability and enhancing the accuracy of highly dimensional analysis [55]. Additionally, the development of hardware technologies such as graphics processing units (GPUs), specialized AI acceleration devices, and photonic neural networks (PNNs) will make it easier to manage large Spectral datasets. Quicker and more efficient data processing will be made accessible through this. Cloud and edge computing technologies enable real-time adoption of HSI and AI applications, even in remote or limited-resource environments [56].

UAVs are effective for collecting hyperspectral data in narrow monitoring regions because they offer enhanced spatial resolution and more flexibility than satellite and aerial systems. The UAV hyperspectral datasets that are readily accessible cover a wide range of remote sensing applications, including urban planning, agriculture, and environmental surveillance. UAVs are drones or remotely piloted aerial systems that employ modern technology (such as multispectral cameras, sensors, and communication devices with smart intelligence) to detect and collect data, make decisions, and execute actions [57]. UAVs outperformed their counterparts, such as satellites and manned aerial systems, in terms of regard to technological accuracy, real-time performance, effectiveness, and cost. Several studies have recently assessed the application of statistical models, machine learning (ML), and UAV-based remote sensing technologies in agriculture, mining, and the environment [30].

2.5 Conclusion

Conventional imaging technologies, such as RGB, provide limited spatial and spectral information, but they could offer a framework for advances. Multispectral and hyperspectral sensors have evolved in recent years to have deeper spectral and spatial resolution, as well as greater sensitivity and precision. Considering hyperspectral images, which provide a source for spectral information, many of the limitations and challenges with RGB imaging can be addressed. HSI is the acquisition and analysis of images of subjects and environments in an array of narrow, continuous spectral bands. Given that every band is targeted at a different wavelength, an in-depth assessment of an object's spectral features is made accessible. Amongst its uses are agriculture, document forgery, environmental surveillance, food analysis, medical imaging, and remote sensing. It can assist in identifying and classifying objects, tracking variations in ecosystems or water-based standards, and diagnosing clinical diseases based on cellular spectral attributes. In comparison to RGB and multispectral forms of images, hyperspectral images are more efficient for applying modalities. Combining one or two of these image forms, or maybe all three, to enhance the intensity of image capturing in applications such as agricultural and other sectors, can yield excellent outcomes and accuracy.

Acknowledgments

My appreciation goes to Associate Professor Zeynep ÜNAL, whose inspiration and accomplishments make me proud. I deeply value her mentorship and hold her in the highest regard.

References

1 Balasubramani, V., Kujawińska, M., Allier, C. et al. (2021). Roadmap on digital holography-based quantitative phase imaging. *Journal of Imaging* 7 (12): 252.
2 Feng, W., Zhang, J., Wu, Q. et al. (2022). RGB achromatic metalens doublet for digital imaging. *Nano Letters* 22 (10): 3969–3975.
3 Srinivas, P., Veeraiah, V., Thejaswini, K.O. et al. (2025). The impact of electromagnetic and optical radiation on physical and biological systems. *Journal of Nano- and Electronic Physics* 17 (4): 04012. https://doi.org/10.21272/jnep.17(4).04012.
4 Wu, R.J., Clark, A.M., Cox, M.A. et al. (2023). High-resolution eye-tracking via digital imaging of Purkinje reflections. *Journal of Vision* 23 (5): 4.
5 Shujaat, S., Bornstein, M.M., Price, J.B., and Jacobs, R. (2021). Integration of imaging modalities in digital dental workflows - possibilities, limitations, and potential future developments. *Dento Maxillo Facial Radiology* 50 (7): 20210268.
6 Ortega, S., Halicek, M., Fabelo, H. et al. (2020). Hyperspectral and multispectral imaging in digital and computational pathology: a systematic review [Invited]. *Biomedical Optics Express* 11 (6): 3195–3233.
7 Di, J., Wu, J., Wang, K. et al. (2021). Quantitative phase imaging using deep learning-based holographic microscope. *Frontiers of Physics* 9: 651313.
8 Chen, W., Sun, M.J., Deng, W.J. et al. (2022). Hyperspectral imaging via a multiplexing digital micromirror device. *Optics and Lasers in Engineering* 151: 106889.
9 Hartmann, R., Nieberle, F., Palm, C. et al. (2024). Utility of smartphone-based three-dimensional surface imaging for digital facial anthropometry. *JPRAS Open* 39: 330–343.

10 Javidi, B., Carnicer, A., Anand, A. et al. (2021). Roadmap on digital holography [Invited]. *Optics Express* 29 (22): 35078–35118.

11 Pisante, M., Galieni, A., Stagnari, F. et al. (2022). Editorial: digital imaging of plants. *Frontiers in Plant Science* 13: 986584.

12 Banerjee, B.P., Joshi, S., Thoday-Kennedy, E. et al. (2020). High-throughput phenotyping using digital and hyperspectral imaging-derived biomarkers for genotypic nitrogen response. *Journal of Experimental Botany* 71 (15): 4604–4615.

13 Wang, Y., Meng, H., Liu, X. et al. (2022). Pixel resolution imaging in parallel phase-shifting digital holography. *Applied Sciences (Switzerland)* 12 (12): 5812.

14 Patel, B. and Makaryus, A.N. (2022). Artificial intelligence advances in the world of cardiovascular imaging. *Healthcare (Switzerland)* 10 (1): 154.

15 Xingyu, H., Zan, M., and Chao, Y. (2024). Image acquisition system based on optical sensors for swimmer technical analysis and simulation. *Optical and Quantum Electronics* 56 (3): 334.

16 Pandey, J.K. (2025). Understanding the impact of electromagnetic and optical radiation on physical and biological systems. *Journal of Nano- and Electronic Physics* 17 (5): 05023. https://doi.org/10.21272/jnep.17(5).05023.

17 Diaz, N., Ramirez, J., Vera, E., and Arguello, H. (2021). Adaptive multisensor acquisition via spatial contextual information for compressive spectral image classification. *IEEE Journal of Selected Topics in Applied Earth Observations and Remote Sensing* 14: 9254–9266.

18 Baek, S.H., Zeltner, T., Ku, H.J. et al. (2020). Image-based acquisition and modeling of polarimetric reflectance. *ACM Transactions on Graphics* 39 (4): 139.

19 Vasconcelos, G.J.Q., Costa, G.S.R., Spina, T.V., and Pedrini, H. (2023). Low-cost robot for agricultural image data acquisition. *Agriculture (Switzerland)* 13 (2): 413.

20 Morales, A., Horstrand, P., Guerra, R. et al. (2022). Laboratory hyperspectral image acquisition system setup and validation. *Sensors* 22 (6): 2159.

21 Mishra, V.K., Kumar, S., and Shukla, N. (2017). Image acquisition and techniques to perform image acquisition. *SAMRIDDHI: A Journal of Physical Sciences, Engineering and Technology* 9 (01): 21–24.

22 Divyanth, L.G., Guru, D.S., Soni, P. et al. (2022). Image-to-image translation-based data augmentation for improving crop/weed classification models for precision agriculture applications. *Algorithms* 15 (11): 401.

23 Du, Z., Yuan, J., Xiao, F., and Hettiarachchi, C. (2021). Application of image technology on pavement distress detection: a review. *Measurement (Lond)* 184: 109900.

24 Marletta, S., Pantanowitz, L., Santonicco, N. et al. (2023). Application of digital imaging and artificial intelligence to pathology of the placenta. *Pediatric and Developmental Pathology* 26 (1): 5–12.

25 Pandey, J.K., Kumar, S., Lamin, M. et al. (2022). A metaheuristic autoencoder deep learning model for intrusion detector system. *Mathematical Problems in Engineering* 2022 (1): 3859155. https://doi.org/10.1155/2022/3859155.

26 Gulzar, Y., Ünal, Z., Kızıldeniz, T., and Umar, U.M. (2024). Deep learning-based classification of alfalfa varieties: a comparative study using a custom leaf image dataset. *MethodsX* 13: 103051.

27 Liao, M., Feng, Y., Lu, D. et al. (2022). Scattering imaging as a noise removal in digital holography by using deep learning. *New Journal of Physics* 24 (8): 083014.

28 De Oliveira, L.M.A., dos Santos, V.B., da Silva, E.K.N. et al. (2020). An environment-friendly spot test method with digital imaging for the micro-titration of citric fruits. *Talanta* 206: 120219.

29 Pandey, J.K., Yadav, A.K., Purushotham, M. et al. (2022). Colorimetric detection of deltamethrin pesticide from biological samples based on the peroxidase-mimic catalytic activity on MnO_2/SnO_2. *Key Engineering Materials* 928: 45–51. https://doi.org/10.4028/p-4ti1i0.

30 Talukdar, V., Veeraiah, V., Roy, D.R.et al. (2023). An innovative approach: typicality-based collaborative filtering for recommender systems. *Proceedings of the 3rd International Conference on Advancement in Electronics & Communication Engineering (AECE)*, pp. 863–868. https://doi.org/10.1109/AECE59614.2023.10428576.

31 Fan, K.J. and Su, W.H. (2022). Applications of fluorescence spectroscopy, RGB-and multispectral imaging for quality determinations of white meat: a review. *Biosensors-Basel* 12 (2): 76.

32 Beemkumar, N., Gupta, S., Bhardwaj, S. et al. (2023). Activity recognition and IoT-based analysis using time series and CNN. In: *Handbook of research on machine learning-enabled IoT for smart applications across industries* (ed. N. Goel and R.K. Yadav), 350–364. IGI Global https://doi.org/10.4018/978-1-6684-8785-3.ch018.

33 Umar, U.M., Ünal, Z., Gulzar, Y. et al. (2025). Deep learning-based classification of lavender for enhanced cultivation and essential oil quality management. *Journal of Essential Oil Bearing Plants* 28 (5): 1063–1085.

34 Aggarwal, 1.L.P. and Papay, F.A. (2022). Applications of multispectral and hyperspectral imaging in dermatology. *Experimental Dermatology* 31 (8): 1128–1135.

35 Huang, F., Chen, Y., Wang, X. et al. (2023). Spectral clustering super-resolution imaging based on multispectral camera array. *IEEE Transactions on Image Processing* 32: 1257–1271.

36 Huo, D., Wang, J., Qian, Y., and Yang, Y.H. (2023). Glass segmentation with RGB-thermal image pairs. *IEEE Transactions on Image Processing* 32: 1911–1926.

37 Brachmann, E. and Rother, C. (2022). Visual camera re-localization from RGB and RGB-D images using DSAC. *IEEE Transactions on Pattern Analysis and Machine Intelligence* 44 (9): 5847–5865.

38 Lopes, A., Souza, R., and Pedrini, H. (2022). A survey on RGB-D datasets. *Computer Vision and Image Understanding* 222: 103489.

39 Liu, Z., Wu, J., Fu, L. et al. (2020). Improved kiwifruit detection using pre-trained VGG16 with RGB and NIR information fusion. *IEEE Access* 8: 2327–2336.

40 Dhiman, P., Kaur, A., Balasaraswathi, V.R. et al. (2023). Image acquisition, preprocessing, and classification of citrus fruit diseases: a systematic literature review. *Sustainability (Switzerland)* 15 (12): 9643.

41 Haavardsholm, T.V., Skauli, T., and Stahl, A. (2020). Multimodal multispectral imaging system for small UAVs. *IEEE Robotics and Automation Letters* 5 (2): 1039–1046.

42 Crucil, G. and Van Oost, K. (2021). Towards mapping of soil crust using multispectral imaging. *Sensors* 21 (5): 1850.

43 Ma, F., Yuan, M., and Kozak, I. (2023). Multispectral imaging: review of current applications. *Survey of Ophthalmology* 68 (5): 889–904.

44 Liu, Y. (2021). Development of hyperspectral imaging remote sensing technology. *National Remote Sensing Bulletin* 25 (1): 4069.

45 Kang, Z., Zhao, Y., Chen, L. et al. (2022). Advances in machine learning and hyperspectral imaging in the food supply chain. *Food Engineering Reviews* 14 (4): 596–616.

46 Ma, J., Zheng, B., and He, Y. (2022). Applications of a hyperspectral imaging system used to estimate wheat grain protein: a review. *Frontiers in Plant Science* 13.

47 Albarrak, K., Gulzar, Y., Hamid, Y. et al. (2022). A deep learning-based model for date fruit classification. *Sustainability* 14 (10): 6339.

48 Gulzar, Y. (2023). Fruit image classification model based on MobileNetV2 with deep transfer learning technique. *Sustainability* 15 (3): 1906.

49 Zhang, T., Fu, Y., Huang, L. et al. (2024). RGB-guided hyperspectral image super-resolution with deep progressive learning. *CAAI Transactions on Intelligence Technology* 9 (3): 679–694.

50 Tlahig, S., Karmous, I., Ayeb, N. et al. (2024). Forage quality profiling of pre-selected alfalfa genotypes: towards genetic enhancement and adaptation for arid dryland farming. *Euro-Mediterranean Journal for Environmental Integration* 9: 1311–1323.

51 Aktaş, H., Kızıldeniz, T., and Ünal, Z. (2022). Classification of pistachios with deep learning and assessing the effect of various datasets on accuracy. *Journal of Food Measurement and Characterization* 16 (3): 1983–1996.

52 Tan, M. and Le, Q.V. (2019). EfficientNet: Rethinking Model Scaling for Convolutional Neural Networks, pp. 6105–6114.

53 Alom, M., Ali, M.Y., Islam, M.T. et al. (2023). Species classification of Brassica napus based on flowers, leaves, and packets using deep neural networks. *Journal of Agriculture and Food Research* 14: 100658.

54 Chaity, M.D. and van Aardt, J. (2024). Exploring the limits of species identification via a convolutional neural network in a complex forest scene through simulated imaging spectroscopy. *Remote Sensing* 16 (3): 498.

55 Gulzar, Y. (2025). PapNet: an AI-driven approach for early detection and classification of papaya leaf diseases. *Applied Fruit Science* 67 (4): 1–11.

56 Pan, X., Jia, N., Mu, Y., and Gao, X. (2023). Survey of small object detection. *Journal of Image and Graphics* 28 (9): 2587–2615.

57 Phang, S.K., Chiang, T.H.A., Happonen, A., and Chang, M.M.L. (2023). From satellite to UAV-based remote sensing: a review on precision agriculture. *IEEE Access* 11: 127057–127076.

3

Preprocessing and Enhancement Techniques for Nutritional Feature Extraction

Sima Tahmouzi[1], Farhang Hameed Awlqadr[3], Javaneh Karimi[2], and Neda Mollakhalili Meybodi[3]

[1] Research Center for Food Hygiene and safety, School of Public Health, Shahid Sadoughi University of Medical Sciences, Yazd, Iran
[2] Department of Food Science and Technology, School of Public Health, Shahid Sadoughi University of Medical Sciences, Sulaymaniyah, Iraq
[3] Department of Food Science and Quality Control, Halabja Technical College, Sulaimani Polytechnic University, Sulaymaniyah, Iraq

3.1 Introduction

Today, digital imaging is widely used in the food industry (e.g., agricultural products) quality by monitoring their biochemical characterization. Hyperspectral imaging (HSI) and multispectral imaging (MSI) are known as new, nondestructive platforms able to extract the spatial-spectral data from agricultural products to estimate pigment concentration, maturity, moisture content, and physiological stress parameters [1]. However, their preprocessing is needed as [2] proved that uncorrected image analysis in agricultural phenotyping pipelines can reduce predictive accuracy by about 15%.

Despite the scalable nutritional assessment across RGB cameras, multispectral sensors, and hyperspectral systems, machine learning automation and rigorous imaging uniformity are prerequisites for their successful analysis. Also, variation in imaging conditions due to sensor variations, calibration inconsistencies, and environmental effects compromises data consistency and consequently neglects integrated consistency as reported by [3]. Standardization steps, including radiometric calibration, denoising, geometric alignment, and color correction, are reported to eliminate objects affecting downstream efficiency. As reported by [4], using Contrast Limited Adaptive Histogram Equalization (CLAHE) and color registration are able to improve the support vector machine (SVM) accuracy from 78% to 97.9% in unmanned aerial vehicle (UAV)-based plant classification.

Raw agricultural images suffer from several degradation issues, including varying levels of sunlight, plant shadows in some images that affect brightness levels, and camera noise. Environmental conditions of dust, humidity, and winds create blur, which affects sharpness and color precision. These factors can alter important light measurements associated with plant health by 25% [5]. Plant variability adds complexity to the previous problems, especially differing shapes, colors, and leaf structures between species. Natural variability makes it difficult to compare images across time periods and locations. Ref. [6] showed that plant variability reduced computer performance by 10% points on a test of images captured at different times.

Often, these problems are addressed through a considerable amount of preprocessing to build confidence in the data for the same location and spectra over different acquisition events. Radiometric correction, geometric distortion removal, spectral matching, and noise reduction should ensure that at least some degree of uniformity (and meaning) is retained in the data [2]. Some industrial examples make this point well; [7] found >85% accuracy in an automated sorting version of their research on sun-burned and rotten apples utilizing HSI and machine learning to reduce their systematic rejection of product, and increase the ability to separate product.

Normalized reflectance ensures stable vegetation indices like the normalized difference vegetation index (NDVI), under variable lighting conditions [8], while accurate geometric registration becomes essential when combining MSI and HSI data [2]. Without these steps, models risk overfitting to artifacts rather than biological signals.

This chapter aims to investigate the imaging and reliable image-based nutritional analysis in the agriculture and food sector using image preprocessing and enhancement techniques.

3.2 Image Quality Challenges and Correction Methods

3.2.1 Common Sources of Distortion

Obtaining high-quality nutritional data from agricultural images is much more complicated than in controlled laboratory settings. Field conditions offer environments that are always changing and affect data quality in multiple ways, where changes in lighting are the most pronounced. Weather changes, time of day, and plant shadows can all impact the brightness and hue of images, and these changes can greatly impact measurements. Ref. [9] shows examples of how lighting influences color images; it also shows that weather changes will influence the measurement of plant health. Lighting control is much less likely in the field, but it is a critical requirement for accurate assessments.

Even identical-appearing RGB cameras have substantial internal differences in lens specifications, color filters, and processing systems, causing the same plant to appear different across camera systems. Ref. [10] compared a variety of plant health indices, including NDVI, among several brands of cameras, and they clearly showed differences, with reports generated by the MicaSense RedEdge and Parrot Sequoia drones for the same crops being markedly different. This necessitates empirical correction protocols for each camera system. More recently, work by [11] gave us some promise on how we might improve some of the issues we have created by showing how they tied the variability in color across cameras using a frequency-separation approach in their application, though each camera system still required separate mapping.

The conversion of RGB photos to hyperspectral images increases quality concerns as these processes are reliant upon quality source images. Ref. [12] helped explain this concern, indicating that small color incorrectness in the RGB image would inherently reproduce incorrectness calculated from the RGB image that is used to estimate plant nutrients. Weather complicates these effects by inducing additional blur via wind or lens contamination via dust or humidity, combined with temperature-induced spectral effects. Ref. [13] carried out an inclusive study with a fixed camera system, in-field research devices positioned in farm fields, and noted NDVI measurement discrepancies for plant health as large as 79% implicated these weather extractions, and employed artificial intelligence (AI) techniques to mitigate weather effects. Ref. [5] reviewed many studies and concluded that atmospheric influences can change the light reflected by the plant structure by 10–25%.

Different crops exhibit varying shapes, textures, and colors that affect light interaction patterns. As reported by [14], light and reflectance imperatives change the ability to properly evaluate specific crops from general methods more broadly across the uncontrolled input parameters of diverse crop types or complex growing conditions. Ref. [15] reported that to produce reliable nutrition imaging, the camera and atmospheric confounding variables need to be fixed, emphasizing these corrections as requirements rather than options. Machine learning models have been found to sensitively react to brightness, type, or texture changes, which necessitates comprehensive error correction. It is worth noting that reliable models are not achievable without the imaging variations like lighting, camera type, weather effects, and crop variations.

3.2.2 Camera Noise

All cameras innately create "noise" during their imaging, which is induced by physical and electronic processes. To develop an effective correction strategy, it is necessary to understand how different cameras make distinct noise (Table 3.1). Multiple noise sources are reported in digital cameras, i.e., thermal, shot, fixed-pattern, and quantization noises, each prone to degrade the image quality differently. While shot noise stems from position distribution on the light intensity reaching the sensor, the thermal and readout noises are induced by camera electronics and amplifiers. Unique noise patterns are achieved by the complementary metal oxide semiconductor (CMOS) and a charge-coupled device (CCD) sensors, which become distinct when exposed to low-light situations or extended exposure time. The recent study by [16] has revealed decisive knowledge to design better noise reduction systems and determined what these patterns are.

In a study conducted by [17], extensive detail about the noise of remote sensing and scientific imaging has been investigated, and shot noise, fixed pattern noise, and stripe artifacts are known as the main ones. Understanding the noise characteristics is described as essential for feature extraction.

Table 3.1 Comparison of dominant noise types across RGB, MSI, and HSI modalities.

Modality	Noise type	Sources	Impacts	Mitigation
RGB	Shot noise	Photon statistics	Grainy texture	Temporal averaging
MSI	Spectral misreg	Lens displacement	Spectral band distortion	Band registration
HSI	Stripe noise	Focal plane array variability	Column artifacts, loss of fidelity	Tensor decomposition

Thermal noise is considered the main noise source in RGB cameras, which is generally induced by electronic fluctuations, shot noise owing to photon detection, and quantization errors in analog-to-digital conversion. These difficulties create low-light situations and grainy and/ or blurred areas, which compromise color precision and fine detail [18] (Table 3.1).

Multispectral cameras are prone to displaying different noise levels across spectral bands, especially in lower-quality systems containing multiple lenses. As reported by [5], potential brightness variations of about 25% have been observed, with 10–15 dB of noise between the red, green, and near-infrared bands when photographing identical objects. This variation leads to two critical problems: degradation of direct image quality and reduced accuracy in vegetation health metrics, e.g., NDVI. The resulting "band weighting bias" occurs when computational analysis incorrectly weights spectral bands, compounding errors in multi-band analyses and requiring either independent band adjustment or extensive camera-specific calibration (Table 3.1).

Despite significant noise concerns, exceptional details have been provided by hyperspectral images containing hundreds of narrow spectral bands (Table 3.1). Considering the complete spectral signature of each pixel, minor electronic issues become readily apparent. Most recently, [14, 19] noted that striping is the greatest concern for hyperspectral data, requiring specialized computational correction methods that add processing costs. Each noise type compromises spectral accuracy and impacts subsequent data analysis quality.

3.2.3 Standardization of Images Before Analysis

Image standardization is critical for reliable remote sensing in precision agriculture, as variations in brightness, lighting, or sensor response reduce machine learning models accuracy [9]. Radiometric normalization adjusts brightness differences from lighting or camera settings, often converting raw multispectral/hyperspectral values to true reflectance using white panels and light sensors (Table 3.2). Ref. [20] achieved $R^2 \approx 0.99$ for red-edge and NIR bands despite changing daylight by calibrating multiple times daily. Ref. [21] found the Empirical Line Method with multiple targets minimized bias vs. single grey panels, though the latter sufficed for many farm uses.

Geometric correction becomes crucial for drone imagery, where motion between captured images requires proper registration. To enable accurate pixel comparisons, all images

Table 3.2 Summary of image standardization techniques across modalities.

Type	Modalities	Software	Method	Goal
Radiometric	MSI, HSI	ENVI, Pix4D	White panel + irradiance	Reflectance calibration
Geometric	UAV it RGB/MSI	QGIS, Pix4D	GCP + RTK/PPK	Orthorectified images
Color calibration	RGB	OpenCV, ENVI	Color checker + ΔE_LAB	Device-independent color
Spectral alignment	MSI, HSI	MATLAB, Python	Mutual info/ correlation	Band-wise registration

need to be transformed to the same coordinate space by registration and rectification processes. These corrections are facilitated by software packages like ENVI, Pix4D, and QGIS, as well as radiometric calibration as reported in Table 3.2. A significantly enhanced orthophoto reliability has been observed in RTK/PPK positioning and ground control points, achieving the geometric accuracies below 1 m (RMSE <0.7 m) [22].

RGB color calibration using the conversion of pixel values to canonical color spaces like CIE Lab is necessary for colors to match real-world object colors. In a study conducted by [23], photographing color checker charts under identical lighting conditions and computing transformation matrices has been used for color calibration. Results indicated a ΔElab score of less than 1.8, which indicated an acceptable color accuracy in farm applications. Controlled lighting and color standards have also been used by [24] to calibrate their images, which show more accurate measurements of color characteristics in plants. Spectral band alignment is critically important for multispectral and hyperspectral systems, as minor misalignments can cause spatial variations. Sub-pixel level band correspondence is observed by image registration algorithms using mutual information or cross-correlation techniques, as reported in Table 3.2 [25].

Standardizing the information about the image, known as sensor type, acquisition angle, timing, and environmental conditions, is also critically important to provide replication. Ref. [26] states that agriculture-focused communities need to agree on ag-specific metadata standards like AgMES, enabling universal data usability across systems and machine learning pipelines. Ref. [27] mentions that using FAIR metadata frameworks is a great initiative, allowing agricultural research datasets to be expressed in discoverable and relatable ways across varying studies. Without proper standardization encompassing radiometric, geometric, color, and metadata corrections, nutrient estimation models face significant challenges when applied across different farms or conditions, making these preprocessing steps fundamental for reliable agricultural image analysis.

3.2.4 Importance of Nutritional Feature Extraction

Combined distortions, noise, and inconsistencies significantly alter measurements of crucial nutritional traits, including color (indicating ripeness or pigment concentration), texture (associated with freshness or firmness), and shape (used for defect detection or variety classification). These issues cause feature extraction algorithms to learn biased patterns, misinterpret variability due to poor signal-to-noise ratios, or to fail. Ref. [28] found that systematic misclassification of traits can originate from radiometric and sensor biases in high-throughput plant phenotyping workflows, which made heavy use of color- and texture-based attributes for classification tasks [29]. Similarly, deep learning models focused on stress detection often learn features from illumination or acquisition artifacts rather than physiologically meaningful plant traits when proper preprocessing is absent.

Therefore, image quality processing steps are fundamental for reliable, scalable, and accurate nutrition-focused imaging pipelines. Proper corrections enable comparability across different platforms, seasons, and environmental conditions, which becomes increasingly critical as datasets accumulate diversity in data provenance, sensor types, and acquisition conditions.

3.3 Image Denoising and Enhancement

Denoising (reducing noise and clarifying key details) is considered critically important in the improvement of data quality in images captured in agricultural and food systems. Artifacts observed in raw images may hide important details about the nutritional state. Main noise reduction techniques [30] are discussed in detail in the next section. Also, methods for increasing the contrast, normalizing color, and sharpening features, which help improve the accuracy of subsequent segmentation and analysis (Figure 3.1), are described.

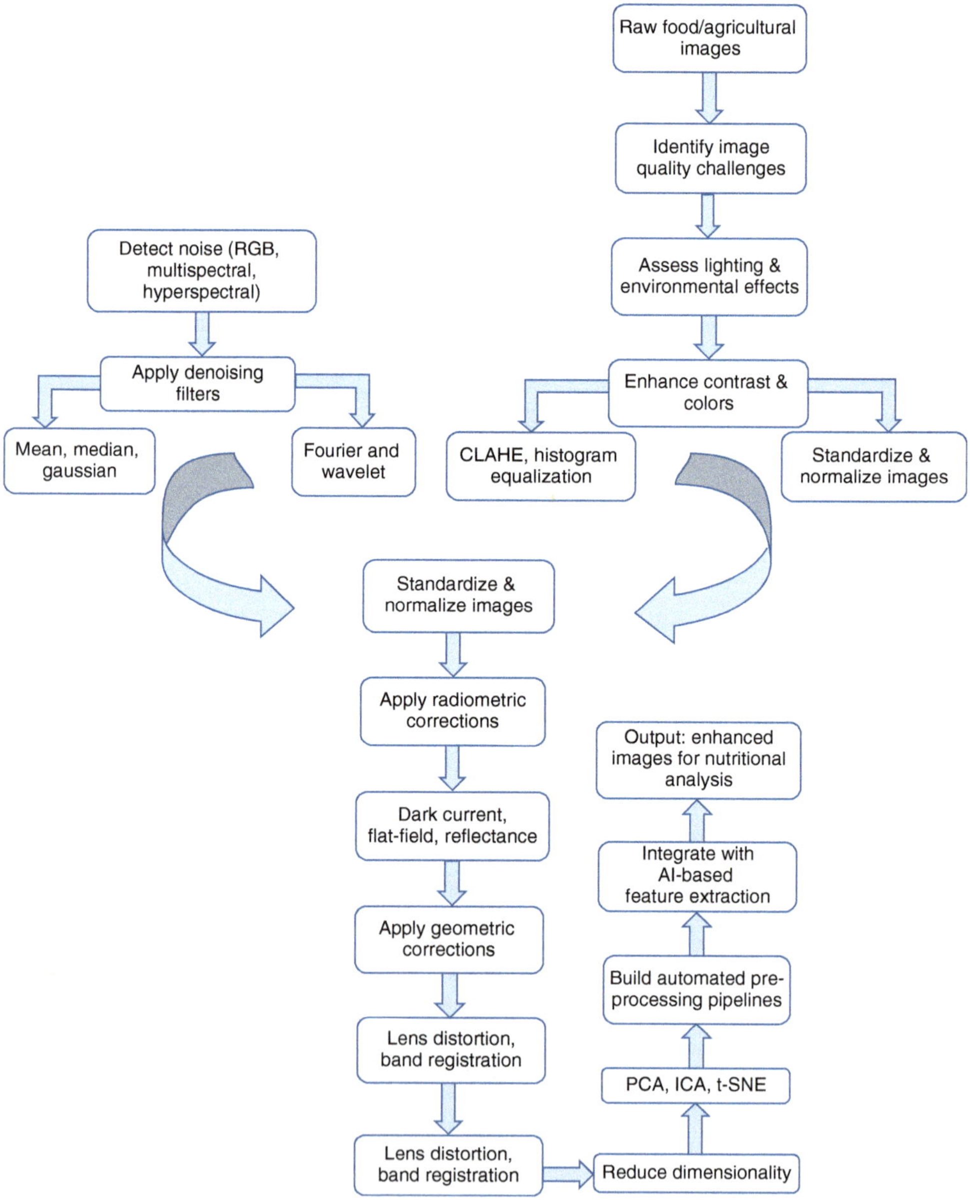

Figure 3.1 Image preprocessing workflow.

3.4 Spatial and Frequency Domain Noise Reduction

Noise reduction is generally aimed at removing undesirable random fluctuations or structured intrusions in the image while keeping the morphological details necessary for nutrient identification [31].

- **Spatial domain filtering** works directly on pixel regions to replace the noisy ones [32]. The most common spatial filters are mentioned in Table 3.3.
  - **Mean Filter:** Calculates the average intensity of pixels within a window and replaces the center pixel. Despite the efficiency of this technique to remove Gaussian noise, it can blur edges and reduce fine textures, which consequently hinders the detection of subtle signs of nutrient deficiencies.
  - **Median Filter:** It is an effective strategy to remove salt-and-pepper noise by replacing a pixel with the median of its neighboring ones. It enables precise segmentation of plant arts by keeping the sharpness of boundaries [33].
  - **Gaussian Filter:** It reduces high-frequency noise by making a smart shadow on the image to remove fine-grained noise. Consequently, it keeps the sharpness of important edges and smooth transitions, which are important for precisely observing the shape/structure of plants.

Table 3.3 Noise reduction techniques in the spatial and frequency domains.

	Domain	How it works	Advantages	Limitations
Mean filter	Spatial	Average pixel intensities within a neighborhood	Reduces Gaussian noise	Blurs edges, loses fine nutrient textures
Median filter	Spatial	Replaces pixel with the median of neighboring values	Removes salt-and-pepper noise without blurring edges	Less effective for Gaussian noise
Gaussian filter	Spatial	Weigh pixels by a Gaussian function for smoothing	Reduces high-frequency noise while preserving gradual transitions	May still blur important edges
Low-pass filtering	Frequency	Removes high-frequency components (noise)	Smooths the image and reduces noise	Overly aggressive filtering blurs key features
Bandpass filtering	Frequency	Keeps only specific frequency ranges	Preserves features relevant to nutrients while suppressing noise	Requires careful tuning; may lose important info
Wavelet denoising	Spatial + frequency	Decomposes the image into multiscale wavelet coefficients; thresholds the noise.	Maintains spatial detail, effective for hyperspectral images	Computationally intensive
Non-local means (NLM)	Spatial	Uses patch similarity across the image to average noise selectively	Preserves fine textures and details	Computationally expensive

- **Frequency domain filtering** involves transforming the image with the Fourier transform to break it down into its frequency components [34]. In this domain, noise is typically linked to high-frequency signals, but nutrient characteristics can be observed across a range of signals.
- **Low-pass filtering** removes high-frequency noise, but when overused, it can blur important edges and textures vital for recognizing nutrient patterns.
- **The bandpass filtering** process then selectively preserves frequencies related to nutritional attributes while removing redundant parts.
- **Wavelet denoising** is especially useful in hyperspectral imagery by breaking an image down into multiple scales and spatial locations of wavelet quantities. Then, noise is removed, and spatial details are preserved [35]. This technique improves the signal-to-noise ratio, which is critically important for identifying spectral features in nutritional images.

These filters need a careful parameter alteration, as nutritional indicators often overlap with noise in frequency bands.

More advanced techniques like non-local means (NLM) give even more power to patch similarity across the image, allowing for selective smoothing of noise while preserving texture and small nutrient details [36]. This means that the best noise reduction technique depends on the type of sensors, the noise characteristics, and the features of the crop or food of interest. Methods used in the spatial and frequency domains often provide better results when combined, as each method has complementary strengths.

3.5 Contrast Enhancement and Histogram Equalization (Including CLAHE)

Poor contrast can be problematic because nutritional factors, often pigment distributions, ripening gradients, or disease symptoms, may not be distinctly visible under poor field lighting or sensor performance. Increasing contrast accentuates these elusive variations and makes them easier to detect by either automated systems or human observers.

- **Histogram equalization (HE)** reorders the pixel values to span the entire range [37]. HE enhances contrast globally by increasing the intensities of both darkness and brightness. Consequently, the hidden details that may be under- or over-exposed will appear. The overamplification of the background with blurring of the edges is considered the main drawback of HE.
- **Contrast Limited Adaptive Histogram Equalization (CLAHE)** prevents over-processing by improving the contrast of the image in its small sections separately [38]. This localized method is spatially adaptive because it responds to spatially varying light conditions (often seen in field photos with highlights and shadows), helping retain fine details and reduce noise exaggeration. Applying CLAHE enhances nutrient-related features, such as chlorophyll concentration gradients or lesion borders, which simplifies their subsequent segmentation or categorization.

3.6 Color Normalization, Sharpening Filters, and Visibility Improvement for Segmentation

Edge sharpness and color fidelity are critical to clearly delineate nutrient characteristics, which, in the case of agricultural and food imagery, more frequently appear as discrepancies in texture or color (Table 3.4).

- **Color Normalization:** Variations in illumination spectrum, camera sensor sensitivity, and other environmental factors can lead to differences in color representation across images [39]. Color normalization involves adjusting image colors to match a reference image, making the images more comparable.
  - **White balance** adjustments eliminate lighting distortions and will result in a more accurate display of colors related to nutrient content.
  - **Reference-based Calibration:** Many calibration methods map color to standard color spaces using color charts or targets that are co-acquired in the image, along with relational environmental factors to green, such as images of the same co-acquired location with variations in relation
- **Nutrition sharpening filters** also help add detail and structure to edges and textures to improve nutrient feature delineation.
  - An **unsharp mask** differs from the original by using a blurred (smoothed) copy of the image, which enhances high-frequency elements that represent edges and other small details such as spots of nutrient deficiencies, areas of ripeness, or blemishes of spoilage [40].
  - **Laplacian sharpening** accentuates sudden intensity changes to create sharp boundaries, resulting in more accurate segmentation masks.

Although sharpening improves the visibility of features, caution should be taken to prevent amplifying noise or creating artifacts that could mislead the algorithm used for nutrient classification. Overall, color normalization combined with sharpening significantly

Table 3.4 Color normalization and sharpening for segmentation.

Technique	Category	Purpose	Advantages	Limitations
White balance adjustment	Color Normalization	Corrects color cast from lighting	Improves true color representation	Dependent on lighting consistency
Reference-based calibration	Color Normalization	Maps colors to a standard reference (e.g., color charts)	Ensures comparability across datasets	Requires reference targets
Unsharp masking	Sharpening	Subtracts blurred version to emphasize edges	Highlights nutrient spots, ripening zones	Can amplify noise if overused
Laplacian sharpening	Sharpening	Highlights abrupt intensity changes	Improves edge clarity for segmentation	May create artifacts in noisy images

Table 3.5 Summary of studies that used image denoising and enhancement for nutritional imaging.

Category	Key methods	Main benefit	References
Noise reduction	Mean, Median, Gaussian, Wavelet, NLM	Removes unwanted noise while preserving nutrient details	[41]
Contrast enhancement	HE, CLAHE	Enhances visibility of subtle features	[42]
Color normalization	White Balance, Calibration	Standardizes color representation across images	[43]
Sharpening	Unsharp Masking, Laplacian	Improves edge definition for segmentation	[44]

boosts the visual and quantitative contrast of nutrient-related areas, which directly enhances the effectiveness of image segmentation, defect detection, and quantitative nutritional analysis.

To summarize, the nutritional analysis through image could be more effective if denoising and enhancement of images had occurred. While median, Gaussian, and wavelet filters are influential in reducing noise, they can preserve the nutrient characteristics. Histogram equalization techniques, such as CLAHE, which improve contrast, can expose low-level nutritional gradients and consequently better extract the features. Color normalization keeps the images more consistent and improves the accuracy, robustness, and simplicity of nutritional characteristic extraction in complex matrices when applied carefully (Table 3.5).

3.7 Radiometric and Geometric Corrections

In food quality assessment based on images, reliable results have been obtained by proper data preprocessing, especially radiometric and geometric corrections. Radiometric correction focuses on the problems induced by sensor variability, object shape, and fluctuating illumination conditions. In contrast, geometric correction concerns distortions from camera movement or lens issues [45, 46].

3.7.1 Importance of Calibration for Quantitative Imaging

Radiometric calibration, which is essential for accurate and consistent food quality analysis, changes the raw image data from sensors into measurable units like radiance or reflectance [47, 48]. Proper calibration guarantees that results obtained from different sensors, lighting conditions, and sessions are comparable [49]. Without it, errors in detecting or classifying food characteristics are caused by changes in lighting, sensor response, or atmosphere [50].

In UAV-based imaging, data reliability may be induced by changing light and sensor instability. However, using reflectance calibration is efficient to obtain consistent results for crop health and nutrient analysis. Without such corrections, mosaicked images may provide uneven brightness in images and consequently lower the accuracy [51].

However, traditional calibration methods are best suited to controlled labs and often ineffective during fluctuating outdoor conditions. Environmental factors, e.g., changes in sunlight intensity, airborne particulates, temperature, and humidity, can reduce the image accuracy, especially in UAV or mobile systems, which need real-time results [52]. To solve this, new adaptive calibration techniques have been developed.

3.7.2 Radiometric Corrections (Dark Current Subtraction, Flat-field, Reflectance Calibration)

The accuracy of hyperspectral data analysis rigorously depends on radiometric preprocessing, which removes noise induced by sensors and lighting conditions [53]. Radiometric correction techniques, including dark current subtraction, flat-field correction, and reflectance calibration, are critical for ensuring the accuracy of data [54].

Dark current subtraction is aimed at eliminating baseline noise induced by sensor electronics. This is typically performed by capturing dark images with the lens covered and subtracting their average from the raw data. This step confirms that the image reproduced the true radiance or reflectance of the sample [55].

Flat-field correction is done to fix uneven brightness and sensor sensitivity by dividing the raw image by a reference image of a uniformly lit white surface. Consequently, the brightness across all pixels will be standardized [46].

Reflectance calibration, as the final step in radiometric preprocessing, converts raw sensor data into definite surface reflectance values through calibration targets with known properties (e.g., the Empirical Line Method). This makes the data across sessions and instruments comparable [56]. The comparative overview of commonly used radiometric correction techniques in HSI is summarized in Table 3.6. Additionally, Figure 3.2 demonstrates the radiometric correction workflow, used for food and agricultural applications, commonly employed in HSI systems.

However, the use of HSI in food quality testing and agriculture is growing; the lack of standardized methods for radiometric correction is still known as the main challenge [57]. Different researchers and manufacturers use different correction techniques (i.e., dark current subtraction, flat-field correction, and reflectance calibration), which consequently lead to inconsistent spectral data across labs and devices. As a result, comparing the results,

Table 3.6 Common radiometric correction techniques used in hyperspectral imaging (food and agricultural applications).

Technique	Input required	Environment	Strengths	Limitations
Dark current subtraction	Dark frames	Lab/Field	Reduces baseline noise	Needs control frames
Flat-field correction	Uniform target	Lab	Normalizes illumination	Sensitive to angle
Reflectance calibration	Reference target	Field	Enables comparability	Requires accurate targets

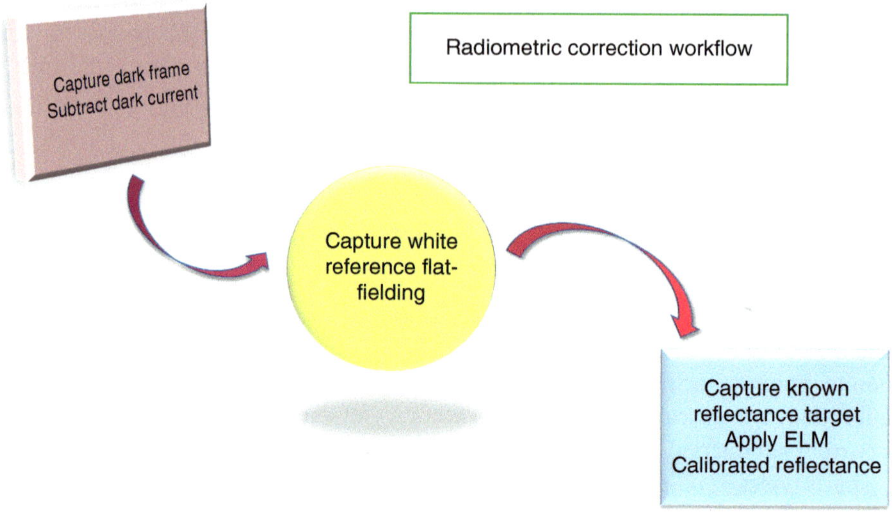

Figure 3.2 Radiometric correction workflow used in hyperspectral imaging for food and agricultural applications.

reproducing the studies, and building large, shared spectral databases for food quality assessment will be hard [58].

Despite the difficulty of using new automated and data-driven calibration methods (e.g., bidirectional reflectance distribution function [BRDF] correction), demonstrating capacity, they are not widely used because of their technical complexity and lack of clear guidelines. Radiometric corrections are also influenced rigorously by sensor type and environmental conditions, which limit their reliability in real-world settings. Sensor-specific factors like viewing geometry and spectral sensitivity further reduce the transferability of correction models between systems [29, 59].

3.8 Geometric Corrections (Lens Distortion, Sensor Geometry) and Image Registration Across Spectral Bands

Geometric correction is critically important for food quality and agricultural imaging to mitigate the shape alterations from cameras, sensors, or terrain. Common lens distortions, such as barrel and pincushion effects, are amended by considering the intrinsic camera parameters like focal length, principal point, and distortion coefficients commonly through tools such as OpenCV [60]. External factors like satellite motion, Earth's rotation, or topographic variations may also lead to geometric errors. However, these distortions can also be fixed by ground control points (GCPs) and polynomial transformations to align images with real-world coordinates [61]. Through HSI and MSI, slight misalignments between spectral bands can significantly affect the accuracy of classification, defect detection, and biochemical analysis. To ensure the precise pixel matching across wavelengths, different registration techniques align these bands either using feature-based (e.g., SIFT, SURF, Harris corner)

Table 3.7 Comparison of image registration methods used in hyperspectral and multispectral imaging for agricultural applications.

Method type	Techniques	Input needed	Advantages	Disadvantages
Feature-based	SIFT, SURF, Harris	Key points	Accurate for textured scenes	Poor in low-texture areas
Intensity-based	Mutual Info, NCC	Full image	Works on low-texture images	Computationally expensive
Deep learning-based	MU-Net, Siamese CNN	Image Pairs	Learns robust matching	Needs training or pretraining

or intensity-based (e.g., mutual information) methods [62, 63]. The summary of common image registration methods in HSI and MSI is provided in Table 3.7.

However, fully automated registration methods remain challenging under noisy or low-texture conditions. Along with the necessity of manual adjustments for traditional algorithms, struggling with sensor drift or lighting changes also exists. To address this, newer methods like the Hyperspectral Stacked Feature Detector (HSI-SFD) also exist, which use redundant spectral data to improve feature matching, while 3D convolutional Siamese networks increase registration in low-contrast or repetitive-texture scenes [64, 65].

Crop fields often have repetitive pattern scenes and few discrete features, which reduces the performance of standard methods, e.g., SIFT and SURF. To improve this, enhanced SIFT-based algorithms have been developed by researchers, which add gradient-based filtering and Gradient Location and Orientation Histogram (GLOH) descriptors and consequently increase the robustness and speed in complex agricultural environments [66]. 3D Siamese networks have also shown better accuracy in aligning low-resolution hyperspectral images under difficult field conditions [65].

3.8.1 Data Reduction and Automated Pipelines

The growing imaging technologies have facilitated the collection of more detailed nutritional characterization in the agri-food industry. HSI is notable in this regard, as it captures continuous spectral information across hundreds of wavelengths. Despite the efficiency of this method in providing rich data for nondestructive analysis of nutrients and food quality, it is also challenging, induced by the high dimensionality of the data [67]. Too many spectral bands may also lead to computation time and, consequently, add noise or redundancy, which hides the important nutritional details [68]. In order to make processing faster and more accurate, data reduction is essential.

As the enhanced volume and velocity of data achievement imaging systems create larger and faster data streams, automated preprocessing pipelines have become necessary. These systems are generated to handle tasks such as noise reduction, calibration, enhancement, geometric correction, and registration, which convert raw images into more reliable nutritional information [69]. Automation ensures consistency, scalability, and real-time performance in agro-food industry applications. Particularly, AI and machine learning approaches for analyzing nutrition in detail are increasingly complemented by automated processes. Data reduction, combined with AI-driven feature extraction, improves the precision in nutrient recognition, model adaptability to various conditions, and facilitates faster decision-making. This chapter reviews the main data reduction techniques crucial for hyperspectral nutritional data, the design and challenges of automated pipelines, the application of AI, and future developments anticipated to shape this evolving field [70].

3.8.2 Dimensionality Reduction Techniques for Hyperspectral Data

- **Principal Component Analysis (PCA)**

  One fundamental dimensionality reduction technique used in HSI for nutritional analysis is PCA. It works by applying a linear transformation that maps high-dimensional spectral data to new coordinates with orthogonal principal components (PCs), which account for decreasing amounts of the dataset's variance [71]. PCA effectively compresses data while minimizing information loss because it retains the main components that explain the majority of the variance (Table 3.8).

  In nutritional imaging, PCA reduces computational complexity and storage needs, enabling much faster processing pipelines. This is essential for real-time nutritional monitoring, such as continuous field monitoring [72]. Notably, PCA can also be used to eliminate noise, as signal-heavy components can be distinguished from those that contain noise but are mainly dominant in low-variance PCs. The noise reduction provides a clearer view and analysis of nutrient-related spectral patterns associated with biochemical components, such as chlorophyll, carotenoids, or moisture content.

  In addition to compression, the main components are often linked to physiological or chemical features, making them easier to interpret in a nutritional context. For example, some of those PCs could correspond to spectral bands sensitive to nitrogen or protein levels in crops, which could help in selecting features for future predictive models. However, the PCA method requires a linear approach, which limits its ability to capture the complex, nonlinear relationships in biological systems; in such cases, more advanced or alternative methods may be necessary.
- **Independent Component Analysis (ICA)**

  ICA extends dimensionality reduction by also identifying sources, specifically statistically independent source signals, from observed mixed spectral data. Unlike PCA, which emphasizes maximizing variance, ICA is used to decompose mixed hyperspectral data into independent component signals, ideally corresponding to pure substances of biochemical or physical nature within the sample [73]. Using ICA to estimate nutrient content offers significant advantages in separating overlapping spectral signals in plant and food materials. For example, spectral responses influenced by a combination of chlorophyll, water, and anthocyanins can be divided into individual sources, enabling more precise determination of specific nutrients. This separation reduces the interference caused by spectral mixing, which can disrupt traditional analysis methods. In nutrition imaging, ICA is practically useful because it can detect biochemical indicators related to nutrient deficiencies or decomposition chemicals.

Table 3.8 Dimensionality reduction techniques for hyperspectral data.

Technique	Type	How it works	Advantages	Limitations	Practical applications
Principal component analysis (PCA)	Linear, variance-based	Computes eigenvectors of the covariance matrix and projects data onto principal components ordered by explained variance. Retains the first few PCs to represent the majority of information.	Reduces high-dimensional spectral data to a manageable size. Removes redundant bands. Segregates noise-dominated low-variance components. Speeds up downstream AI models.	Linear assumption cannot capture complex biological variability. PCs are abstract and may lack a clear physical meaning. Sensitive to scaling and outliers.	Identifying nutrient-related spectral features (e.g., chlorophyll, carotenoids). Preprocessing before classification of crop nutrient status. Field-based real-time nutrient monitoring.
Independent component analysis (ICA)	Linear, statistical independence	Decomposes hyperspectral data into independent source signals based on higher-order statistical independence.	Separates mixed signals, e.g., overlapping spectra from water, pigments, and structural components. Reveals hidden biochemical constituents. Improves nutrient quantification accuracy.	Requires assumptions about data distribution. Sensitive to noise and initial conditions. May yield nonunique solutions without constraints.	Isolating nutrient deficiency markers in leaves. Detecting spoilage compounds in food quality analysis. Differentiating biochemical pathways in plant physiology.
t-Distributed stochastic neighbor embedding (t-SNE)	Nonlinear manifold learning	Maps pairwise similarity of points from high-dimensional to low-dimensional space, preserving local structure for visualization.	Reveals clusters, outliers, and natural groupings. Highlights subtle spectral differences between nutrient states. Useful for exploratory data analysis.	Not scalable for massive datasets. Nondeterministic, requires careful parameter tuning. Cannot directly produce features for modeling.	Visualizing nutrient clusters in hyperspectral space. Differentiating healthy vs. stressed crops. Guiding feature selection for classification models.

The ability of ICA to detect hidden spectral factors improves the quality of nutrient signals, interpretation, and the accuracy of quantitative analysis. However, ICA algorithms can be prone to incorrect assumptions about data distribution and may require proper preprocessing to ensure convergence and meaningful results [74].

- **t-Distributed Stochastic Neighbor Embedding (t-SNE)**
  t-SNE is also a nonlinear dimensionality reduction tool. However, it is designed to work well when data are visualized in two or three dimensions [75]. Unlike PCA or ICA, where feature extraction is often the main focus, t-SNE is particularly useful for identifying very local structures and clusters in various spectral data. This makes it highly valuable in exploring data in nutritional research. t-SNE helps reveal existing groupings or even anomalies based on nutrient content, crop type, or food quality category by transforming hyperspectral data into two- or three-dimensional space. The resulting clustered spectral profiles may show trends that can be used to develop classifications or identify outlier samples for further study [76].

  Although successful as a visualization and clustering tool, t-SNE is not very scalable. When working with large hyperspectral datasets in fully automated pipelines, t-SNE cannot always be used. Its nonparametric nature and lack of explicit feature transformations also make it unsuitable for direct use in predictive modeling pipelines to build feature extractors. As a result, t-SNE is typically used as a complementary tool rather than as a primary reduction technique in operational nutritional imaging systems.

3.9 Designing Automated Preprocessing Workflows for High-throughput Imaging

- **Components of Automated Workflow Pipelines**
  Modern nutritional imaging applications require automation to meet the demands of throughput and reproducibility in preprocessing [77]. These automated pipelines consist of modules with linear components that process raw images in sequential steps. Common parts include (Table 3.9):
- **Noise Removal:** Applying spatial and frequency filters to eliminate noise without losing nutrient features.
- **Calibration:** SENSOR outputs across different sessions and devices will be standardized through radiometric calibration methods such as dark current subtraction, flat-field correction, and reflectance scaling.
- **Enhancement:** Standardize colors and contrast adjustments to improve visibility and ensure feature consistency.
- **Geometric corrections** involve adjusting for optical distortions in lenses and aligning images across spectral bands or modalities.
- **Registration:** Accurate co-registration of multispectral or hyperspectral data cubes to ensure spatial alignment of spectral data.

The modular design offers flexibility in terms of sensor types, imaging conditions, and the objectives of specific nutrient testing. Depending on deployment needs, these pipelines can be implemented as sequential scripts, integrated software platforms, or cloud services. It is

Table 3.9 Automated preprocessing workflow components and challenges.

Workflow component	Detailed function	Challenges and considerations	Practical examples
Noise reduction	Removes random noise and environmental interference while preserving spectral details relevant to nutrients. Uses spatial filters (Median, Gaussian) or frequency-domain filters.	Noise profiles vary across sensors and bands. Must balance noise removal with detail preservation. Hyperspectral data has high spectral correlation, requiring advanced denoising like wavelets.	Removing salt-and-pepper noise in UAV crop images to detect nutrient stress.
Calibration	Includes dark-current subtraction, flat-field correction, and reflectance standardization to remove sensor-specific biases.	Calibration targets may not always be available in field conditions. Sensor drift and varying illumination complicate consistency.	Correcting raw hyperspectral cubes for reflectance calibration in nitrogen analysis.
Enhancement	Improves visibility of subtle nutrient features via contrast stretching, CLAHE, and color normalization.	Over-enhancement can introduce artifacts. Needs adaptive approaches for variable field lighting.	Enhancing leaf chlorophyll gradients to detect nitrogen deficiency.
Geometric corrections	Corrects lens distortion, perspective shifts, and aligns images from different spectral bands.	UAV movement and vibration cause misalignment. Requires precise ground control points for correction.	Aligning multispectral drone images for precise nutrient mapping.
Registration	Ensures precise spatial alignment between multiple modalities (e.g., RGB, hyperspectral, and thermal) for accurate fusion.	Complex for large scenes with parallax. Misregistration leads to inaccurate feature fusion.	Aligning hyperspectral and thermal imagery for ripeness or spoilage detection.

essential to maintain adjustable parameters that accommodate various data collection conditions and dataset attributes, and to incorporate logging, quality control, and error identification functions [78].

- **Challenges in Automation for Agri-food Systems**
  There are specific issues related to automation in agri-food nutritional imaging that require careful resolution. The main challenge is the changing environment, where variations in brightness, weather, and leaf quality affect image quality and sensor responses unpredictably [79]. These factors challenge fixed-parameter pipelines and highlight the need for adaptive preprocessing that can respond flexibly to input variations. Additionally, differences in sensor hardware, such as spectral sensitivity, spatial resolution, and noise characteristics, complicate standardization and require the use of modular calibration and normalization modules designed specifically for each device.

The large amounts of data generated by HSI pose challenges related to extensive storage, transmission, and slow processing times. Field monitoring or industrial food lines based on high-throughput imaging require pipelines that are efficient and robust, capable of operating without human intervention [80]. Fault tolerance and error management strategies, along with automatic quality control measures, are essential to keep the pipeline running 24/7. These include procedural error detection, remedial proposals, and foolproof stopping mechanisms to prevent erroneous records from contaminating the study process.

- **Scalability and Real-time Processing**
  Scalability and real-time capability are crucial for applications such as precision agriculture decision support and real-time food quality monitoring [81]. Some methods used to achieve these needs include:
  - **Parallel processing and hardware acceleration** involve using multicore CPUs, GPUs, or dedicated processing units to speed up computationally intensive tasks that occur before the primary process in the preprocessing stage, such as denoising, calibration, and dimensionality reduction.
  - **Edge Computing:** Performing computations close to imaging sensors, especially in field environments with limited connectivity, to reduce latency and data transfers.
  - **Internet of Things (IoT)-enabled Sensor Networks:** Combining sensor arrays, real-time data integration, and preliminary processing to support end-to-end, automated monitoring of nutritional programs.

Together, these methods minimize bottlenecks caused by data size and complexity in processes, ensuring that pipelines can manage the input/output characteristics of entire farm imaging systems both in terms of data rates and computational loads that produce insights, often within minutes.

3.9.1 Integration with AI-Based Feature Extraction

- **Preprocessing Impact on AI Model Performance**
  High-quality, preprocessed data with built-in cleaning features is essential for successfully implementing AI or deep learning techniques to extract nutritional information [82]. Raw data can be highly inaccurate due to noise and inconsistencies, which can greatly reduce model accuracy and lead to incorrect nutrient estimates or misclassifications. Preprocessing also normalizes input, minimizes irrelevant variation, and highlights important aspects, thereby improving training efficiency and prediction reliability.

  The redundancy method of data reduction, such as PCA, helps manage the curse of dimensionality by removing redundant and noisy data. This prevents oversaturated training and reduces computation time in machine learning algorithms. The resulting smaller feature spaces enable faster convergence and improved generalization across datasets with different acquisition conditions. Postprocessing and applying AI models to these results can then more accurately identify nutrient-specific areas and measure biochemical components, whether they are deficient or spoiled. For example, convolutional neural networks (CNNs) trained on enhanced hyperspectral cubes can accurately and efficiently classify nutrients at both pixel and object levels (Table 3.10).

Table 3.10 AI integration and multimodal data fusion.

Aspect	Detailed description	Benefits	Real examples
Impact of preprocessing on AI	Noise removal and dimensionality reduction produce cleaner input data, improving AI model accuracy and reducing overfitting.	More robust nutrient classification. Faster model convergence. Better generalizability across conditions.	CNNs for predicting leaf nitrogen after PCA-based reduction.
Feedback loops with AI	AI models dynamically adjust preprocessing parameters (e.g., adaptive denoising) based on prediction accuracy.	Adaptive optimization of pipelines. Continuous improvement with minimal human intervention.	Reinforcement learning for fine-tuning CLAHE parameters in field imaging.
Multimodal data fusion	Combines hyperspectral with RGB, thermal, and gas sensing for richer nutrient profiling.	Enhances detection sensitivity. Compensates for missing info in a single modality. More robust to environmental noise.	Fusion of thermal + hyperspectral for fruit ripeness monitoring.
Edge computing and IoT integration	Processing data directly at the sensor level using low-power AI accelerators.	Reduces data transfer costs. Enables real-time decision support. Ideal for remote agricultural sites.	IoT-enabled hyperspectral cameras on drones for on-site nutrient stress detection.

- **Feedback Loops between AI and Preprocessing**
  Starting with fourth-generation nutritional imaging pipelines, feedback systems are becoming more widespread, with AI output providing real-time adjustments to preprocessing parameters. These adaptable workflows enhance image quality and reduce noise based on performance metrics, aiming to maximize detection sensitivity [83].

  To some extent, parameter tuning may be driven by techniques such as reinforcement learning or active learning, which continually improve noise levels, contrast enhancements, or registration settings based on AI feedback. This collaboration has enabled ongoing improvements in identifying nutrient characteristics, adapting to environmental variations, and compensating for sensor drift throughout the process. Case studies demonstrate the effectiveness of active pipelines in increasing classification accuracy and reducing false detections over time, suggesting that an autonomous nutritional monitoring system could achieve self-optimization.
- **Multimodal Data Integration and Fusion**
  Imaging methods in dietary assessment can be combined since each provides unique information. With a focus on the accuracy of hyperspectral data, automated systems can merge that information with RGB images, thermal maps, or gas sensor readings to provide a more complete, multidimensional view of nutrient levels, maturity, and spoilage.

  Uninterrupted integration requires that preprocessing steps are automated and designed to match different spatial resolutions, spectral bands, and acquisition times by implementing proper registration, normalization, and synchronization [84]. Such multisource data are then fused using AI-based fusion frameworks to enhance the sensitivity and specificity of detection. Multimodal solutions in this way will increase resilience to changing environmental factors, make extensive food quality and safety assessments more accessible, and support more advanced decision-making capabilities in agricultural management and food processing.

3.9.2 Future Trends in Data Reduction and Automated Pipelines

- **Advances in Dimensionality Reduction Algorithms**
  The so-called nonlinear and deep learning methods are introducing new dimensionality reduction techniques among contemporary approaches. Autoencoders and variational autoencoders, along with manifold learning, provide scalable frameworks that can learn spectral features from complex nutrient signals. These tools enhance interpretability through domain knowledge and domain-specific feature selection, and can outperform supervised methods since they are entirely unsupervised, unlike PCA and ICA, which have linear limitations. Real-time adaptability and noise robustness are open research areas, enabling the development of practical algorithms suited for embedded and edge devices.
- **Autonomous End-to-End Nutritional Imaging Systems**
  Older nutritional surveillance systems are being improved by integrating robotics, IoT, AI, and advanced imaging sensors to make them fully automated. These systems involve collecting information, preprocessing data, extracting features, and making decisions within consolidated systems that can operate continuously on-site. In precision agriculture, examples include hyperspectral imagers mounted on drones or vehicles, which enable real-time onboard processing and localized nutrient management. In food sorting, automated lines incorporate both visual and chemical sensors along with adaptive AI for spoilage detection and quality grading. These advanced systems hold promise as operational tools that can transform nutritional analytics into resources for increasing efficiency, reducing waste, and enhancing food quality and safety (Table 3.11) [85].
- **Open Challenges and Research Directions**
  Although significant progress has been made, serious obstacles remain to fully realizing the potential of automated nutritional imaging pipelines.
- Obscure benchmark databases and evaluation indicators hinder objective comparison and validation of preprocessing and reduction techniques across various crops, foods, and imaging situations.
- Environmental and sensor variability present a major challenge. On-demand, learning-informed algorithms should be versatile and not require repeated re-evaluation or manual adjustment [86].
- Deep learning and complex reduction algorithms are computationally expensive and were previously limited by the computing power of low-power edge and IoT devices, which requires further innovation in both algorithm efficiency and hardware-software co-design.

Table 3.11 Future trends and open challenges.

Future trend	Detailed description	Potential impact	Challenges to overcome
Advanced nonlinear reduction	Deep autoencoders and manifold learning for spectral feature extraction.	Captures complex nonlinear nutrient signals beyond PCA/ICA.	Requires large labeled datasets and computational resources.
Autonomous end-to-end systems	Full integration of sensors, AI, and robotics for closed-loop nutrient monitoring.	Enables continuous, on-site precision agriculture and food quality monitoring.	Needs reliable power, connectivity, and adaptive models for dynamic conditions.
Benchmarking and standardization	Shared datasets and evaluation protocols for comparing algorithms.	Improves reproducibility and accelerates adoption.	Requires collaboration among academia, industry, and policymakers.
Robust edge AI	Optimized AI models for low-power devices in the field.	Enables affordable, scalable deployment in agriculture.	Balancing model accuracy vs. computational constraints.
Explainable AI (XAI)	Transparent AI decisions for regulatory and consumer trust.	Increases adoption in sensitive food supply chains.	Striking a balance between model complexity and interpretability is a challenging task.

- Transparency and explainability of an AI model are essential for model approval by regulators and building consumer trust when implementing nutritional assessment technologies. This often involves researching interpretable models and quantifying uncertainty.

Such challenges should be addressed through coordinated efforts in data sharing, algorithm development, and system design to unlock the full potential of image-based nutritional analysis at scale.

3.10 Conclusion and Future Directions

Image enhancement and preprocessing form the basis of accuracy, effectiveness, and reliability of nutritional imaging techniques used in agricultural and food systems. Sensor noise, changing illumination, and optical distortions, as well as the benefits of the environments, can degrade the raw data obtained in such natural conditions, which may obscure or distort nutritional information. Systematic noise reduction methods, including median, Gaussian, wavelet, and NLM filtering, allow subtle nutrient-related signals to be preserved but also reduce artifacts associated with acquisition [87]. Likewise, computational modalities on contrast and color enhancement, such as CLAHE and calibrated color normalization, also enhance the visibility of features, involving greater attention to low-intensity spectral differences, which relate to potential biochemical characteristics.

In addition, radiometric and geometric corrections are essential to maintain analytical accuracy by compensating for sensor drift, variations in illumination, and spatial distortions. Coupled with strong band registration and metadata standardization, these will allow reproducibility and platform-, environmental-, and dataset interoperability. Since hyperspectral data are characterized by a high-dimensional space, dimensionality reduction techniques such as PCA and ICA, besides minimizing computational costs, enhance the robustness of the models by noise filtering and preservation of physically/physiologically meaningful features.

It is possible to see fully automated preprocessing pipelines that can deploy a variety of sensor modalities (hyperspectral, RGB, thermal) and feature extraction using AI. The modular workflow designs provide flexibility to various crops, imaging platforms, and environmental variations, whereas parallel processing, edge computing, and integration of IoT provide real-time analysis in field and industrial applications [88]. The so-called feedback loops between the AI models and preprocessing modules hold the promise of adaptive optimization that would give us self-calibrating systems that will be able to perform well under changing environmental conditions or even changes in the hardware [89].

Further studies need to focus on the creation of uniform calibration guidelines, automated correction systems that could scale, and standards that could be used to compare the algorithms. Fusion of multiple data sets will also improve detection accuracy through a combination of complementary data sets to aid with more extensive quality and nutritional analysis. Besides, it is necessary to implement explainable AI models to develop trust and gain regulatory approval in food supply chain software. The development of embedded calibration sensors, miniaturized imaging hardware, and lightweight, efficient AI models will allow a slower pace of deployment in precision agriculture and food processing.

Through the combination of novel preprocessing algorithms, real-time automation, and adaptive AI integration, next-generation nutritional imaging systems have the potential to provide precise, reproducible, and actionable data, transforming the agricultural decision-making process and directly supporting sustainable global food production systems by reducing food waste and improving food quality [90].

References

1 Dung, C.D., Trueman, S.J., Wallace, H.M. et al. (2023). Hyperspectral imaging for estimating leaf, flower, and fruit macronutrient concentrations and predicting strawberry yields. *Environmental Science and Pollution Research* 30 (53): 114166–114182.

2 Shin, H.-S., Go, S.-H., and Park, J.-H. (2024). Robust radiometric and geometric correction methods for drone-based hyperspectral imaging in agricultural applications. *Korean Journal of Remote Sensing* 40 (3): 257–268.

3 Beemkumar, N., Gupta, S., Bhardwaj, S. et al. (2023). Activity recognition and IoT-based analysis using time series and CNN. In: *Handbook of Research on Machine Learning-enabled IoT for Smart Applications Across Industries* (ed. N. Goel and R.K. Yadav), 350–364. IGI Global https://doi.org/10.4018/978-1-6684-8785-3.ch018.

4 Ojo, M.O. and Zahid, A. (2023). Improving deep learning classifiers' performance via preprocessing and class imbalance approaches in a plant disease detection pipeline. *Agronomy* 13 (3): 887.

5 Haque, M.A., Reza, M.N., Ali, M. et al. (2024). Effects of environmental conditions on vegetation indices from multispectral images: a review. *Korean Journal of Remote Sensing* 40 (4): 319–341.

6 Prieur, C., Braham, N.A.A., Tresson, P. et al. (2024). Prospects for mitigating spectral variability in tropical species classification using self-supervised learning. *2024 14th Workshop on Hyperspectral Imaging and Signal Processing: Evolution in Remote Sensing (WHISPERS)*. IEEE.

7 Khort, D.O., Kutyrev, A., Smirnov, I. et al. (2024). Enhancing sustainable automated fruit sorting: hyperspectral analysis and machine learning algorithms. *Sustainability* 16 (22): 10084.

8 Zhu, H., Huang, Y., An, Z. et al. (2024). Assessing radiometric calibration methods for multispectral UAV imagery and the influence of illumination, flight altitude, and flight time on reflectance, vegetation index, and inversion of winter wheat AGB and LAI. *Computers and Electronics in Agriculture* 219: 108821.

9 Wang, J., Wang, Y., Li, G., and Qi, Z. (2024). Integration of remote sensing and machine learning for precision agriculture: a comprehensive perspective on applications. *Agronomy* 14 (9): 1975.

10 Slade, G., Fawcett, D., Cunliffe, A.M. et al. (2023). Optical reflectance across spatial scales—an intercomparison of transect-based hyperspectral, drone, and satellite reflectance data for dry season rangeland. *Drone Systems and Applications* 11: 1–20.

11 Juárez-Trujillo, I.A., Zavala-de Paz, J.P., Isaza, C. et al. (2024). Methodology for multispectral camera calibration using frequency component separation. *Multimedia Tools and Applications* 83 (8): 22327–22346.

12 Ahmed, M.T., Monjur, O., Khaliduzzaman, A., and Kamruzzaman, M. (2025). A comprehensive review of deep learning-based hyperspectral image reconstruction for agri-food quality appraisal. *Artificial Intelligence Review* 58 (4): 96.

13 Talukdar, V., Veeraiah, V., Roy, D.R. et al. (2023). An innovative approach: typicality-based collaborative filtering for recommender systems. *Proceedings of the 3rd International Conference on Advancement in Electronics & Communication Engineering (AECE)*, pp. 863–868. https://doi.org/10.1109/AECE59614.2023.10428576.

14 Li, C., Czyż, E.A., Halitschke, R. et al. (2023). Evaluating potential of leaf reflectance spectra to monitor plant genetic variation. *Plant Methods* 19 (1): 108.

15 Konara, B., Krishnapillai, M., and Galagedara, L. (2024). Recent trends and advances in utilizing digital image processing for crop nitrogen management. *Remote Sensing* 16 (23): 4514.

16 Joglekar, M. and Deshpande, A.M. (2025). A comprehensive review of hyperspectral image denoising techniques in remote sensing. *International Journal of Remote Sensing* 46 (16): 5961–5995.

17 Rasti, B., Chang, Y., Dalsasso, E. et al. (2021). Image restoration for remote sensing: overview and toolbox. *IEEE Geoscience and Remote Sensing Magazine* 10 (2): 201–230.

18 Pandey, J.K., Yadav, A.K., Purushotham, M. et al. (2022). Colorimetric detection of deltamethrin pesticide from biological samples based on the peroxidase-mimic catalytic activity on MnO_2/SnO_2. *Key Engineering Materials* 928: 45–51. https://doi.org/10.4028/p-4ti1i0.

19 Zhang, Z. and Yang, F. (2022). Hyperspectral image denoising and destriping based on sparse representation, graph Laplacian regularization, and stripe low-rank property. *EURASIP Journal on Advances in Signal Processing* 2022 (1): 96.

20 Swaminathan, V., Thomasson, J.A., Hardin, R.G. et al. (2024). Radiometric calibration of UAV multispectral images under changing illumination conditions with a downwelling light sensor. *The Plant Phenome Journal* 7 (1): e70005.

21 Daniels, L., Eeckhout, E., Wieme, J. et al. (2023). Identifying the optimal radiometric calibration method for UAV-based multispectral imaging. *Remote Sensing* 15 (11): 2909.

22 Atik, M.E., Arkali, M., and Atik, S.O. (2025). Impact of UAV-derived RTK/PPK products on geometric correction of VHR satellite imagery. *Drones* 9 (4): 291.

23 Mohammadi, V., Ansari, K., Gouton, P., and Attig, H. (2024). Developing a new method of transformation for obtaining XYZ color values from RGB images for agricultural applications. *Sensors* 24 (23): 7728.

24 Conley, M.M., Hejl, R.W., Serba, D.D., and Williams, C.F. (2024). Visualizing plant responses: novel insights possible through affordable imaging techniques in the greenhouse. *Sensors* 24 (20): 6676.

25 Caporaso, N., Whitworth, M.B., and Fisk, I.D. (2017). Application of calibrations to hyperspectral images of food grains: example for wheat falling number. *Journal of Spectral Imaging* 6: 1–15.

26 Urdu, D., Goense, D., Booij, J. et al. (2023). Analysis of metadata standards for the exchange of image datasets and algorithms in the agricultural domain: a metadata-oriented approach to identify minimum interoperability mechanisms for image data and deep learning algorithms that are used for vision-based applications in agriculture. Sprint Robotics Project PL4. 0 WP7.

27 Bahlo, C. (2024). Advancing FAIR agricultural data: the AgReFed FAIR assessment tool. *Data Science Journal* 23 (1).

28 Gill, T., Gill, S.K., Saini, D.K. et al. (2022). A comprehensive review of high-throughput phenotyping and machine learning for plant stress phenotyping. *Phenomics* 2 (3): 156–183.

29 Mishra, P. (2021). Chemometric approaches for calibrating high-throughput spectral imaging setups to support digital plant phenotyping by calibrating and transferring spectral models from a point spectrometer. *Analytica Chimica Acta* 1187: 339154.

30 Rojanaphan, P. (2024). Automated nutrient deficiency detection and recommendation systems using deep learning in nutrition science. Available at SSRN 5032705.

31 Janani, M. and Jebakumar, R. (2023). Detection and classification of groundnut leaf nutrient level extraction in RGB images. *Advances in Engineering Software* 175: 103320.

32 Jia, H., Yin, Q., and Lu, M. (2022). Blind-noise image denoising with block-matching domain transformation filtering and improved guided filtering. *Scientific Reports* 12 (1): 16195.

33 Shah, A., Bangash, J.I., Khan, A.W. et al. (2022). Comparative analysis of median filter and its variants for removal of impulse noise from gray scale images. *Journal of King Saud University, Computer and Information Sciences* 34 (3): 505–519.

34 Lin, S., Zhang, Z., Huang, Z. et al. (2023). Deep frequency filtering for domain generalization. *Proceedings of the IEEE/CVF Conference on Computer Vision and Pattern Recognition.*

35 Boyer, M., Bouyer, L., Roy, J.S., and Campeau-Lecours, A. (2023). Reducing noise, artifacts, and interference in single-channel EMG signals: a review. *Sensors* 23 (6): 2927.

36 Dhillon, D. and Chouhan, R. (2023). Edge-preserving image denoising using noise-enhanced patch-based non-local means. *Multimedia Systems* 29 (3): 1025–1041.

37 Park, G.-H., Cho, H.-H., and Choi, M.-R. (2008). A contrast enhancement method using dynamic range separate histogram equalization. *IEEE Transactions on Consumer Electronics* 54 (4): 1981–1987.

38 Thepade, S.D. and Pardhi, P.M. (2023). Mammographic image enhancement using a linear weighted fusion of contrast improvement technique and CLAHE. *2023 IEEE 3rd Mysore Sub Section International Conference (MysuruCon)*. IEEE.

39 Mazdeyasna, S., Arefin, M.S., Fales, A. et al. (2025). Evaluating normalization methods for robust spectral performance assessments of hyperspectral imaging cameras. *Biosensors* 15 (1): 20.

40 Pandey, J. K., Kumar, S., Lamin, M., Gupta, S., Dubey, R.K. and Sammy, F. (2022). A metaheuristic autoencoder deep learning model for an intrusion detector system. *Mathematical Problems in Engineering*, 2022, 3859155, 11. https://doi.org/10.1155/2022/3859155

41 Pandey, D., Kapoor, I., Singh, I. and Sharma, V.K. (2025). Food label analysis and ingredient safety. *2025, the 3rd International Conference on Communication, Security, and Artificial Intelligence (ICCSAI)*. IEEE.

42 Pandey, J.K. (2025). Understanding the impact of electromagnetic and optical radiation on physical and biological systems. *Journal of Nano- and Electronic Physics* 17 (5): 05023. https://doi.org/10.21272/jnep.17(5).05023.

43 Li, X. and Plataniotis, K.N. (2015). A complete color normalization approach to histopathology images using color cues computed from saturation-weighted statistics. *IEEE Transactions on Biomedical Engineering* 62 (7): 1862–1873.

44 Liu, W., Zhang, P., Lei, Y. et al. (2022). A generalized framework for edge-preserving and structure-preserving image smoothing. *IEEE Transactions on Pattern Analysis and Machine Intelligence* 44 (10): 6631–6648.

45 Nocerino, E., Dubbini, M., Menna, F. et al. (2017). Geometric calibration and radiometric correction of the Maia multispectral camera. *The International Archives of the Photogrammetry, Remote Sensing and Spatial Information Sciences* 42 (3W3): 149–156.

46 Lian, X., Li, J., and Sun, D. (2025). Adaptive nonuniformity correction methods for push-broom hyperspectral cameras. *IEEE Journal of Selected Topics in Applied Earth Observations and Remote Sensing* 18: 7253–7263.

47 Castro, W., Mejía, J., de-la-Torre, M. et al. (2023). Radial grid reflectance correction for hyperspectral images of fruits with rounded surfaces. *Computers and Electronics in Agriculture* 213: 108179.

48 Lawrence, K.C., Park, B., Windham, W.R. et al. (2003). Reflectance calibration of focal plane array hyperspectral imaging system for agricultural and food safety applications. In: *Remote Sensing for Agriculture, Ecosystems, and Hydrology IV* (ed. M. Owe), 133–140. Bellingham, WA: SPIE.

49 Im, J. and Jensen, J.R. (2008). Hyperspectral remote sensing of vegetation. *Geography Compass* 2 (6): 1943–1961.

50 Khorram, S., van der Wiele, C.F., Koch, F.H. et al. (2016). *Principles of Applied Remote Sensing*. Cham: Springer.

51 Jiang, J., Zhang, Q., and Gao, S. (2025). Enhancing multi-flight unmanned-aerial-vehicle-based detection of wheat canopy chlorophyll content using relative radiometric correction. *Remote Sensing* 17 (9): 1557.

52 Park, B. and Yoon, S.-C. (2015). Real-time hyperspectral imaging for food safety. In: *Hyperspectral Imaging Technology in Food and Agriculture* (ed. B. Park and R. Lu), 305–329. New York, NY: Springer, Springer.

53 Anusha, R. Prediction of attrition using classification algorithm in data mining. In: *Global Perspectives on Multidisciplinary Research & Development*, 179. India: MeriMax Publications.

54 Qian, Y., Li, Q., Liu, H. et al. (2014). A preprocessing algorithm for hyperspectral images of vessels. *2014 7th International Conference on Biomedical Engineering and Informatics*. IEEE.

55 Zhang, H., Li, C., Xue, H. et al. (2023). Dark background correction of the infrared detector for hyperspectral remote sensing application. *Sensors and Actuators A: Physical* 349: 114088.

56 Kizel, F., Benediktsson, J.A., Bruzzone, L. et al. (2018). Simultaneous and constrained calibration of multiple hyperspectral images through a new generalized empirical line model. *IEEE Journal of Selected Topics in Applied Earth Observations and Remote Sensing* 11 (6): 2047–2058.

57 Markelin, L., Honkavaara, E., Näsi, R. et al. (2017). Radiometric correction of multitemporal hyperspectral UAV image mosaics of seedling stands. *The International Archives of the Photogrammetry, Remote Sensing and Spatial Information Sciences* 42: 113–118.

58 Köppl, C.J., Malureanu, R., Dam-Hansen, C. et al. (2021). Hyperspectral reflectance measurements from UAS under intermittent clouds: correcting irradiance measurements for sensor tilt. *Remote Sensing of Environment* 267: 112719.

59 Elmasry, G., Kamruzzaman, M., Sun, D.W., and Allen, P. (2012). Principles and applications of hyperspectral imaging in quality evaluation of agro-food products: a review. *Critical Reviews in Food Science and Nutrition* 52 (11): 999–1023.

60 Srinivas, P., Veeraiah, V., Thejaswini, K.O. et al. (2025). The impact of electromagnetic and optical radiation on physical and biological systems. *Journal of Nano- and Electronic Physics* 17 (4): 04012. https://doi.org/10.21272/jnep.17(4).04012.

61 Tawfeik, M., Elhifnawy, H., Hamza, E., and Shawky, A. (2016). Determination of suitable requirements for geometric correction of remote sensing satellite images when using ground control points. *International Research Journal of Engineering and Technology* 3 (10): 54–62.

62 Zhou, Y., Rangarajan, A., and Gader, P.D. (2019). An integrated approach to registration and fusion of hyperspectral and multispectral images. *IEEE Transactions on Geoscience and Remote Sensing* 58 (5): 3020–3033.

63 Dalmiya, C. and Dharun, V. (2015). A survey of registration techniques in remote sensing images. *Indian Journal of Science and Technology* 8 (26): 1–7.

64 Yan, Z. and Wu, Z. (2020). Stacked local feature detector for hyperspectral image. *The International Archives of the Photogrammetry, Remote Sensing and Spatial Information Sciences* 43: 495–500.

65 Perera, C.J., Premachandra, C., and Kawanaka, H. (2023). Enhancing feature detection and matching in low-pixel-resolution hyperspectral images using 3D convolution-based siamese networks. *Sensors* 23 (18): 8004.

66 Wu, P., Liu, H., Yi, X. et al. (2024). Research on the registration of aerial images of *cyclobalanopsis* natural forest based on optimized fast sample consensus point matching with SIFT features. *Forests* 15 (11): 1908.

67 Ghanghas, S., Kumar N., Singh V. K., Kumar S., Birania S., Kumar A., Image processing technology, imaging techniques, and their application in the food processing sector. *Nonthermal Food Engineering Operations*, 2024: p. 193–223, Wiley.

68 Borah, K. et al. (2024). A review on advancements in feature selection and feature extraction for high-dimensional NGS data analysis. *Functional & Integrative Genomics* 24 (5): 139.

69 Bilal, M., Ali, G., Iqbal, M.W. et al. (2022). Auto-prep: efficient and automated data preprocessing pipeline. *IEEE Access* 10: 107764–107784.

70 Sagan, V., Maimaitijiang, M., Paheding, S. et al. (2021). Data-driven artificial intelligence for calibration of hyperspectral big data. *IEEE Transactions on Geoscience and Remote Sensing* 60: 1–20.

71 Garg, N. and Dwivedi, P. (2024). A novel approach for exploring data-driven nutritional insights using clustering and dimensionality reduction techniques. *SN Computer Science* 5 (8): 1019.

72 Kannan, R.J. (2024). Advances in digital image processing techniques for microscopy and their applications in biomedical research and material science. *Parkavi and K, Madhumitha, Advances in Digital Image Processing Techniques for Microscopy and Their Applications in Biomedical Research and Material Science (November 28, 2024).*

73 Judith, A.M., Priya, S.B., and Mahendran, R.K. (2022). Artifact removal from EEG signals using regenerative multi-dimensional singular value decomposition and independent component analysis. *Biomedical Signal Processing and Control* 74: 103452.

74 Kumar, A., Kulshreshtha, S., Shrivastava, A., and Saini, A. (2024). Biosensors for food spoilage detection: a comprehensive review of current advances. *Journal of Food Chemistry and Nanotechnology* 10: S73–S82.

75 Anvari, M.A., D. Rahmati, and S. Kumar, t-Distributed stochastic neighbor embedding, in J. A. Rad, S. Chakraverty, & K. Parand (Eds.) *Dimensionality Reduction in Machine Learning.* 2025, Cambridge, MA, USA: Morgan Kaufmann (Elsevier). p. 187–207.

76 Luo, N., Yang, X., Sun, C. et al. (2021). Visualization of vibrational spectroscopy for agro-food samples using t-Distributed stochastic neighbor embedding. *Food Control* 126: 107812.

77 Pérez-Sanpablo, A.I., Quinzaños-Fresnedo, J., Gutiérrez-Martínez, J. et al. (2025). Transforming medical imaging: the role of artificial intelligence integration in PACS for enhanced diagnostic accuracy and workflow efficiency. *Current Medical Imaging* 21 (1): e15734056370620.

78 Patro, R.N., Subudhi, S., Biswal, P.K. et al. (2021). Error removal by energy scaling from hyperspectral images for performance improvement of spectral classifiers. *International Journal of Remote Sensing* 42 (20): 7863–7879.

79 Renda, A. (2019). The age of foodtech: optimizing the agri-food chain with digital technologies. In: *Achieving the Sustainable Development Goals Through Sustainable Food Systems* (ed. C.M. Galanakis), 171–187. Cham, Cham: Springer.

80 Adão, T., Hruška, J., Pádua, L. et al. (2017). Hyperspectral imaging: a review on UAV-based sensors, data processing and applications for agriculture and forestry. *Remote Sensing* 9 (11): 1110.

81 Mehedi, I.M., Hanif, M.S., Bilal, M. et al. (2024). Remote sensing and decision support system applications in precision agriculture: challenges and possibilities. *IEEE Access* 12: 44786–44798.

82 Lin, Y. et al. (2023). Applications of machine learning techniques for enhancing nondestructive food quality and safety detection. *Critical Reviews in Food Science and Nutrition* 63 (12): 1649–1669.

83 Chakilam, C. (2023). Next-generation healthcare: merging AI, ML, and big data for accelerated disease diagnosis and personalized treatment. *American Online Journal of Science and Engineering (AOJSE)* 1 (1): 1–4. (ISSN: 3067-1140).

84 Ferraz, M.A.J., Barboza, T.O.C., Arantes, P..S. et al. (2024). Integrating satellite and UAV technologies for maize plant height estimation using advanced machine learning. *AgriEngineering* 6 (1): 20–33.

85 Sharma, K. and Shivandu, S.K. (2024). Integrating artificial intelligence and Internet of Things (IoT) for enhanced crop monitoring and management in precision agriculture. *Sensors International* 5: 100292.

86 Tank, K.H., Ghanem, M.C., Gasiorowski, P. et al. (2025). Synchronization, optimization, and adaptation of machine learning techniques for computer vision in cyber-physical systems: a comprehensive analysis. *IET Cyber-Physical Systems: Theory & Applications* 10 (1): e70031.

87 Mitchell, S. and Remmel, T. (n.d.). Impacts of sensor noise on land cover classifications: sensitivity analysis using simulated noise.

88 Li, F., Wu, Z., Li, H. et al. (2024). A frozen suspension external calibration strategy for elemental quantitative imaging of fresh plant soft tissues by LA-ICP-MS with cryogenic ablation cell. *Analytical and Bioanalytical Chemistry* 416 (20): 4519–4529.

89 Xie, P., Ma, Z., Du, R. et al. (2023). Generating high-quality 3DMPCs by adaptive data acquisition and NeREF-based radiometric calibration with UGV plant phenotyping system. arXiv preprint arXiv:2305.06777.

90 Chi, B., Zhang, J., Huang, G. et al. (2025). An iterative adaptive polarization calibration method independent of corner reflectors. *IEEE Journal of Selected Topics in Applied Earth Observations and Remote Sensing* 18: 5255–5270.

4

Segmentation and Object Detection in Crop and Food Images

Sagnik Bachhar[1], Sumanta Bhattacharjee[2], and Biswarup Yogi[2]

[1] *Adamas University, Kolkata, West Bengal, India*
[2] *Brainware University, Kolkata, West Bengal, India*

4.1 Introduction

Agriculture and food quality assurance lie at the base of health, nutrition, and economic stability globally. With increasing populations and growing concerns over food security, modern farming and food systems are being increasingly tested to see if they can produce more output while ensuring minimum wastage and quality checks up and down the value chain. This has forced a gradual transition from manual labor into AI-driven automation and computer vision-enabled analytics, especially in developing economies. However, among these tools, image segmentation and object detection are the most powerful, since they allow machines to "comprehend" and "interpret" visual data at scale [1].

Computer vision facilitates real-time monitoring, analysis, and decision-making by performing tasks that are otherwise complex to automate, such as crop monitoring, disease detection, yield estimation, and quality inspection. Such tasks, in the most part, involve subjectivity, have the potential to go wrong, and are time-consuming. An image-processing AI presents precision, impeccable repeatability, and scalability to execute in seconds what else would otherwise prove hard to carry out. Segmentation involves breaking an image into meaningful regions like leaves, soil, or fruit, whereas object detection becomes a case of detecting and localizing objects like apples, pest spots, or packing defects [2]. Figure 4.1 shows the illustration of the difference between segmentation (pixel-level) and object detection (bounding-box level) on a crop image.

Segmentation and object detection combine to offer automation across the agriculture and food processing pipeline from drone crop monitoring to robotic sorting of fruits on a conveyor belt. These methods, meanwhile, intend to improve the accuracy of autonomous systems wherein complex, real-world conditions prevail [3].

Precision agriculture works in site-specific management and intervention based on data. Here, segmentation algorithms are used for mapping the leaf canopy, weed identification, crop row identification, or nutrient stress detection from leaf color or shape. Object detection tasks include fruit counting, disease lesion localization, or invasive species. The computer vision in the food sector ensures standard grading, surface defect detection, contamination

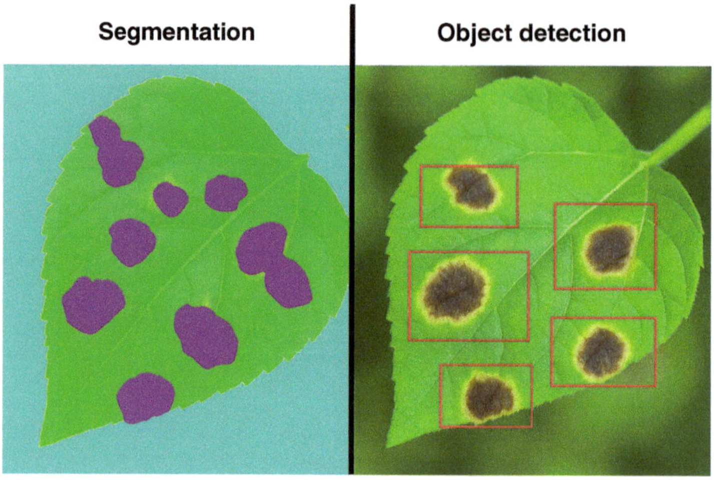

Figure 4.1 Illustration of the difference between segmentation (pixel-level) and object detection (bounding-box level) on a crop image.

inspection, and even ripeness classification. Hyperspectral imaging, coupled with multimodal fusion – RGB, IR, and depth data, can capture the very faint and extremely minute differences related to the classification of food items [4]. For example, fungal spots on the apples can be segmented as an object for semantic segmentation, while the rest are graded by object detectors, greatly automating their packaging and sorting [5].

Thus, some traditional segmentation methods like thresholding, region growing, or edge-based methods could never guarantee satisfactory results under difficult or noisy scenarios. Currently, deep learning is mainly preferable for modern applications, with major prominence to CNNs, especially U-Net, Mask R-CNN, DeepLabV3+, Transformer-like Segmenter, and Swin Transformer [6]. Regarding the object detection task, R-CNN → Fast R-CNN → Faster R-CNN → YOLOv5-v10, SSD, EfficientDet, and DETR are popular options, with YOLOv8 striking a fine balance among speed, size, and accuracy, which is suitable for real-time applications onboard farming drones and embedded devices [7]. These models are trained on annotated datasets and are fine-tuned after training using loss functions like cross-entropy, Dice loss, or IoU-based metrics.

In real-life settings, however, hurdles and limitations come into play:

a) **Scarcity of Data:** Labeled datasets for minor crops or rare diseases are difficult to obtain [8].
b) **Variability:** Because of illumination, occlusion, and background diversity, outdoor environments differ.
c) **Low-power Deployment:** Such models must be compressed by pruning or quantization for deployment to low-power farm robots.
d) **Generalization Will Cease:** A model trained in one region and applied to another crop field, or factory setup, may not work well there.

Explainability is a very important aspect. Wrong detection may imply financial or environmental losses in agriculture. Farmers and plant pathologists need interpretable models.

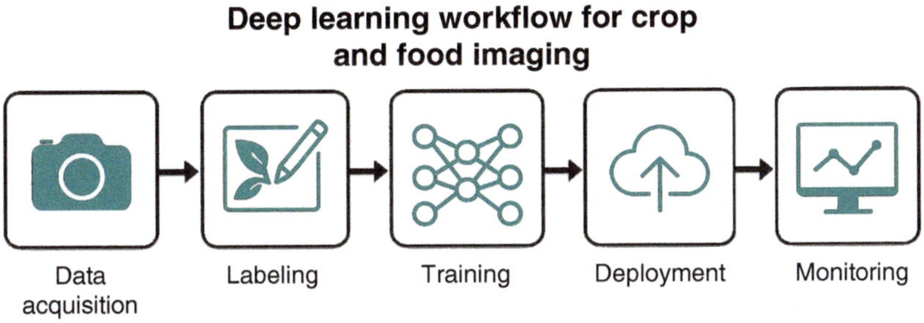

Figure 4.2 Deep learning workflow for crop and food imaging.

Methods, such as Grad-CAM, SHAP, and attention heatmaps, are being extensively used to interpret decisions of deep-learning-based methods [9]. And here are some currently very popular emerging technologies and research directions:

a) **Multimodal Fusion:** The combination of RGB, thermal, depth, and hyperspectral data leads to higher accuracy in detection even in challenging situations.
b) **Transfer Learning and Domain Adaptation:** The process of pretraining results in a huge drop in the number of data instances.
c) **Few-shot and Zero-shot Learning:** It detects extremely rare diseases with only a few labeled examples [10].
d) **Federated Learning:** Different farms can participate in training together while keeping their data private.
e) **Lightweight Models:** MobileNet, TinyYOLO, and EfficientNet are some examples of models that are designed to run on mobile devices and drones [11].

This chapter mainly focuses on the topics of the mathematical basis–related techniques, architecture, metrics (mAP, IoU, etc.), and their practical application in agriculture and the food industry. AI and image processing have helped not only to increase productivity but also to set the way for sustainable agriculture and smart food systems. Figure 4.2 shows the deep learning workflow for crop and food imaging.

4.2 Fundamentals of Image Segmentation

Image segmentation is considered one of the leading tasks in the areas of computer vision and digital image processing. The procedure involves decomposing an image into different and nonoverlapping parts reflecting various objects or regions of interest. The agricultural and food sectors highly benefit from the segmentation technique since it allows, among other things, the thorough inspection of visual data, the separation of crops from soil, the uncovering of the diseased spots on leaves, or even counting fruits on one tree. Segmentation is a prerequisite for many operations like classification, counting, or quality assessment [12].

4.2.1 Types of Segmentation

According to the criterion of scale and semantic considerations, image segmentation falls into the following categories:

4.2.1.1 Semantic Segmentation

Semantic segmentation works by providing a class label to every single pixel in the image (e.g., "fruit," "leaf," "stem," or "background"). It doesn't separate different objects belonging to the same category. Thus, in a photograph, all apples will be tagged "apple," no matter if they are individual or not. The technique of semantic segmentation finds its application in field mapping, canopy estimation, and large-scale agriculture monitoring.

4.2.1.2 Instance Segmentation

Instance segmentation represents a more complex approach that not only detects and outlines the objects but also establishes the individuals of the same class as well. For example, a picture with several overlapping tomatoes will get assigned different IDs for each tomato through the instance segmentation process. This technique is used in fruit counting, defect detection, and robotic harvesting systems, where the precise position of the objects is essential [13].

4.2.1.3 Panoptic Segmentation

Panoptic segmentation represents a fresh method that merges semantic and instance segmentation techniques, providing total comprehension of the scene through pixel-level class labels and object instance IDs. Panoptic segmentation can be used in complex agricultural scenarios such as heavy crop clusters and mixed-species fields.

4.2.2 Traditional Ways

Pre-deep learning, classical image processing techniques considered pixel-level statistics and simple heuristics to achieve segmentation. Such traditional methodologies are simple, meaningful, and computationally efficient, but they are found lacking in the realms of a more complex natural scene, varying lighting, and overlapping objects [14].

4.2.2.1 Thresholding

A threshold is one of the simplest ways for segmenting an image. Based on the intensities, an image is divided. For example, in fruit detection under uniform illumination, the pixels with intensities above a particular threshold would be considered as fruit. Adaptive thresholding can compensate for local variations in illumination. This, of course, alone cannot do a good job when illumination is not homogeneous, or when the object and background share similar intensities.

4.2.2.2 Edge Detection

The segmentation is performed by edge-based methods by detecting discontinuities in the pixel discourse. The commonly used edge detectors are Sobel, Prewitt, Canny, and Laplacian. They work well when objects have clearly defined borders. Still, edge detection is noisy; hence, the resulting edges often require some additional processing, such as contour joining or region filling, to form coherent segments [15].

4.2.2.3 Region-Based Methods

The basic idea behind a region-growing algorithm is that a few seed points grow into all foreground regions by including neighboring pixels of similar properties (intensity, texture, etc.). Region splitting and merging try to improve the aforesaid approach by recursively partitioning or merging regions based on heterogeneity criteria. Segmentation can also be achieved by methods that cluster feature points. K-means and mean-shift are examples of such algorithms.

4.2.2.4 Watershed Algorithm

In the watershed algorithm, a grayscale image is treated as a topographic surface where intensity is taken as elevation. The method floods the image starting from low-gradient regions, and the watershed lines serve to separate the respective catchment basins. While it succeeds in separating touching objects, watershed segmentation suffers from over-segmentation, which can be reduced through gradient filtering or marker-controlled watershed.

4.2.3 Evaluation Metrics

Several quantitative metrics are used to assess the performance of segmentation methods in agricultural and food imaging:

4.2.3.1 Pixel Accuracy

Pixel accuracy is the total number of correctly classified pixels divided by the total number of pixels. Easy to interpret, yet PA may yield a misleading notion in an imbalanced scenario, say a few pixels representing small diseases on huge leaves, whereas the count of pixels of the background makes up the majority in the pixel count [16].

4.2.3.2 Mean Intersection over Union (mIoU)

Intersection over union (IoU) measures the amount of overlap existing between the predicted segments and the ground truth segments. It is then averaged over all classes to obtain the mean IoU to provide an unbiased evaluation, more so for multi-class implementations. This is the metric used in the PASCAL VOC and COCO benchmarks.

4.2.3.3 Dice Coefficient

The Dice coefficient is a similarity measure between two sets: it is more sensitive to small objects and is generally defined as follows:

$$\text{Dice} = \frac{2 \times |\text{A} \cap \text{b}|}{|\text{A}| + |\text{B}|} \tag{4.1}$$

where A is the predicted region, and B is the ground truth. It is widely used in medical and agricultural imaging, where detecting small abnormalities is critical.

4.3 Object Detection: Concepts and Techniques

Object detection is one of the core tasks in computer vision. This involves detecting the objects of interest, localizing them within an image using bounding boxes, and assigning categories to each bounding box. In agricultural and food imaging, object detection is used in the precise identification of fruits of interest, pests, diseases, and quality anomalies, thereby allowing for automation of sorting, grading, and monitoring.

Discussed here are classical approaches to object detection through to the latest deep learning detectors, as well as the prominent metrics used to assess how well a model performs.

4.3.1 Traditional Techniques

Before deep learning, object detection was based on hand-crafted feature extraction combined with classical machine learning algorithms. These methods, though far from being adaptable, enriched the field with the foundational work.

4.3.1.1 Haar Cascades

In this method, Haar-like rectangular features are calculated with the help of integral images for speed, first described by Viola and Jones. The cascade classifier architecture allows real-time object detection. Working in agri-food imaging, it is utilized to detect simple shapes such as fruit clusters, but it is not robust under occlusion or changes in lighting.

4.3.1.2 HOG

Histogram of oriented gradients (HOGs) by Dalal and Triggs keeps track of the shape of an object by considering edge formulas within images. An approach that, when plugged with SVMs (support vector machines), works wonderfully in very structured environments such as food packaging lines under human control, yet finds difficulties working in cluttered agricultural environments.

4.3.1.3 Limitations

Under diverse backgrounds, scale variations, and real-world noise, these image-based detection techniques fail to deliver due to their limited ability to generalize situations – limited generalization. Thus, deep learning succeeded.

4.3.2 Deep Learning-Based Techniques

The rise of convolutional neural networks (CNNs) and large annotated datasets gives a new meaning to object detection. Deep learning methods automatically extract features, always aiming for higher accuracy and better generalization ability.

4.3.2.1 R-CNN and Its Varieties

R-CNN: The Selective Search generates region proposals, followed by the application of a CNN (say AlexNet) to each of these regions to perform classification and bounding box regression.

Fast R-CNN: Further enhancement of R-CNN: features are computed just once per image, and then RoI pooling is applied.

Faster R-CNN: The Region Proposal Network (RPN) is introduced for the generation of proposals, whereupon end-to-end training is performed.

Use in Agriculture: Faster R-CNN is a very accurate approach for such problems as plant disease detection and fruit localization; however, it may be computationally expensive for real-time applications.

4.3.2.2 YOLO (You Only Look Once)

YOLO treats detection as a regression problem by dividing the image into a grid and simultaneously predicting bounding boxes and class probabilities in one forward pass. YOLOv5 and YOLOv8 are used widely to perform fast, real-time agricultural duties such as fruit counting or pest detection on embedded devices.

4.3.2.3 SSD (Single Shot MultiBox Detector)

It improves the detection of small objects by the use of multi-scale feature maps and anchors. Striking a balance between speed and accuracy, it has been used in ripeness detection, fruit grading, and pest localization.

4.3.2.4 DETR (Detection Transformer)

DETR turns object detection into a set prediction problem through Transformer attention. It is free from anchors or region proposals and is suited for cluttered, dense crop imagery. With the downside of being computationally heavier and more time-consuming than CNN-based models, DETR provides an excellent alternative.

4.3.3 Performance Metrics

Several standard evaluation metrics are used to assess object detection models:

4.3.3.1 Mean Average Precision (mAP)

mAP@IoU = 0.5: Average precision score corresponding to an IoU criterion of 0.5.

mAP@[0.5:0.95]: Average mAP over several IoU criteria (0.5–0.95), thus providing a more stringent evaluation.

4.3.3.2 Precision and Recall

Precision: That proportion of all objects detected that were actually predicted correctly.

Precision = TP / (TP + FP): How many predicted boxes are correct.
Where TP is True Positive, and FP is False Positive.

$$\text{Precision} = \frac{\text{TP}}{(\text{TP} + \text{FP})} \tag{4.2}$$

Recall: The proportion of true objects detected by the model. It is useful for trade-offs between under-detection and false positives.

Recall = TP / (TP + FN): How many ground-truth objects are detected.
Where TP is True Positive, and FN is False Negative.

$$\text{Recall} = \frac{\text{TP}}{(\text{TP} + \text{FN})} \tag{4.3}$$

4.3.3.3 Intersection over Union (IoU)

IoU measures how well the predicted box matches the ground truth. In many cases, the detection is termed a true positive if IoU ≥ 0.5.

$$\text{IoU} = \frac{\text{Area of Overlap}}{\text{Area of Union}} \tag{4.4}$$

At every stage of the object detection pipeline, constant innovation is seen in many systems, mostly the Agri-on-cloud systems. The traditional detectors like Haar and HOG have disappeared, while the topmost detectors are YOLO and DETR. Numerous detection systems exist, and, besides accuracy and speed, they can be classified according to their adaptability to real-world setups. The right model is chosen depending on the factors of environment complexity, hardware limitations, and output speed, which are all seen as very important for the sustainability of agri-tech operations.

4.4 Applications in Crop Imaging

The usage of computer vision methods like object recognition and segmentation has now changed agricultural practices greatly. Automation of visual inspection tasks in the field or greenhouse not only gives better results but also cuts down on the need for human work, thus making it easier for farmers to rely on data for their decisions. We shall indicate the main uses of these technologies in crop imaging in the coming paragraphs.

4.4.1 Plant Disease Detection

Diseases of plants like blight of leaves, rust, and powdery mildew severely diminish yield and reduce the quality of a crop. Early and correct detection is, therefore, very important to reduce its spread.

Semantic segmentation is a crucial technique that allows for the separation of infected regions at the pixel level. Furthermore, the network is presented with annotated images as training data, for instance, U-Net, DeepLab, or Swin-based segmenters, to divide the healthy tissues from those that are diseased. These techniques detect lesions, spots, or wilting and, in some cases, do so even earlier than a human would be able to notice the signs. The generation of high-definition segmentation masks makes it possible to evaluate the disease severity very accurately by calculating the ratio of the affected leaf area.

For instance, a U-Net-based model trained on a tomato leaf blight dataset achieves greater than a 90% Dice score in separating lesion boundaries, even under varied illumination and leaf angles. By integrating it with smartphone-based capture systems, farmers get near-real-time scenarios of disease development as it spreads throughout large fields. The disease severity indices obtained through segmentation maps allow agronomists to determine the thresholds for treatments, plan fungicide applications, and eliminate unnecessary spraying.

4.4.2 Fruit Counting and Yield Estimation

Accurate yield estimation helps in logistics planning, forecasting the supply in the market, and scheduling harvest. Object detection models like YOLOv8, SSD, or Faster R-CNN are generally used to count fruits such as apples, mangoes, citrus, and grapes.

These models are trained on an annotated dataset having bounding boxes surrounding visible fruits. In the orchards, the images get captured either via drones or tractor-mounted cameras and then get fed into the detection model. The algorithm returns bounding boxes and class labels (e.g., "ripe mango") that can be aggregated per tree or block to yield an estimation.

High-speed detectors such as YOLOv8 process images faster than at an average speed of 30 frames per second, allowing a drone to cover several hectares in a single flight. To deal with occlusion (fruits hidden behind foliage), advanced paradigms resort to instance segmentation or depth sensing to infer if any fruits are partially visible. Empirical research shows that the accuracy of counting via automation lies within ±5% of manual counting, thus lessening the labor cost and variance in prediction.

Furthermore, the estimated yield can become input data for precision harvest scheduling and supply chain coordination, thereby cutting down on spoilage and improving quality control for perishables.

4.4.3 Weed Detection

The weeds compete with crops for nutrients, water, and sunlight. Traditionally, herbicides have been used indiscriminately, causing environmental destruction. Computer vision allows instance segmentation to distinguish weed species from crops so that site-specific weeding can be carried out.

Instance Segmentation networks (e.g., Mask R CNN) label every pixel; they differentiate among plant instances even when weed and crop overlap. In precision-agriculture robots, real-time segmentation maps are used as guidance for mechanical or chemical weeders, which apply selective treatment of weeds in the preservation of crop plants.

For instance, trained on soybean- or maize-field data containing multiple types of weeds, models are capable of weed segmentation with precision around 85–90%. Indeed, by way of robotic trials, selective herbicide application has diminished the quantity of chemicals used by about 60–70%, maintaining yields at equivalent levels. This environmental advantage and cost saving make weed segmentation a prime candidate for object detection in sustainable crop management.

4.4.4 Crop Phenotyping

Phenotyping is the quantitative assessment of traits such as plant height, leaf area, canopy density, and leaf branching, which are important aspects in plant breeding and genomics. High-throughput detection and segmentation tools for objects allow phenotyping to be done at scale.

Segmentation differentiates individual leaves or whole plant architecture from metadata-received imagery, either aerial or side-view. Object detection will localize leaf tips, fruits, or flowers. Metrics sort through leaf area index, mean leaf length, canopy cover percentage, and plant height with a pixel-to-metric conversion based on camera calibration. Such traits are key to the assessment of genetic variants within breeding programs and to the selection of crop lines performing well.

Estimation of leaf area index with less than 5% error against manual measurements is achieved, for example, with segmentation using U-Net. Complemented with object detectors for flower or pod counting enables analysis of hundreds of genotypes in a field season, greatly speeding up data-driven breeding for drought tolerance, nutrient efficiency, or disease resistance, and, thereby, a step toward more resilient crop varieties.

4.4.5 Integration and Deployment Considerations

Most crop applications combine segmentation and detection in hybrid pipelines: segmentation first isolates relevant plant regions (e.g., foliage), then the detection stage counts the fruits or disease spots. Such integrated systems are deployed on drones, ground vehicles, or stationary cameras connected to edge computing nodes for near-real-time inference.

Key Considerations During Deployment Are

a) **Concerning the Lighting Changes:** The lighting conditions vary in different seasons and also depend on the weather and the type of trees. Thus, it is highly recommended to invest a lot of time in training with an augmented dataset that will contain various lighting changes, including brightness shifts and shadows.
b) **Occlusions are Overcome:** Sometimes, fruits are only partially recognized automatically based on how hidden they are – a detection method based on the shape or context has to be used.
c) **Edge Computing Restriction:** Lightweight models (YOLO-based MobileNet, pruned Mask R-CNN) are required to achieve a real-time performance level on devices with low power supply and hence limited energy.
d) **Labeling and Dataset Diversity:** Appropriate model training would require high-quality datasets with a wide variety of crops, seasons, and phenological stages.

4.5 Food Imaging Operations

The combination of smart food systems and Industry 4.0 has put image processing in the limelight for monitoring the quality, safety, and traceability of food products. The joining together of segmentation technologies and object detection in the food imaging process yields accurate analysis of visual attributes, which helps to maintain uniformity, reduce waste, and automatically verify. Their use is widespread from food processing to supply chains and retail, with a few cases being very close to consumers, like diet monitoring and smart kitchens.

This section will cover the use of segmentation and object detection in food imaging, along with operational importance, recognition of performance, and scenarios of the deployment of these technologies.

4.5.1 Quality Grading

Computer vision plays a pivotal role in the food processing industries, and among its various applications, food-grade inspection is the most prominent one. The edible quality of food items like fruits, vegetables, and packaged ones is assessed. Quality control using the conventional methods is totally dependent on the inspectors, who are subjective, inconsistent, and the process takes a lot of time. But image-based systems can offer an automated, repeatable, and objective method for conducting this task.

Segmentation is the predominant technique for inspecting surfaces. For instance, when semantic segmentation is applied to the likes of apples or bananas, the edible part will be separated from the background, and the areas of specific defects like bruising, rotting, cracking, or fungal spots will be marked. Different algorithms, such as U-Net, PSPNet, or Mask R-CNN, can be used for training to demonstrate the localization areas with pixel-level detail.

Utilizing object-detection models such as YOLOv5 and Faster R-CNN, along with these methods, enables defect identification and severity assessment. For instance, a rectangular box can be drawn around a rotting or puncture mark, and the area can then be classified as "minor bruise," "mold patch," or "deep crack."

These methods are also applied in the case of packaged food products, where they ensure that proper sealing has been done, missing labels are found, and discrepancies in the appearance of baked goods or dairy products are detected. With the combination of high-speed cameras and GPU-powered detectors, it is possible to perform these operations in real-time on the production lines, which not only helps in cutting down labor costs but also minimizes the occurrence of false positives.

4.5.2 Ripeness Estimation

The maturity of fruits and vegetables is among the main problems that the imaging technique for food will encounter. The maturity of fruits and vegetables is one of the most important factors affecting their quality, the length of time they can be stored, and the price at which they will be sold. Up to now, it was thought that the ripeness assessment was done through very subjective techniques, i.e., color and feel. The computer vision technique, however, looks at the same product attributes through a quantitative lens and produces a corresponding ripeness grade.

In color-based segmentation, the image is segmented into parts representing the various maturity levels, and the parts can be subjected to further analysis. For example, the segmentation of green-to-yellow gradients in bananas or the red blush on apples is done using color spaces like HSV or LAB. Local binary patterns (LBPs) and Gabor filters are the two most widely applied texture descriptors for assessing the texture (softness or roughness) of the surface.

The deep learning model can be a CNN or a lighter MobileNet variant that has been trained on images with labels indicating different ripeness stages, such as underripe, ripe, and overripe. The models will rely on global shape features and local pixel-level features and combine them to produce a ripeness score or class as an output.

When this technique is put to use in an industrial setting, it significantly helps in sorting out the different grades of produce for packaging. In the case of supermarkets, an automated kiosk comprising a camera and an object detector is used to assist customers in getting the best option, mainly in the fresh produce department. In the case of a warehouse, technology for ripeness tracking is used in real-time, and hence the inventory is turned over more rapidly and, consequently, food waste is reduced.

4.5.3 Food Item Recognition

With the rise of personalized nutrition, diet tracking, and calorie counting, food recognition is turning into an essential application area for consumers. In these systems, the automatic identification of food carried out either by smartphone cameras or smart kitchen sensors would be for nutrition logging and dietary monitoring.

High-level object detection models like YOLOv8 or EfficientDet are used to picturize and measure the different food items in one's picture. All items that are recognized are drawn around with the bounding boxes and labeled, e.g., "rice," "grilled chicken," "broccoli," or "french fries." After that, the nutritional database is used to get an estimate of the calorie intake and the amount of macronutrients based on those detections.

In such a situation, instance segmentation is particularly required when items overlap or merge visually, such as in curries with rice or mixed salads. For volume and portion estimations, the model-Mask R-CNN separates instances of food from each other.

The present technology is seen embedding itself in hospital meal monitoring systems, canteen billing, and dietary compliance platforms for patients with diabetes or cardiovascular diseases, beyond calorie counting. Recently, there has been some research also integrating food recognition with vision-language models (VLMs) for zero-shot detection, effectively allowing for the identification of new dishes without retraining.

4.5.4 Foreign Object Detection

Food safety is an overwhelming priority for regulators and consumers. Contamination by plastic, metals, glass, or wood during processing or packaging can seriously harm health. Foreign object detection using image-based systems plays a critical role in safeguarding.

X-ray or metal detectors are still mostly used in the industry, yet with the advancements of vision inspection via object detection methods, they serve as a remote and inexpensive redundancy. Vision systems trained to detect abnormalities in shape, color, and reflectance can pick out plastic shards from bread, wood pieces in processed meat, or insects in bulk grain.

It is feasible to modify an object detection algorithm such as YOLOv5 or RetinaNet to work with datasets containing both clean and contaminated samples. The use of models along with multi-spectral or hyperspectral imaging shows that the materials have very little spectral difference, which implies that detection sensitivities are increased even when the lighting or packaging conditions are changed.

These detection systems, which are part of the HACCP regulations, are incorporated into smart conveyor belts fitted with automatic rejection systems that ensure the removal of contaminated units not only instantly but also from the supply chain, thereby maintaining customer goodwill or regulatory acceptance.

4.5.5 Key Benefits

- Large-scale standardized food grading
- Visual features-based quantitative ripeness monitoring
- Automated food recognition for dietary reporting
- Very precise detection of contamination in packaged goods

4.6 Deep Learning and Model Architectures

Deep learning is a game-changer for the image analysis of agriculture and food systems, and for the manual feature extraction process; it can even predict symptoms through various images and generalize perfectly over them, no matter if they are of different varieties. On the one hand, the human-centric approach to image analysis consists of creating features through engineering and then teaching the machine to recognize patterns in those features. On the other hand, deep learning is finding hierarchical representations from the input raw data, thus bringing the models to be used for complicated visual tasks like crop segmentation, disease detection, fruit counting, and food quality inspection.

This section presents a review of the latest architectures in deep learning, such as CNNs, transformers, transfer learning, and ensemble learning, with crop and food image analysis as the discussed application domains.

4.6.1 CNN-Based Architectures

4.6.1.1 U-Net

U-Net is a convolutional network architecture originally developed for biomedical image segmentation, but has become widely applied in agriculture for capturing fine-grained spatial information. As an encoder-decoder network, U-Net obtains contextual features through encoding and reconstructs into pixel-wise predictions during decoding; hence, skip connections between corresponding layers help with spatial resolution preservation.

In agricultural practice, it has found massive usage for segmenting leaf diseases, analyzing root system architecture, and delineating canopy structure. The pixel-level accuracy and small size of U-Nets allow them to be used in embedded devices for field-monitoring systems.

4.6.1.2 Mask R-CNN

Mask R-CNN improves Faster R-CNN with the addition of a parallel branch for segmentation mask prediction along with the object detection task. It performs instance segmentation by object classification, generating bounding boxes, and pixel-wise masking of each detected instance.

In crop imaging, Mask R-CNN is known to give very precise instance segmentation of fruits, especially in cases involving overlapping or clustered objects such as those of grape, apple, or tomato. Its multi-task learning nature makes it a perfect candidate for cases where localization and fine boundary delineation are both required.

4.6.2 Transformer-Based Architectures

Transformer architectures learned first for natural language processing have been successfully transferred to vision tasks because of their ability to model long-range dependencies and contextual relationships.

4.6.2.1 Vision Transformers (ViTs)

ViTs take an image and split it into fixed-size patches, which are then linearly embedded and processed by the standard Transformer encoders that work on sequences of patches. Unlike CNNs, ViTs do not work with convolutional filters; hence, they rely purely on attention mechanisms.

ViTs are used in agricultural imaging for semantic segmentation of crop fields and distinguishing weeds from plants, as well as for identifying pests. The global receptive field assists in handling large-scale scenes where objects appear at various scales and under variable lighting conditions.

4.6.2.2 Swin Transformer

Swin Transformer is a better version of the ViT, introducing a hierarchical framework and shifted windows for attention computation, organizing things such that it can handle very high-resolution images efficiently and actually balances between local and global context.

Swin Transformers are very effective tools in the high-resolution agricultural imaging segmentation area, with the ability to pinpoint even the minutest diseased areas on a leaf and fine root structures, as some examples. Thus, their friendship and the ability to adapt have opened the way for their introduction as a rival to the traditional CNNs in the precision farming applications domain.

4.6.3 Transfer Learning

Creating a deep learning model from scratch for AI necessitates an enormous amount of labeled data, which, most of the time, is a constraint in the agricultural sector. But, transfer learning is here to help by adopting a pretrained model (typically trained over ImageNet) either as a static feature extractor or as a network that is fine-tuned.

Typical backbone architectures include:

a) **ResNet:** Characterized by its use of residual connections and ability to train very deep networks for feature learning.
b) **VGG:** Simple and deep, and often used as a feature extractor.
c) **MobileNet:** Intended to be deployed on mobile and edge devices due to its lightweight nature.

In agriculture and food analysis, transfer learning brings benefits to fruit classification, disease detection, and vegetable grading, all the while suffering from the drawback of limited labeled datasets. The fine-tuning of pretrained models considerably reduces the time of training while retaining accuracy for converting to new crops, new imaging conditions, or sensor types.

Besides, lightweight pretrained models such as MobileNetV2 and EfficientNet-Lite are used for on-device inference on smartphones, drones, and robotic harvesters.

4.6.4 Ensemble Learning

Ensemble learning refers to combining multiple models to ensure prediction stability, accuracy, and generalization. It is an attempt to aggregate the strengths of varying architectures or model checkpoints, thereby counteracting an individual weakness.

In the agri-food vision systems, ensembles can be formed by:

a) **Model Variants:** U-Net-like CNNs combined with transformer models like Swin Transformer.
b) **Data Augmentation Ensembles:** Training the same model on differently augmented datasets.
c) **Cross-domain Ensembles:** Combining perhaps three models based on RGB, hyperspectral, and thermal modalities.

For example, an ensemble of YOLOv8 and Mask R-CNN can provide real-time fruit detection and high-precision segmentation for post-processing. In food inspection pipelines, ensembles also reduce the risk of false positives during foreign object detection, thereby ensuring greater compliance with safety standards in those operations.

However, the computational expense of these techniques is greater, but they are very indispensable in the applications where the utmost precision and dependability are required, for instance, food safety, crop disease outbreak prevention, and yield forecasting.

4.7 Datasets and Tools

The successful application of segmentation and object detection based on deep learning in crop and food imaging mainly depends on high-quality datasets and tools as the primary factors. In agriculture or food systems, data is usually collected in uncontrolled conditions, and therefore variations in lighting, occlusion, cluttered backgrounds, and scale differences make the choice of datasets and tools even more important. The major datasets, annotation platforms, and deep learning frameworks that are highly regarded and have a considerable impact on the research and application of vision modeling in these areas are discussed in this part of the article.

4.7.1 Popular Datasets

4.7.1.1 PlantVillage Dataset

The PlantVillage dataset stands among the most prominent open-source datasets for plant disease classification and segmentation. The database consists of 50,000 highly detailed leaf images from over 14 crop species and 26 diseases, along with healthy classes. It provides a real version as well as one with the background removed, allowing it to be used either for training or transfer learning.

- **Use Case:** Disease classification, semantic segmentation.
- **Format:** RGB images, labeled per class.
- **Notable Application:** Early detection of tomato leaf mold, corn rust, and grape blight.

4.7.1.2 Fruits-360 Dataset

More than 90,000 images of fruits and vegetables in a controlled environment are included in this dataset. Being preprocessed to be centered and free of background, they help with object classification and early model prototyping.

- **Use Case:** Fruit classification, training lightweight detection models.
- **Format:** 100 × 100 RGB images with class labels.
- **Limitations:** It cannot be used for segmentation or natural scene detection because of the uniform background.

4.7.1.3 DeepWeeds

It is a field-collected database of 17,000 images of eight invasive weed species in agricultural landscapes in Australia. Due to changes in illumination, backgrounds, and weed morphology, the dataset presents a challenge.

- **Use Case:** Object detection and weed-crop segmentation.
- **Format:** Annotated images with bounding boxes.
- **Noble Feature:** Presence of GPS and sensor metadata, which helps in context-aware detection models.

4.7.2 Annotation Tools

Precise labeling is necessary in the training of robust segmentation and detection models. The various open-source annotation tools have been developed according to the diverse annotation requirements, from bounding boxes to pixel-level masks.

4.7.2.1 LabelImg

LabelImg is a straightforward annotation tool based on Python for labeling with bounding boxes. It supports PASCAL VOC and YOLO formats and is suitable for object detection workflows.

- **Advantages:** It is light, fast, and easy to install.
- **Limitations:** It is not the best choice for segmentation or polygon annotation.

4.7.2.2 VGG Image Annotator (VIA)

VIA was developed by the Visual Geometry Group at the University of Oxford. It is a powerful tool for annotation of bounding boxes, polygons, and pixel-level masks. It works in a browser and hence does not require any installation.

- **Use Case:** Tasks involving semantic and instance segmentation.
- **Features:** Supports attribute tagging and relationships among image regions.
- **Format:** JSON export, and should be supported by most deep learning frameworks.

4.7.3 Frameworks and Libraries

Building and deploying segmentation and object detection models requires a set of strong frameworks that provide modular design, pretrained weights, and ease of experimentation. These are the commonly used libraries in agri-food imaging projects:

4.7.3.1 TensorFlow and Keras

TensorFlow (by Google) – while Keras is its high-level API – provides an extensive ecosystem for setting up CNN and Transformer-based architectures. Keras provides a simple interface for defining and training models, while TensorFlow enables deployment on cloud, edge, and mobile platforms.

- **Use Case:** U-Net, Mask R-CNN, MobileNet segmentation.
- **Strengths:** Excellent documentation, support for TensorFlow Lite on the edge.
- **Visualization:** TensorBoard support is there for visualization and tracking of training metrics.

4.7.3.2 PyTorch

Meta AI's PyTorch is loved for its dynamic computation graph, ease of debugging, and strong community support. It enables quick prototyping and has become the de facto implementation in academic research.

- **Use Case:** ViTs, Swin Transformer, YOLOv5/YOLOv8 models.
- **Strengths:** Native GPU acceleration, ONNX support.
- **Tooling:** Perfect work with Jupyter, WandB, and other ML pipelines.

4.8 Challenges and Limitations

Computer vision and deep learning are very powerful technologies nowadays. Nevertheless, segmentation and object detection models are still hard to apply in agriculture and food systems because of a variety of challenges. Such challenges are caused by the difficulties in nature scenes, the scarcity of data, the need for instantaneous processing, and the inability to generalize. The primary problems and restrictions, which are going to be addressed in this part of the paper, are the very ones that have to be recognized by every single researcher and practitioner intending to build a solid and widely used vision-based solution.

4.8.1 Environmental Variability

Among the other challenges, environmental variability is one of the greatest that affect object detection and segmentation in the food and crop imaging domain. Unlike in laboratory experiments, agricultural fields and food processing zones are very dynamic and unpredictable.

a) **Lighting Conditions:** Along with sunlight exposure, the existence of shadows and variations due to time of day alter pixel intensities and imply color features that eventually hamper the generalization capability of segmentation models conventionally trained under set lighting.

b) Occlusion and overlapping objects can be fruits being completely hidden by leaves, vegetables in trays being partly overlapped, or some weeds being partially obscured. Difficulties in seeing clearly the borders between the objects and their astonishingly creating problems, for instance, segmentation models like Mask R-CNN and panoptic segmenters in general are the consequences that the situations bring along.

- **Mitigation Approaches:**

a) Advanced data augmentation techniques (e.g., brightness/contrast jitter, shadow simulation).

b) Domain adaptation and fine-tuning using target-specific data.

4.8.2 Limited Annotated Data

A major bottleneck that continues to arise in agricultural and food imaging is still the dearth of annotated datasets. Most of the large image datasets (e.g., COCO, ImageNet), being generic in nature and urban-centric in their domains, lack the fine-grained, domain-oriented information required in agri-food work.

a) **Time-Intensive Labeling:** Labeling at a pixel level for leaves, pests, or food contaminants is tedious and requires domain expertise.

b) **Under-Represented Geographic Diversity:** The datasets capture specific crops from specific regions, specific seasons, or specific soil conditions, limiting their generalization.

- **Mitigation Approaches:**

a) Synthesizing data via GANs or simulation engines.

b) Weak supervision, where models get trained with only coarse labels or partial annotations.

4.8.3 Real-Time Constraints

Latency can vary from critical for inference, exercising several agricultural and food processing conditions. Autonomous tractors and drone surveillance for quality control on conveyor belts need the least possible latency.

a) **High Computational Load:** Fast and accurate, deep-learning models such as Faster R-CNN or DETR are computationally heavy and GPU-intensive.

b) **Deployment on Edge Devices:** Some agri-food systems may be based on embedded systems such as a Raspberry Pi, NVIDIA Jetson, or microcontrollers with available memory and low processing capacity.

- **Mitigation Approaches:**

a) **Model Compression:** Methods such as pruning (eliminating redundant weights), quantization (using low-precision arithmetic), and distillation (transferring knowledge to smaller models).

b) **Hardware Acceleration:** Using specialized chips such as TPUs or FPGAs to speed up inference.

4.8.4 Generalization Across Contexts

One critical obstacle in vision-based agricultural and food applications is restricted generalization. Models trained on one dataset or in one region may fail when applied under different circumstances.

a) **Regional Variation:** Same crop species may differ in appearance from one geographic region to another based on varying climate, soil type, irrigation, and cultivation practices.
b) **Seasonal Changes:** Changes in the morphological features of plants and fruit occur with their growth stages and thus across seasons. A model that has been trained on mature mangoes would not be able to detect the unripe or ripening stages effectively.

- **Mitigation Approaches:**

a) Cross-domain transfer learning and continual learning to build an adaptation to new data without initially having to retrain the model from the start.
b) Meta-learning and few-shot learning methods that facilitate quick adaptation to previously unseen categories with as little data as possible.

4.9 Future Directions

In the agricultural automation and food quality monitoring domains, we are witnessing a paradigm shift, mainly driven by the rapid growth in deep learning, computer vision, and sensor technologies. Nonetheless, to fully realize segmentation and object detection techniques in practical applications, future research must target the challenges of deployment, scalability, transparency, and data diversity. This section describes some of the key future directions set to shape the next generation of intelligent agri-food imaging systems.

4.9.1 Edge AI and IoT Integration

An elaborate move is on to shift AI processing from cloud servers into edge devices, also known as Edge AI. Thus, a real-time decision can be made on-site without incurring the time delay for data to be transferred over the Internet.

Devices such as NVIDIA Jetson Nano, Google Coral, and Raspberry Pi with deep learning accelerators allow lightweight deployment of segmentation and detection models in tractors, drones, or handheld tools. Fruit-counting systems with YOLOv5-nano may now run on low-power cameras, carrying out harvest monitoring in actual fields.

Further, combining AI with IoT sensors allows continuous monitoring. AI-enabled cameras on drones and autonomous robots inspect the large fields autonomously, detect the onset of symptoms of diseases, and initiate timely responses such as pesticide spraying or irrigation.

This merging of AI and IoT will be key in the evolution of scalable, energy-efficient, and autonomously functioning precision agriculture.

4.9.2 Synthetic Data Generation

The lack of richly annotated datasets is a common bottleneck in the development of agricultural vision applications, especially in the case of rare diseases or unusual crops. Alternatively, synthetic data generation of agricultural images by GANs might be able to fill this void.

GANs can generate crop images that look real, simulate diverse lighting conditions, or generate composite images of leaves affected by diseases that are rarely found in real life. Unrealistic or Unity-based simulators can take fields, weather effects, and growth stages into consideration to create a labeled dataset that reflects true-to-life variability.

Such synthetic datasets not only increase the training amount but can also be used for data balancing and domain adaptation so that models can be generalized to different locations and seasons.

4.9.3 Explainable AI (XAI)

With the continuous integration of AI in critical tasks of the food and agriculture sector, the requirement for explainability is increasing. The farmers, agronomists, and policymakers need to know the reasoning behind the model's decision if there is a case of the detection of disease or the classification of fruit as overripe.

Image areas that had the most impact on the decision can be visualized through the use of explainable AI techniques such as Grad-CAM, attention heatmaps, and saliency maps. The trust for the models is not only built up as a result of this process, but it also helps in debugging and improving the models.

In cases where there might be food contamination detection or plant disease diagnosis, the explainability should not only be a characteristic of AI systems but also a reason for making human-in-the-loop systems indispensable so that the decisions made by AI are more accountable and trustworthy.

4.10 Conclusion

Object detection and image-segmentation methods are among the most systematic and intelligent solutions to make agriculture and food processing activities more sustainable, smart, and data-driven. Having gained more ground in modern agricultural applications, these techniques are chiefly responsible for the detection of plant diseases and estimation of fruit yields while being involved in food quality grading and implementation of food safety as well. The systems that allow machines to see, group, and assess visual objects with an extremely high degree of accuracy have already started to change the whole process of crop monitoring and food assessment throughout the value chain.

At the start of the chapter, we looked at segmentation and object detection through the lens of traditional methods and then moved on to deep learning methods such as U-Net, YOLO, Mask R-CNN, and ViTs. Each of these techniques has its specific advantages depending on the agri-food context characterized by such factors as accuracy, speed, and the available resources for operations. We also talked about the need for the practical tools, datasets, and workflows that facilitate the training, evaluation, model building, and deployment processes in the real world.

Technological evolution has not eliminated all problems. The models' variances in performance depending on the input, which is due to changes in lighting, obstructions, and different types of crops, are among the problems. Then we have the lack of annotated datasets, especially for crops grown in specific regions or for rare diseases, which is another issue that hinders model performance. Moreover, real-time systems need to be made ready for rural or edge environments with limited resources by going through quantization, pruning, and transfer learning.

Future research directions are bright and promising, and they propose solutions that could effectively tackle the problems mentioned above. Edge AI combined with IoT will allow real-time, on-site analyses to take place at both the farm and food processing unit levels. The use of GANs in the generation of synthetic data could result in creating more training data for the many scenarios that are underrepresented. Furthermore, multimodal learning and sensor fusion could be a new path in developing richer, more context-aware insights. Explainable AI techniques might be used to win over users by disclosing the reasoning behind the model's decision and thus making the decision more acceptable.

In conclusion, the applications of segmentation and object detection have evolved from being purely academic disciplines to becoming the very groundwork for the agricultural and food future that was promised. The world would only be able to approach food production that is smart, safe, and sustainable enough to meet the demands of both 21st-century producers and consumers if continuous improvements in these tools are made, along with overcoming current limitations.

Acknowledgment

The authors want to express their utmost gratitude to all the people who were in some way or other responsible for the successful completion of the project. Their help, in terms of stimulating ideas and giving constructive criticism, as well as motivation, was great throughout the writing of the chapter, and we owe much to the mentors, colleagues, and peers mentioned above.

Our deepest gratitude goes to our beloved Institutions for the provision of resources, facilities, and a research-conducive environment that were necessary. The broader research community also has a right to our thanks, as their papers and datasets have greatly contributed to the establishment and validation of our findings.

We would like to thank our families and friends for being so patient, understanding, and supportive during the research process, which was indeed very necessary.

References

1 Patrício, D.I. and Rieder, R. (2018). Computer vision and artificial intelligence in precision agriculture for grain crops: A systematic review. *Computers and Electronics in Agriculture* 153: 69–81.

2 Kamilaris, M. and Prenafeta-Boldú, F.X. (2018). Deep learning in agriculture: a survey. *Computers and Electronics in Agriculture* 147: 70–90.

3 Sladojevic, S., Arsenovic, M., Anderla, A. et al. (2016). Deep neural networks based recognition of plant diseases by leaf image classification. *Computational Intelligence and Neuroscience* 2016: 1–11.

4 dos Santos Ferreira, J.P., Matte Freitas, D., Gonçalves da Silva, G. et al. (2017). Weed detection in soybean crops using ConvNets. *Computers and Electronics in Agriculture* 143: 314–324.

5 Mishra, P. (2019). Recent trends in food quality evaluation using computer vision. *Trends in Food Science & Technology* 85: 263–276.

6 Ronneberger, O., Fischer, P., and Brox, T. (2015). U-net: Convolutional networks for biomedical image segmentation. In: International Conference on Medical image computing and computer-assisted intervention, 234–241. Cham: Springer International Publishing.

7 Jocher, G. (n.d.). YOLO by Ultralytics: Real-time Object Detection for Industry and Research.

8 Too, R. (2019). A comparative study of fine-tuning deep learning models for plant disease identification. *Computers and Electronics in Agriculture* 161: 272–279.

9 Selvaraju, R. (2017). Grad-CAM: visual explanations from deep networks via gradient-based localization. *ICCV*.

10 Wang, L. (2022). Few-shot learning for plant disease recognition. *Proceedings of the AAAI Conference on Artificial Intelligence*.

11 Beemkumar, N., Gupta, S., Bhardwaj, S. et al. (2023). Activity recognition and IoT-based analysis using time series and CNN. In: Handbook of Research on Machine Learning-enabled IoT for Smart Applications Across Industries (ed. N. Goel and R.K. Yadav), 350–364. IGI Global https://doi.org/10.4018/978-1-6684-8785-3.ch018.

12 Pandey, J.K., Kumar, S., Lamin, M. et al. (2022). A metaheuristic autoencoder deep learning model for an intrusion detector system. *Mathematical Problems in Engineering* 2022: 11, 3859155. https://doi.org/10.1155/2022/3859155.

13 Talukdar, V., Veeraiah, V., Roy, D.R. et al. (2023). An innovative approach: typicality-based collaborative filtering for recommender systems. *Proceedings of the 3rd International Conference on Advancement in Electronics & Communication Engineering (AECE)*, pp. 863–868. https://doi.org/10.1109/AECE59614.2023.10428576.

14 Srinivas, P., Veeraiah, V., Thejaswini, K.O. et al. (2025). The impact of electromagnetic and optical radiation on physical and biological systems. *Journal of Nano- and Electronic Physics* 17 (4): 04012. https://doi.org/10.21272/jnep.17(4).04012.

15 Pandey, J.K., Yadav, A.K., Purushotham, M. et al. (2022). Colorimetric detection of deltamethrin pesticide from biological samples based on the peroxidase-mimic catalytic activity on MnO_2/SnO_2. *Key Engineering Materials* 928: 45–51. https://doi.org/10.4028/p-4ti1i0.

16 Pandey, J.K. (2025). Understanding the impact of electromagnetic and optical radiation on physical and biological systems. *Journal of Nano- and Electronic Physics* 17 (5): 05023. https://doi.org/10.21272/jnep.17(5).05023.

5

Hyperspectral Image Analysis for Nondestructive Estimation of Crop Nutrient Composition

Devanakonda Venkata Sai Chakradhar Reddy[1], *Selvaprakash Ramalingam*[1], *and Divya Dharshini Saravanan*[2]

[1] *Division of Agricultural Physics, ICAR - Indian Agricultural Research Institute, New Delhi, India*
[2] *Department of Agricultural Meteorology, Chaudhary Charan Singh Haryana Agricultural University, Hisar, India*

5.1 Introduction

The main objective of efficient and sustainable nutrient management in today's agriculture is to enhance crop production, improve resource utilization, and contribute to global food security [1]. The demands being placed on agriculture by an increasing world population, expected to grow to approximately 10 billion people by 2050 [2], will require the industry to increase food production using fewer resources than ever before. Increasing environmental restrictions on the amount of available arable land mean there has never been a greater need for new technology to accurately assess crop nutritional status over large areas. Traditional nutrient diagnostic procedures rely upon the collection of samples from crops, followed by laboratory analysis of the collected samples. While traditional methods provide accurate and consistent measures of nutrient status, they are time-consuming, expensive, and labor intensive. Most importantly, traditional methods do not lend themselves well to frequent sampling across a large area of crops, nor do they allow for quick decision-making during rapidly changing conditions found in fields [3]. Sample collection, laboratory analysis, and ultimately transforming the results into useful decisions create a delay. Furthermore, traditional methods are destructive; thus, repeated sampling is impossible [4]. Hyperspectral imaging (HSI), a type of remote sensing technology, has become increasingly recognized as a viable alternative to traditional methods for continuous crop monitoring. HSI integrates two concepts, spectroscopy and digital imaging, to collect reflected light at numerous wavelengths. HSI captures the spectral reflectance at hundreds of discrete and contiguous wavelength bands in a single image, providing a wealth of biochemical and physiological information about the crop canopy [5]. Due to hyperspectral images' ability to identify minute differences in plant physiology, hyperspectral images are especially appropriate for detecting nutrient deficiencies. Nutrient deficiencies at early stages of development cause biochemical and morphological changes in plants, such as reduced chlorophyll content, changes in the water balance of plants, and disruption to cellular structures, which are detected in the spectral reflectance characteristics of plants [6].

Nitrogen-deficient plants typically result in reduced chlorophyll content, resulting in increased reflectance in the visible spectrum (notably in the red part of the spectrum) and a change in the red edge (near 700 nm) [7]. Phosphorus and potassium-deficient plants also alter specific spectral characteristics associated with photosynthesis, stomatal conductance, and cell wall integrity [8]. When hyperspectral data is combined with advanced analytic techniques such as machine learning (ML) and deep learning (DL) models, the utility of hyperspectral data is greatly increased. ML techniques such as partial least squares regression (PLSR), support vector regression (SVR), and random forest (RF) are capable of modeling the complex, nonlinear relationships between hyperspectral features and nutrient concentrations in plants. For example, [9] demonstrated how SVR was able to surpass PLSR when estimating rice leaf nitrogen content via hyperspectral LiDAR data. Also, recent studies have shown that the use of unmanned aerial vehicle (UAV) hyperspectral imagery combined with DNNs and PLSR has led to increased accuracy in the mapping of rice leaf nitrogen content [10]. Additionally, hybrid models incorporating physically based PROSAIL-PRO models along with ML methods (Gaussian process regression) have been effective in mapping nitrogen concentrations aboveground in agricultural environments [11]. DL models, specifically convolutional neural networks (CNNs), can also add value by identifying the spatial and contextual information contained in hyperspectral cubes. Such analytic approaches allow for predictive models that can accurately estimate nutrient levels with high accuracy, robustness, and generalization [12]. An additional advantage of HSI is its ability to collect data from various spatial scales and platforms. HSI systems can be mounted on the ground, UAVs, manned aircraft, or satellites [13]. Ground systems can provide the highest resolution and, therefore, are most often used in calibration/validation studies. Systems mounted on UAVs can provide a good trade-off between resolution and coverage and are most appropriately used to monitor an area within an individual field. Although the spatial resolution of HSI systems mounted on satellites will be less than that of those mounted on ground systems/UAVs, they do offer a means of conducting regional/global assessments of agriculture, which are critical for developing plans for large-scale agricultural production and monitoring food security [14]. Preprocessing of hyperspectral data is necessary to ensure that the data collected is reliable and accurate. Raw hyperspectral data is often corrupted by noise, variability in lighting conditions, sensor errors, and atmospheric effects. Therefore, to improve the quality of signals from hyperspectral images and remove redundant data, various preprocessing methods like radiometric calibration, geometric correction, noise filtering (e.g., Savitzky-Golay smoothing), and dimensionality reduction (e.g., principal component analysis [PCA]) are used [15]. As well as to increase the efficiency of subsequent analytical models, these preprocessing techniques ensure the compatibility of results obtained using different equipment and under different conditions. With the ability to analyze a wide variety of agricultural products, including cereals (wheat, rice, maize), oilseed crops (canola, sunflower), legume crops, and horticulture crops (tomato, lettuce, grapevine), HSI has gained significant recognition in recent years. It was shown by numerous studies conducted in field conditions that HSI can be used to determine the content of nitrogen (N), phosphorus (P) and potassium (K) in the soil under different agroclimatic conditions [6]. The use of HSI in combination with other types of sensing technologies (thermal imaging, LiDAR, RGB imaging) expands the scope and reliability of the diagnosis of nutrients by means of data fusion [16]. Using multiple sensing technologies allows for an integrated evaluation of the health status of plants by combining physiological, structural, and temperature-related information. Although it has great potential, there are some barriers that may limit the practical application of HSI in commercial agriculture [17]. First, the high price of hyperspectral sensors and specialized hardware for processing hyperspectral image data creates a barrier for small-scale producers and countries or regions with limited resources. Second, due to the fact that the response of plants to light varies depending on species, stage of growth, and environment, the development of reliable and portable models is necessary. Current research is focused on creating large databases of spectra, the implementation of transfer learning methods to allow for adaptation of models between domains, and the creation of low-weight systems that enable real-time processing of hyperspectral data directly at the farm [18]. Some emerging trends in hyperspectral research relate to the miniaturization of hyperspectral sensors, the connection of the sensors to IoT, and their integration into smart-farming systems. All these new trends are expected to make hyperspectral technology more accessible and available to farmers worldwide, which will enable them to apply it in precision agriculture [19]. For instance, edge computing solutions that enable on-board processing of hyperspectral data can significantly shorten the time required to process the data, and thus, provide real-time decision-making support for management of nutrient availability. Cloud-based platforms and decision-support systems that combine hyperspectral data with other environmental and agricultural inputs are also being developed to provide farmers, agronomists, and policymakers with actionable recommendations [20]. In summary, HSI in the context of plant nutrition represents a revolution in diagnostics of plant nutrition from traditional approaches based on destructive sampling, to modern, data-driven and intelligent approaches to agriculture. Through providing a nondestructive, spatially resolved, and real-time assessment of nutrient availability in plants, HSI contributes to the optimization of fertilizer usage, minimizing the negative impact of agriculture on the environment, and increasing crop yields. Therefore, HSI is positioned as one of the cornerstones of the next generation of precision farming systems that are sustainable, efficient in terms of resources, and adaptable to changing climatic conditions.

5.2 Fundamentals of Hyperspectral Imaging

The primary way HSI works is to measure how plants reflect various wavelengths of electromagnetic energy over a very wide spectrum (from 400 to 700 nm for the visible and from 700 to 2500 nm for the near infrared), and the hundreds of bands it measures provide very specific and precise spectral signatures about each plant's physiological state [5]. Multispectral images have fewer, wider bands of light; hyperspectral images have many more, narrower bands of light, allowing for identification of slight differences in a plant's physiological state [5]. Hyperspectral signatures are directly related to the physical characteristics of plants, including concentration of pigments (chlorophyll, carotenes, etc.), concentration of water, and structural properties of the plant [21]. Deficiencies of nutrients in plants cause physiological changes, such as chlorotic leaf color, stunted growth, or alterations in water balance, and these changes result in distinct spectral patterns [22].

Nitrogen-deficient plants generally have less chlorophyll than nutrient-sufficient plants. This reduction in chlorophyll content results in a decrease in reflectance in the red-edge region (approximately 680–750 nm), where most of the photosynthetic activity takes place [23].

5.3 Nutrient Detection and Spectral Sensitivity

5.3.1 Macronutrient Estimation

The HSI technology offers very detailed measurements of the macronutrient content in plants using wavelength-based absorption and scattering of light according to the chemical composition of each macronutrient (Table 5.1). The relationship between the macronutrient nitrogen (N), which is part of chlorophyll and proteins, and plant response to light is directly related to the light absorption in the red-edge (680–750 nm) and near-infrared (750–1300 nm) parts of the spectrum. Nitrogen deficiency in plants generally results in less chlorophyll absorbing light, thus creating a measurable difference in the red-edge area of the spectrum; the use of first-derivative spectral indices enhances the sensitivity of the detection of small variations in the spectral information [27, 28]. Phosphorus (P) deficiency can also cause some visible effects on plants as it affects both the pigment ratio within the plant and the transfer of energy in photosynthesis processes, resulting in differentiable features in the blue (430–470 nm) and green (530–570 nm) areas of the spectrum. The use of the photochemical reflectance index (PRI) helps to differentiate between phosphorus-stressed plants [29]. Potassium (K) deficiencies are primarily related to the ability of the plant to regulate water and maintain turgor pressure, which affects mesophyll scattering properties. The reflectance patterns of potassium-deficient plants are different in the short-wave infrared (SWIR) portion of the spectrum, specifically at the 1500–1800 nm and 2000–2400 nm ranges, where there is a large amount of absorption of light due to water [28]. Advanced methods, including PLSR, RFs, and hybrid ensemble modeling techniques, allow for better predictions of nutrient levels using full-spectrum hyperspectral imagery [27, 29].

Table 5.1 Spectral sensitivity and deficiency signatures of key macronutrients.

Macronutrient	Physiological role	Deficiency effect	Spectral sensitivity	References
Nitrogen (N)	Chlorophyll and protein synthesis, photosynthetic capacity	Chlorosis, reduced leaf area, lower biomass	Red edge (680–750 nm), NIR (750–1300 nm)	[24]
Phosphorus (P)	Energy transfer (ATP/ADP), root, and reproductive development	Dark green foliage, poor root growth, delayed maturity	Blue (430–470 nm), green (530–570 nm)	[24]
Potassium (K)	Osmotic regulation, enzyme activation, and stomatal conductance	Leaf scorching, poor water balance, weak stems	SWIR (1500–1800 nm; 2000–2400 nm)	[25, 26]

Table 5.2 Spectral sensitivity and deficiency signatures of key micronutrients.

Micronutrient	Physiological role	Deficiency effect	Spectral sensitivity	References
Iron (Fe)	Chlorophyll biosynthesis, electron transport	Interveinal chlorosis, reduced photosynthesis	Blue (400–500 nm), red edge (680–720 nm)	[30]
Zinc (Zn)	Enzyme activation, auxin synthesis, protein metabolism	Stunted growth, small leaves, reduced flavonoids	UV–blue (<450 nm), secondary metabolite bands	[31]
Manganese (Mn)	Photosystem II activity, chlorophyll fluorescence regulation	Leaf speckling, impaired electron transport	Green–red transition, fluorescence indices	[32]

5.3.2 Micronutrient Monitoring

Since micronutrient diagnostics are inherently a lot more complicated, since micronutrients like Iron (Fe), Zinc (Zn), and Manganese (Mn) create secondary, subtle physiological effects, as an example iron (Fe) deficiency impairs chlorophyll biosynthesis, which results in interveinal chlorosis that can be identified by HSI from changes to the blue region (400–500 nm) and minor changes to the red edge (around 680–720 nm) [30]. Zinc (Zn), which has been determined to be vital for activating enzymes and synthesizing auxins, will affect flavonoid and protein metabolism and therefore have detectable shifts in the ultraviolet–blue region (<450 nm) in hyperspectral imagery [31] (Table 5.2). Manganese (Mn), which is necessary for photosystem II activity and electron transport, affects chlorophyll fluorescence; its deficiency produces measurable spectral changes at the green–red transition [32]. To increase the detection of such micronutrient stress requires high signal-to-noise sensors, spectral derivative analysis, and chemometric modeling techniques, such as SVR or CNNs [33]. In addition, combining HSI with either fluorescence or thermal imaging through data fusion frameworks increases the ability to detect micronutrient-specific stress and enables early intervention prior to visible symptoms appearing [32].

5.4 Data Acquisition Platforms

The hyperspectral diagnostics scalability is dependent upon acquisition platforms. Ground-based acquisition systems (i.e., handheld spectroradiometer and tractor-mounted sensor) can acquire high-spectral fidelity and high-spatial resolution (Figure 5.1) and are therefore necessary for calibration, physiological studies, and algorithm validation [22, 24]. They have limitations in terms of coverage and manpower requirements that limit their ability to be used on a field-wide basis. UAVs allow for the flexibility of flight planning, submeter resolution, and rapid data collection of multiple fields (Table 5.3). UAVs are equipped with sensors that range from VNIR (400–1000 nm) to SWIR (up to 2500 nm), which allows for early stress detection and nutrient mapping [14, 36]. UAVs can also integrate GPS, LiDAR, cloud-based analytics, etc., to support the scalable precision agriculture, but there are

Figure 5.1 Different image acquisition platforms and sensors.

Table 5.3 Different image acquisition platforms and sensor details.

Platform	Example sensors	Key specs/features	References
Satellite	PRISMA (Italy), DESIS (DLR), EnMAP (Germany), HyspIRI (NASA, planned)	400–2500 nm, 30 m spatial resolution (typical), global coverage	[5]
Manned Aircraft	AVIRIS (NASA), HyMap, CASI (Itres)	High SNR, flexible campaigns, large-area coverage	[14]
UAV (Drones)	Headwall Nano-Hyperspec, Cubert UHD 185-Firefly, Resonon Pika	400–1000 nm (VNIR) or 400–2500 nm (SWIR), lightweight (<2 kg), portable for precision agriculture	[5, 34]
Ground-based/ Proximal	FieldSpec (ASD Inc.), HySpex (Norsk Elektro Optikk)	Portable spectroradiometers, close-range plant and soil analysis, high spectral fidelity	[14]
Laboratory Imaging	Push-broom line scanners, Snapshot sensors	Used for food quality, plant physiology, medical imaging, and high control over illumination	[35]

several operational barriers to using UAVs, including flight regulations, battery endurance, and data volume [14]. Satellites, such as PRISMA and the upcoming SBG (Surface Biology and Geology) mission by NASA, will allow for large-scale monitoring with broad spectral coverage. Although satellites are subject to coarse resolution (10–30 m) and atmospheric noise, they are essential for regional nutrient dynamics [37, 38]. Ultimately, the most effective strategy is to use a multiscale monitoring framework that uses ground calibration, UAV precision mapping, and satellite synoptic coverage.

5.5 Preprocessing and Data Handling

Preprocessing is a necessity for hyperspectral data to ensure that both the signal remains intact and that computation time can be minimized. The radiometric correction process will convert the raw digital number of hyperspectral data into surface reflectance and correct for differences between the sensor used to collect the data and how the atmosphere distorts what we see [39]. The geometric correction process includes orthorectification, using GPS/IMU data, and co-registering the hyperspectral data with a basemap in order to provide an accurate location of the spectral pixels to ground truth [22]. In addition, noise reduction is essential for removing noise from hyperspectral data when the signal-to-noise ratio is low, especially at the band edge and in the SWIR region. Smoothing techniques (Savitzky-Golay), Minimum Noise Fraction (MNF) transformation, and wavelet-based denoising remove spurious noise while preserving diagnostic absorption features [40]. Dimensionality reduction methods such as PCA, ICA, and band selection based on information theory reduce redundant bands in hyperspectral data cubes and, therefore, decrease dataset size and computational requirements by extracting the most biologically important features [41].

5.6 Machine Learning and Analytical Techniques

ML models are capable of modeling linear and nonlinear relationships between hyperspectral data and nutrients as well as exploiting spatial-spectral dependencies within hyperspectral cubes by use of CNNs or hybrid three-dimensional two-dimensional CNN frameworks; therefore they can represent a wide variety of possible reflectance–nutrient interactions [42, 43]; additionally hyperspectral datasets have complex biochemical–spectral interactions, and as such require an advanced computational method for nutrient estimation [44, 45].

5.6.1 Regression Models

PLS regression continues to be a standard tool for hyperspectral nutrient prediction because it can effectively manage the high levels of collinearity present among the hundreds of contiguous bands within hyperspectral data. The use of PLS regression allows for the development of interpretable models as well as for the extraction of meaningful predictive information related to macronutrients [46]. Support Vector Machine (SVM) regression utilizes kernel functions to better represent nonlinear reflectance–nutrient relationships. SVM regression can perform well under conditions where the data are heterogeneous due to the presence of environmental factors that complicate the application of linear models; however, the accuracy of this type of model will depend upon the optimization of the parameters used [47, 48].

5.6.2 Deep Learning Methods

CNNs have been shown to improve the ability to model both spectral and spatial characteristics through the representation of hierarchical features derived from spectral information and/or spatial information. These hierarchical representations allow CNNs to identify subtle absorption patterns as well as canopy texture and are particularly effective when analyzing heterogeneous fields and evaluating multiple nutrients at once [49]. Recurrent Neural Networks (RNNs), such as Long Short-Term Memory networks (LSTMs), can also model temporal dependencies between consecutive bands and therefore provide a means for predicting changes in nutrient content over time [50, 51]. The quality of these models can be evaluated using a variety of metrics (RMSE, R^2, MAE), and using cross-validation methods can help ensure that they apply to a broader set of data. Additionally, the combination of regression and DL approaches (i.e., ensembles) can also increase the overall accuracy of nutrient predictions for practical applications [52, 53] (Table 5.4).

5.7 Case Studies and Applications

5.7.1 Cereal Crops

HSI, especially in the context of early detection of nitrogen deficiency in cereal crops like wheat and rice, is rapidly becoming a valuable tool due to its ability to detect variations in reflectance in various spectral regions. The red edge (680–750 nm) and near infrared (NIR, 750–1300 nm) are particularly relevant as they both respond to changes in leaf chlorophyll and canopy structure that accompany N status changes; thus, many researchers have found strong correlations between these two regions of interest and N status. Researchers have also utilized field-based studies employing UAV-based hyperspectral sensors to demonstrate high levels of predictive accuracy of estimating N content when utilizing ML algorithms such as PLSR and SVR. For example, [62] developed a model that had an $R^2 > 0.85$ in predicting the nitrogen concentration in winter wheat when using optimized regression models in conjunction with UAV-based HSI. Additionally, [63] showed that UAV-based hyperspectral images were able to successfully capture the variation in nitrogen in the canopy of wheat at varying levels of spatial resolution. Maps of nutrients that have been created from hyperspectral image data may be used to integrate with variable rate fertilizer application systems that improve the nitrogen use efficiency (NUE) and minimize nitrogen loss through leaching or volatilization.

5.7.2 Oilseed Crops

In recent years, HSI has been used to monitor levels of phosphorus and potassium in oilseed crops, including canola and sunflowers. Plants deficient in phosphorus have been shown to display changes in their reflectance in the blue-green range of wavelengths (430–570 nm), whereas deficiencies in potassium have been found to affect the water-related scattering signals that occur in the SWIR wavelength ranges (1500–1800 nm and 2000–2400 nm). In addition to being able to identify nutrient deficiencies, recent HSI field trials, along with ML classifiers (including RF and SVM), were able to classify nutrient deficiency levels at approximately 80%. This is expected to enable farmers to make proactive decisions in terms of adjusting their nutrient applications based on real-time information regarding crop nutrient status. For example, [64] demonstrated the use of a three-layer ensemble modeling framework that utilized hyperspectral data to achieve an accuracy of 80.76% when

Table 5.4 Details of DL approaches for abiotic stress.

Technique/model	Description and key characteristics	Example crops/ applications	Model performance/details	References
Partial Least Squares Regression (PLSR)	Classical regression for spectral calibration and prediction; robust baseline	Wheat, apple, citrus, *Macadamia*	$R^2 = 0.75–0.90$ for N, P, K	[54]
Support Vector Regression (SVR)	Nonlinear regression, resistant to overfitting in high-dimensional data	Rice, wheat, strawberry, soybean	$R^2 > 0.85$ for N/P estimation	[55]
Random Forest (RF)	Ensemble learning for feature selection, robust to noise/overfitting	Lettuce, fruit, General crops	Used for N, P, K, pigment, and quality estimation	[56]
Convolutional Neural Networks (CNN/3D-CNN/Hybrid)	Deep learning for joint spatial–spectral feature extraction	Wheat, quinoa, fruits, cowpea, millet	>94% accuracy for N/P stress, yield prediction	[12, 57]
Recurrent Neural Networks (RNN/LSTM/GRU)	Sequence modeling across wavelengths (spectral dependency)	Fruit, cereal grains, vegetables	Used for time-series/long-range spectral learning	[57]
Autoencoders (AE, Stacked AE)	Unsupervised nonlinear feature extraction, data compression	Tomato, general fruits/veg	Improves dimensionality reduction, data denoising	[58]
Generative Adversarial Networks (GANs)	Generate high-quality synthetic leaf images and improve model robustness	Rice	Improved classification accuracy	[59]
Transfer Learning (TL)	Utilizes knowledge from large general datasets for small agrifood datasets	New crops, small-sample yield/N stress	Fast adaptation, strong in limited data regimes	[60]
Active Learning (AL)	Interactive data selection to train ML models efficiently	Soil, disease spotting, remote imaging	Maximizes model improvement with minimal annotation	[61]

classifying stresses caused by nitrogen, phosphorus, and potassium in winter oilseed rape. The development of nutrient mapping and site-specific fertilization strategies is also enabled by these developments and will be expected to improve both seed quality and crop yields while minimizing losses of nutrients into the environment.

5.7.3 Horticulture Crops

High-value horticultural crops, like lettuce, tomatoes, and spinach, can be adversely affected by nutrient imbalance. HSI-based multinutrient monitoring, therefore, has the potential to greatly enhance high-value horticulture through multispectral monitoring. Research within controlled environment agriculture has demonstrated that CNNs can identify hyperspectral data's spatial-spectral characteristics with greater robustness than traditional ML techniques. A deep-attention network applied to hyperspectral data collected during greenhouse production of lettuce was able to achieve an R^2 of approximately 0.75 and an RMSE of approximately 2.02 when predicting lettuce chlorophyll content [65]. Although there are still many limitations associated with developing closed-loop fertigation systems using HSI-based multispectral monitoring, the use of HSI-based multispectral monitoring to provide real-time nutrient delivery adjustments in greenhouse growing conditions offers much promise in enhancing crop uniformity, improving the nutritional quality of crops, and increasing the market value of crops produced.

5.8 Data Fusion and Integration

The ability of HSI to diagnose is enhanced through the use of additional sensor modalities that are complementary to it, which will allow for more richly developed feature extraction and greater robustness to environmental confounders. HSI can be combined with other types of data, including thermal, LiDAR, fluorescent, and RGB images, to enhance the ability to separate nutrient stress from confounding variables such as soil background, illumination changes, and canopy structure. For instance, combining thermal and hyperspectral data can improve the discrimination of nutrient stress vs. drought since thermal data captures the temperature and transpiration of the canopy, and hyperspectral reflectance provides information on biochemical and pigment compounds. The complementary sensing strategy also provides a more reliable means to separate stress factors that have similar spectral manifestations [66]. Jointly estimating LiDAR-derived structural parameters of vegetation (e.g., canopy height, leaf area index) and hyperspectral-derived biochemical properties (chlorophyll, water, and nutrient concentrations) using LiDAR and HSI can be very powerful. The use of these two types of data together has been shown to improve the accuracy with which 3D maps of nutrient and canopy structure are produced; therefore, this fusion method may provide better insight into crop health and productivity than either technique separately [66]. Additionally, the coupling of hyperspectral and fluorescent sensors increases sensitivity to photosynthetic dysfunction, thereby enabling the early detection of micronutrient deficiencies before reflectance changes are evident [67].

5.8.1 HSI + LiDAR

Combining structural elements (canopy height, LAI, biomass density, etc.) from LiDAR data with biochemical properties obtained through HSI provides orthogonal information that may assist in distinguishing between similar spectral signatures when identifying stresses related to the environment or plant physiology. Studies have shown that there are improved estimates of nutrient levels and physiological characteristics in heterogeneous canopies that result from the combination of hyperspectral and LiDAR data using 3D Spectral-Point Cloud Fusion Frameworks. In addition, the integration of hyperspectral data with ensemble learning techniques that utilize multimodal data enables robust crop stress monitoring and diagnosis in field conditions (Fassnacht et al., 2016).

5.8.2 HSI + Thermal Imaging

Thermal imaging is significant for assessing the physiological characteristics of crops through measurement of temperature changes in canopies due to variation in water content in plants and variation in stomatal conductance. When combined with HSI, the physiological assessment of crops based on thermal imaging becomes an efficient method to identify potassium-related stress in terms of alterations in osmoregulation affecting transpiration cooling. Data fusion from thermal and hyperspectral images provides enhanced ability to distinguish between abiotic stresses (i.e., nutrient deficiency or drought), as it links the biochemical information derived from spectral reflectance with physiological indicators from transpiration cooling dynamics. Methods for combining thermal and hyperspectral image data have included pixel-wise combination methods (data level fusion) and decision level ensemble models (model stacking) to enhance early stress detection and improve crop resilience when subjected to varying environmental conditions [67].

5.8.3 HSI + RGB Imagery

While RGB data are limited in terms of spectral data, they provide the highest level of spatial resolution and context as compared to other types of remote sensing data; this is why RGB is so valuable for tasks like segmentation, canopy delineation, and identification of disease and nutrient stress symptoms. The addition of RGB to hyperspectral data can be used to increase the spatial and biochemical representation of features of the objects being analyzed by providing additional information on both spatial and biochemical characteristics. In addition to enabling object-based image analysis, RGB-hyperspectral fusions support multi-stream DL architectures, which use RGB data for spatial features and hyperspectral data for biochemical features. As examples of such applications, [68] created a multi-input CNN that combined RGB and hyperspectral data to identify the quality tier of bananas with an accuracy of 98% through the combination of color, size, and spectral data, while [69] developed a ShuffleNet-COCSP architecture that combined RGB and hyperspectral data to generate super-resolution hyperspectral images and was able to detect soybean kernel damage. These large-scale, multimodal DL frameworks enable precision agriculture applications to be more scalable, robust, and generalized than those using single-modal approaches.

5.9 Challenges and Future Directions

5.9.1 Challenges

There are many operational barriers to deploying HSI in agriculture. One barrier is the high cost of the HSI technology. Another is the demand for computers to process the large amounts of data generated by HSI systems. And finally, the technical nature of the HSI systems makes them difficult to adopt, especially for the smaller-scale farm systems that many farmers have. In addition to the initial costs and technical difficulties, there are also additional barriers, such as the need for specialized expertise to calibrate and interpret the data from HSI systems. Another significant barrier to using HSI is the bulk size of the ground-based systems and the payload limitations of unmanned aerial vehicles (UAVs), both of which can limit the ability to use HSI in a real-time manner for in-field monitoring of nutrient levels and to deploy over an entire field. One of the biggest remaining issues is the lack of standardization for HSI spectral nutrient stress signatures. Since the spectral nutrient stress signature varies among different crops, during different crop growth stages, and in different agro-ecological zones, this reduces the transferability of the spectral index and predictive model development [13]. Further reducing model transferability is the fact that algorithms developed to work with specific cultivars, seasons, or regions do not perform well when used in other locations because of differences in the underlying soils, atmospheric conditions, and plant physiological characteristics [60]. The domain dependency of these models is another example of how limited the ability to scale up HSI-based precision agriculture platforms and is therefore another example of how operationalizing HSI-based precision agriculture platforms may be challenging.

5.9.2 Future Directions

Advancements in future development will be focused on a curated database (spectral library), which will represent the signature of nutrients from all crops at all phases of growth in every environment and thereby allow for universal calibration. Cross-domain generalization will be enhanced through transfer learning, as well as adapting pretrained models of DL to use for new crops or new locations with little to no re-training. The combination of edge computing and lightweight artificial intelligence into UAVs and handheld sensors will provide the ability to perform real-time in-field nutrient analysis without reliance on cloud-based services. These three developments are designed to bring scalable, cost-effective, and operational HSI technology to Precision Agriculture.

5.10 Conclusion

HSI has transformed how we evaluate plant health by allowing us to determine nutrient deficiencies both noninvasively and spatially, while relating to plant physiological responses to the deficiencies. Therefore, HSI is ready for large-scale application of nutritional diagnostics when HSI is integrated with robust preprocessing techniques, advanced ML, and multiple sensor systems for data collection. Additionally, integrating HSI with

LiDAR, thermal, and RGB sensors will further enhance the predictive ability of HSI. New technologies, including spectral libraries, transfer learning, and edge computing, are also addressing many of the limitations associated with current applications of HSI. Overall, HSI represents a significant advancement in technology related to precision nutrient management and can contribute to improved agricultural productivity, resource efficiency, and sustainability.

References

1 Drinkwater, L.E. and Snapp, S.S. (2022). Advancing the science and practice of ecological nutrient management for smallholder farmers. *Frontiers in Sustainable Food Systems* 6: 921216.

2 United Nations, Department of Economic and Social Affairs, Population Division (2019). World population prospects 2019: highlights. United Nations. https://www.un.org/sustainabledevelopment/blog/2019/06/growing-at-a-slower-pace-world-population-is-expected-to-reach-9-7-billion-in-2050-and-could-peak-at-nearly-11-billion-around-2100-un-report.

3 Liu, J., Cai, H., Chen, S. et al. (2023). A review on soil nitrogen sensing technologies: challenges, progress, and perspectives. *Agriculture* 13 (4): 743.

4 Kashyap, B. and Kumar, R. (2021). Sensing methodologies in agriculture for soil moisture and nutrient monitoring. *IEEE Access* 9: 14095–14121.

5 Lu, B., Dao, P.D., Liu, J. et al. (2020). Recent advances in hyperspectral imaging technology and applications in agriculture. *Remote Sensing* 12 (16): 2659.

6 Pandey, P., Ge, Y., Stoerger, V., and Schnable, J.C. (2017). High throughput in vivo analysis of plant leaf chemical properties using hyperspectral imaging. *Frontiers in Plant Science* 8: 1348.

7 Debaeke, P., Rouet, P., and Justes, E. (2006). Relationship between the normalized SPAD index and the nitrogen nutrition index: application to durum wheat. *Journal of Plant Nutrition* 29 (1): 75–92.

8 Chen, C.T., Lee, C.L., and Yeh, D.M. (2018). Effects of nitrogen, phosphorus, potassium, calcium, or magnesium deficiency on the growth and photosynthesis of Eustoma. *HortScience* 53 (6): 795–798.

9 Du, L., Shi, S., Yang, J. et al. (2016). Using different regression methods to estimate leaf nitrogen content in rice by fusing hyperspectral LiDAR data and laser-induced chlorophyll fluorescence data. *Remote Sensing* 8 (6): 526.

10 Ma, X., Chen, P., and Jin, X. (2022). Predicting wheat leaf nitrogen content by combining deep multitask learning and a mechanistic model using UAV hyperspectral images. *Remote Sensing* 14 (24): 6334.

11 Berger, K., Verrelst, J., Féret, J.B. et al. (2020). Retrieval of aboveground crop nitrogen content with a hybrid machine learning method. *International Journal of Applied Earth Observation and Geoinformation* 92: 102174.

12 Okyere, F.G., Cudjoe, D., Sadeghi-Tehran, P. et al. (2023). Modeling the spatial-spectral characteristics of plants for nutrient status identification using hyperspectral data and deep learning methods. *Frontiers in Plant Science* 14: 1209500.

13 Beemkumar, N., Gupta, S., Bhardwaj, S. et al. (2023). Activity recognition and IoT-based analysis using time series and CNN. In: Handbook of Research on Machine Learning-enabled IoT for Smart Applications Across Industries (ed. N. Goel and R.K. Yadav), 350–364. Hershey, PA, USA: IGI Global Scientific Publishing. https://doi.org/10.4018/978-1-6684-8785-3.ch018.

14 Adão, T., Hruška, J., Pádua, L. et al. (2017). Hyperspectral imaging: a review on UAV-based sensors, data processing, and applications for agriculture and forestry. *Remote Sensing* 9 (11): 1110.

15 Yao, H. and Lewis, D. (2010). Spectral preprocessing and calibration techniques. In: D.-W. Sun (Ed.), Hyperspectral Imaging for Food Quality Analysis and Control, 45–78. San Diego, CA, USA: Academic Press.

16 Talukdar, V., Veeraiah, V., Roy, D.R. et al. (2023). An innovative approach: typicality-based collaborative filtering for recommender systems. *Proceedings of the 3rd International Conference on Advancement in Electronics & Communication Engineering (AECE)*, pp. 863–868. https://doi.org/10.1109/AECE59614.2023.10428576.

17 Arias, F., Zambrano, M., Broce, K. et al. (2021). Hyperspectral imaging for rice cultivation: applications, methods, and challenges. *AIMS Agriculture and Food* 6 (1): 273–307.

18 Wang, C., Liu, B., Liu, L. et al. (2021). A review of deep learning used in hyperspectral image analysis for agriculture. *Artificial Intelligence Review* 54 (7): 5205–5253.

19 Zhang, C. and Kovacs, J.M. (2012). The application of small unmanned aerial systems for precision agriculture: a review. *Precision Agriculture* 13 (6): 693–712.

20 Kamilaris, A., Kartakoullis, A., and Prenafeta-Boldú, F.X. (2017). A review of the practice of big data analysis in agriculture. *Computers and Electronics in Agriculture* 143: 23–37.

21 Carter, G.A. and Knapp, A.K. (2001). Leaf optical properties in higher plants: linking spectral characteristics to stress and chlorophyll concentration. *American Journal of Botany* 88 (4): 677–684.

22 Thenkabail, P.S., Lyon, J.G., and Huete, A. (2018). Advances in hyperspectral remote sensing of vegetation and agricultural crops. In: P.S. Thenkabail, J.G. Lyon, & A. Huete (Eds.), Fundamentals, Sensor Systems, Spectral Libraries, and Data Mining for Vegetation, 3–37. Boca Raton, FL, USA: CRC Press.

23 Miphokasap, P., Honda, K., Vaiphasa, C. et al. (2012). Estimating canopy nitrogen concentration in sugarcane using field imaging spectroscopy. *Remote Sensing* 4 (6): 1651–1670.

24 Mahajan, G.R., Sahoo, R.N., Pandey, R.N. et al. (2014). Using hyperspectral remote sensing techniques to monitor nitrogen, phosphorus, sulphur, and potassium in wheat (*Triticum aestivum L.*). *Precision Agriculture* 15 (5): 499–522.

25 Ge, Y., Atefi, A., Zhang, H. et al. (2019). High-throughput analysis of leaf physiological and chemical traits with VIS–NIR–SWIR spectroscopy: A case study with a maize diversity panel. *Plant Methods* 15 (1): 66.

26 Yao, Q., Zhang, Z., Lv, X. et al. (2022). Estimation model of potassium content in cotton leaves based on wavelet decomposition spectra and image combination features. *Frontiers in Plant Science* 13: 920532.

27 Sanaeifar, A., Yang, C., Min, A. et al. (2024). Noninvasive early detection of nutrient deficiencies in greenhouse-grown industrial hemp using hyperspectral imaging. *Remote Sensing* 16 (1): 187.

28 Singh, L., Mutanga, O., Mafongoya, P. et al. (2021). Detecting nutrient deficiencies in Eucalyptus grandis trees using hyperspectral remote sensing and random forest. *South African Journal of Geomatics* 10 (2): 207–222.

29 Misbah, K., Laamrani, A., Dhiba, D. et al. (2023, May). Selection of NPK specific spectral bands using hyperspectral imagery and an ensemble machine learning approach over agricultural lands in Morocco. *EGU General Assembly Conference Abstracts*, pp. EGU-8637.

30 Li, H., Wan, L., Li, C. et al. (2024). Hyperspectral imaging technology for phenotyping iron and boron deficiency in *Brassica napus* under greenhouse conditions. *Frontiers in Plant Science* 15: 1351301.

31 Pandey, J.K., Yadav, A.K., Purushotham, M. et al. (2022). Colorimetric detection of deltamethrin pesticide from biological samples based on the peroxidase-mimic catalytic activity on MnO_2/SnO_2. *Key Engineering Materials* 928: 45–51. https://doi.org/10.4028/p-4ti1i0.

32 Masoni, A., Ercoli, L., and Mariotti, M. (1996). Spectral properties of leaves deficient in iron, sulfur, magnesium, and manganese. *Agronomy Journal* 88 (6): 937–943.

33 Li, C., He, M., Cai, Z. et al. (2023). Hyperspectral imaging with machine learning approaches for assessing soluble solids content of tribute citru. *Food* 12 (2): 247.

34 Pandey, J. K., Kumar, S., Lamin, M., Gupta, S., Dubey, R.K. and Sammy, F. (2022). A metaheuristic autoencoder deep learning model for an intrusion detector system. *Mathematical Problems in Engineering,* 2022, 3859155, 11. https://doi.org/10.1155/2022/3859155

35 Alanazi, A., Wahab, N.H.A., and Al-Rimy, B.A.S. (2024, May). Hyperspectral imaging for remote sensing and agriculture: a comparative study of transformer-based models. *2024 IEEE 14th Symposium on Computer Applications & Industrial Electronics (ISCAIE)*, pp. 129–136. IEEE.

36 Yue, J., Yang, G., Tian, Q. et al. (2019). Estimate of winter-wheat above-ground biomass based on UAV ultrahigh-ground-resolution image textures and vegetation indices. *ISPRS Journal of Photogrammetry and Remote Sensing* 150: 226–244.

37 Cawse-Nicholson, K., Townsend, P.A., Schimel, D. et al. (2021). NASA's surface biology and geology designated observable: A perspective on surface imaging algorithms. *Remote Sensing of Environment* 257: 112349.

38 Cogliati, S., Sarti, F., Chiarantini, L. et al. (2021). The PRISMA imaging spectroscopy mission: overview and first performance analysis. *Remote Sensing of Environment* 262: 112499.

39 Chen, J., Jönsson, P., Tamura, M. et al. (2004). A simple method for reconstructing a high-quality NDVI time-series data set based on the Savitzky–Golay filter. *Remote Sensing of Environment* 91 (3–4): 332–344.

40 Hao, Y., Dongyan, Z., Linsheng, H., and Jinling, Z. (2014). Wavelet-based threshold denoising for imaging hyperspectral data. *International Journal of Agricultural and Biological Engineering* 7 (3): 36–42.

41 Kumar, M. V., & Yarrakula, K. (2017). Comparison of Efficient Techniques of Hyperspectral Image Preprocessing for Mineralogy and Vegetation Studies.

42 Gewali, U.B., Monteiro, S.T., and Saber, E. (2018). Machine learning based hyperspectral image analysis: a survey. arXiv preprint arXiv:1802.08701.

43 Chen, B., Lu, X., Yu, S. et al. (2022). The application of machine learning models based on leaf spectral reflectance for estimating the nitrogen nutrient index in maize. *Agriculture* 12 (11): 1839.

44 Yang, X., Zhang, X., Wang, S., and Yang, W. (2020, April). A hybrid 2D/3D convolutional neural network for hyperspectral image classification. *International Conference on Education, Economics and Information Management (ICEEIM 2019)*, pp. 265–269. Atlantis Press.

45 Guerri, M.F., Distante, C., Spagnolo, P. et al. (2024). Deep learning techniques for hyperspectral image analysis in agriculture: a review. *ISPRS Open Journal of Photogrammetry and Remote Sensing* 12: 100062.

46 Kawamura, K., Watanabe, N., Sakanoue, S., and Inoue, Y. (2008). Estimating forage biomass and quality in a mixed sown pasture based on partial least squares regression with waveband selection. *Grassland Science* 54 (3): 131–145.

47 Li, Z., Nie, C., Wei, C. et al. (2016). Comparison of four chemometric techniques for estimating leaf nitrogen concentrations in winter wheat (Triticum aestivum) based on hyperspectral features. *Journal of Applied Spectroscopy* 83 (2): 240–247.

48 Tian, T., Wang, J., Tao, Y. et al. (2024). Estimating rice leaf nitrogen content and field distribution using machine learning with diverse hyperspectral features. *Agronomy* 14 (12): 2760.

49 Pandey, J.K. (2025). Understanding the impact of electromagnetic and optical radiation on physical and biological systems. *Journal of Nano- and Electronic Physics* 17 (5): 05023. https://doi.org/10.21272/jnep.17(5).05023.

50 Mou, L., Ghamisi, P., and Zhu, X.X. (2017). Deep recurrent neural networks for hyperspectral image classification. *IEEE Transactions on Geoscience and Remote Sensing* 55 (7): 3639–3655.

51 Yuan, H., Yang, G., Li, C. et al. (2017). Retrieving soybean leaf area index from unmanned aerial vehicle hyperspectral remote sensing: analysis of RF, ANN, and SVM regression models. *Remote Sensing* 9 (4): 309.

52 Chen, Y., Wang, Y., Gu, Y. et al. (2019). Deep learning ensemble for hyperspectral image classification. *IEEE Journal of Selected Topics in Applied Earth Observations and Remote Sensing* 12 (6): 1882–1897.

53 Kamilaris, A. and Prenafeta-Boldu, F.X. (2018). Deep learning in agri-culture: a survey. *Computers and Electronics in Agriculture* 147: 70–90.

54 Hapuarachchi, N.S., Trueman, S.J., Kämper, W. et al. (2023). Hyperspectral imaging of adaxial and abaxial leaf surfaces for rapid assessment of foliar nutrient concentrations in Hass avocado. *Remote Sensing* 15 (12): 3100.

55 Srinivas, P., Veeraiah, V., Thejaswini, K.O. et al. (2025). The impact of electromagnetic and optical radiation on physical and biological systems. *Journal of Nano- and Electronic Physics* 17 (4): 04012. https://doi.org/10.21272/jnep.17(4).04012.

56 Belgiu, M. and Drăguţ, L. (2016). Random forest in remote sensing: a review of applications and future directions. *ISPRS Journal of Photogrammetry and Remote Sensing* 114: 24–31.

57 Nie, S., Gao, W., Liu, S. et al. (2024). Hyperspectral imaging combined with deep learning models for the prediction of geographical origin and fungal contamination in millet. *Frontiers in Sustainable Food Systems* 8: 1454020.

58 Windrim, L., Ramakrishnan, R., Melkumyan, A. et al. (2019). Unsupervised feature-learning for hyperspectral data with autoencoders. *Remote Sensing* 11 (7): 864.

59 Zhang, Z., Gao, Q., Liu, L., and He, Y. (2023). A high-quality rice leaf disease image data augmentation method based on a dual GAN. *IEEE Access* 11: 21176–21191.

60 Feng, L., Wu, B., He, Y., and Zhang, C. (2021). Hyperspectral imaging combined with deep transfer learning for rice disease detection. *Frontiers in Plant Science* 12: 693521.

61 Tuia, D., Volpi, M., Copa, L. et al. (2011). A survey of active learning algorithms for supervised remote sensing image classification. *IEEE Journal of Selected Topics in Signal Processing* 5 (3): 606–617.

62 Chen, X., Li, F., Shi, B., and Chang, Q. (2023). Estimation of winter wheat plant nitrogen concentration from UAV hyperspectral remote sensing combined with machine learning methods. *Remote Sensing* 15 (11): 2831.

63 Hassani, K., Gholizadeh, H., Taghvaeian, S. et al. (2023). Assessing the impact of spatial resolution of UAS-based remote sensing and spectral resolution of proximal sensing on crop nitrogen retrieval accuracy. *International Journal of Remote Sensing* 44 (14): 4441–4464.

64 Liu, S., Yang, X., Guan, Q. et al. (2020). An ensemble modeling framework for distinguishing nitrogen, phosphorus, and potassium deficiencies in winter oilseed rape (*Brassica napus* L.) using hyperspectral data. *Remote Sensing* 12 (24): 4060.

65 Ye, Z., Tan, X., Dai, M. et al. (2024). A hyperspectral deep learning attention model for predicting lettuce chlorophyll content. *Plant Methods* 20 (1): 22.

66 Dalponte, M., Bruzzone, L., and Gianelle, D. (2008). Fusion of hyperspectral and LIDAR remote sensing data for classification of complex forest areas. *IEEE Transactions on Geoscience and Remote Sensing* 46 (5): 1416–1427.

67 Zarco-Tejada, P.J., González-Dugo, V., and Berni, J.A. (2012). Fluorescence, temperature, and narrow-band indices acquired from a UAV platform for water stress detection using a micro-hyperspectral imager and a thermal camera. *Remote Sensing of Environment* 117: 322–337.

68 Mesa, A.R. and Chiang, J.Y. (2021). Multi-input deep learning model with RGB and hyperspectral imaging for banana grading. *Agriculture* 11 (8): 687.

69 Zheng, L., Zhao, M., Zhu, J. et al. (2023). Fusion of hyperspectral imaging (HSI) and RGB for identification of soybean kernel damages using ShuffleNet with convolutional optimization and cross-stage partial architecture. *Frontiers in Plant Science* 13: 1098864.

6

Computer Vision for Nutrition Deficiency in Field Crops

Dilbar Hussain[1] *and Fahiza Fauz*[2]

[1] *Department of Computer Science, Faculty of Engineering Science & Technology (FEST), Iqra University, Karachi, Sindh, Pakistan*
[2] *Department of Electrical Engineering, National University of Modern Languages, Islamabad, Pakistan*

6.1 Introduction

The benefits are particularly significant in precision agriculture, where the deficiencies can be identified in time and with precision to allow the optimal utilization of water, fertilizers, and other inputs [1, 2]. It also promotes the economic and social well-being of billions of individuals all over the globe and food security [3]. The field-crop sector, which consists of wheat, maize, rice, and barley, is a valuable source of essential nutrients and contributes greatly to human nutrition. These crops are also important in the provision of the increasing world demand for food due to their large-scale production [4, 5].

6.1.1 Field Crops and Their Importance in Global Food Systems

The domain of food security in the world depends on field crops that provide most of the calories and nutrients in the diet, including wheat, maize, rice, barley, and the soybean [3, 6]. They facilitate the daily life of millions of people and play a major role in the agricultural production and consumption worldwide [7]. These crops are also used to provide food security in addition to supporting rural livelihoods, especially in developing countries where agriculture is the primary income earner.

Field crops also have impacts on everyday life and macroeconomic indicators. They are helping maintain food safety nets around the world, supplying much of the energy consumed, and sustaining massive farming systems, both biofuel and animal food, as well as an assortment of industrial inputs. These are the crops that are needed to maintain adequate, nutritious supply chains in the face of the growing population of the world. This is because some of the complex challenges that the industry experiences and which do not allow it to perform optimally include nutrient shortages, sustainability, and the increasing impacts of climate change on crop production.

6.1.2 Problem Statement: Micronutrient Deficiencies and Hidden Hunger

The outcome of the nutrient deficiency in crops is reduced nutrient density, yields, and quality harvests, which are often caused by the depletion or unavailability of soil nutrients [8]. They are especially devastating in case the adverse health outcomes of the insufficient intake of micronutrients are aggravated by the fact that the person does not consume a sufficiently diverse diet [9, 10]. Although staple crops provide a significant share of calories in the world, they normally do not contain adequate amounts of micronutrients. As an example, zinc, iron, and folate may be low in wheat and maize, which are high in carbohydrates [11]. Since they do not have important micronutrients to support normal growth and development, people might be meeting their calorie requirements, but one may still experience hidden hunger."

6.1.3 Rising Pressures and Research Need

There is still a major concern of micronutrient adequacy despite the improvement in yield, which is helping in satisfying the increasing food demand. This means that farmers and researchers have to enrich the micronutrient levels of the staple crops so that they balance them with nutrition and energy. This requires a multipronged approach that combines both traditional breeding and unusual technologies, such as computer vision (CV) and artificial intelligence (AI), to keep track of crops in the field and enhance the environment in which they grow [12, 13].

The synergistic impacts of climate change and the unreliable agronomic states are complicating the mitigation of these gaps [14]. The current realities complicate the process of sustaining healthy soils, water, and creating optimum conditions of growth, resulting in faster nutrient loss and huge loss of yield and content.

6.2 Chapter Aim and Scope

Newer technologies enable modern comprehensive crop nutrition management to ease these pressures. Of these, CV is a subfield of AI offering an opportunity to create field-scale surveillance (continuous). By integrating CV with machine learning (ML) methods, farmers can monitor fields in real-time, anticipate nutrient deficiencies, and react proactively before local problems turn into system-wide problems.

In this chapter, the authors evaluate how AI-assisted CV can be used to optimize the yield and nutritional quality, promote climate-resilient, sustainable agronomy, and detect and correct the nutrient deficiencies at an early stage. We introduce nutrient-management requirements and climate stresses, reveal precision and digital strategies, overview the types of deficiencies and their effects, introduce CV principles and processes, survey the key applications, discuss prospects, and wrap up with actionable points.

6.2.1 The Growing Need for Crop Nutrient Management

Sustainability is also sought to maintain food production due to the projections that the world population will still be more than nine billion people by the year 2050 [15]. Nutrient availability is currently having a significant influence on the field crop yield and quality. Nitrogen (N), phosphorus (P), and potassium (K) are crucial to the health of plants; hence, when the elements are deficient, plants exhibit external conditions such as yellowing of the leaves, sluggish growth, reduced flowering, and low yield [16].

The common causes of nutrient deficiencies are poor soil quality, high land use, and poorly designed fertilization programs. Early detection is vital so that its effects on productivity are minimized. Common diagnostic methods of visual inspection and laboratory soil testing prove to be very time-consuming, labor-intensive, and often impossible to make timely and field-wide decisions [17].

6.2.2 Challenges in Crop Nutrition and the Impact of Climate Change

The nutritional status of field crops is dependent on various environmental factors, fertilization practices, and the physicochemical characteristics of soil. South Asia and Africa have the highest levels of micronutrient depletion, hence these areas have the highest rates of deficiencies in zinc and iron [18]. Imbalances are widespread due to poor methods of managing the soil and a lack of nutrients or an excess of the same. As an example, the soils that lack micronutrients such as iron or zinc can be used to produce cereals with low concentrations of these factors. This issue is particularly critical in most developing nations where high-quality fertilizers and methods to improve the soil are limited.

Anthropogenic climate change exacerbates nutrient problems. The metabolism and uptake of nutrients are disturbed by flooding, heat stress, and drought [19, 20]. This is because the variability in temperature and precipitation and the more frequent occurrence of extreme events (like floods and droughts) weaken crop performance, leading to stunted phenotypes, low nutrient levels, and low yields. Raised CO_2 and temperatures also have an impact on crop physiology; it is known that CO_2 can decrease the micronutrient levels of grains (like iron and zinc), which puts people with already high risks of malnutrition at greater risk [21].

These stressors are expected to intensify in the coming decades, requiring producers to realign management practices. Farmers need timely technologies and decision support to maintain crop nutritional quality under variable conditions. Advanced precision-farming tools that monitor environmental variables in real time and provide fine-grained information on crop health have therefore become essential.

6.2.3 The Importance of Early Detection and Intervention

The main nagging issue facing the practitioners in the agricultural sector is the capacity to identify deficiencies of nutrients early so that corrective action can be taken before the crops can succumb [22]. Nitrogen, phosphorus, and potassium are considered essential macronutrients to support optimum plant growth, and iron, zinc, and magnesium are considered essential micronutrients, especially in photosynthesis, enzyme reactions, and chlorophyll formation. Detection of the mentioned imbalances at an early stage allows precise nutrient management, accurate fertilizing, and eventually better crop yields.

CV and the introduction of drone technology and satellite imagery have now enabled real-time surveillance of large-scale fields. The high-resolution data not only enables the detection of the lacks on the individual plant level but also provides the decision support on the field-wide level regarding crop management in the field. In addition, the monitoring of pathogen spreading and pest infestation can also be observed using the capacity of CV, thus offering a thorough platform through which plants can be managed and kept healthy [23].

6.2.4 A Data-Driven Approach to Precision Agriculture

Smart farming technologies: The rise of smart farming technologies has encouraged the growth of precision agriculture, a systems-level technology implementation strategy that harnesses and uses technology to track and maximize field operation efficiency [24]. In the sphere, CV has a special role to play, as it supplies the farmers with the tools they can use to monitor crops at all times and identify problematic areas before they become serious [13, 25]. Large-scale farms can utilize that opportunity to exploit automation and the ability to monitor crop health in real-time through CV, thus allowing practitioners to better understand the conditions of plants being treated and administer necessary treatment more precisely and effectively. Digital automation strengthens sustainability in large farming systems [26].

6.2.5 Nutrition Deficiency and Its Impact

Nutrient deficiencies are a burdensome topic in modern farming, particularly since worldwide food security has become an important concern. The availability of fundamental elements like nitrogen (N), phosphorus (P), and potassium (K) is pivotal for plant growth, development, and reproduction [27]. A lack of these factors or reduced availability causes the plants not to perform vital physiological functions and end up developing slowly and with poor yield. This discussion goes through the major nutrient deficiencies, the physiological impacts on healthy plant growth, and their impact on crop productivity and food security overall.

6.3 Types of Nutrient Deficiencies

6.3.1 Nitrogen Deficiency

Nitrogen is one of the major nutrients that is essential in the formation of chlorophyll, amino acids, and the production of proteins. It supports many tissue structures and is paramount in the photosynthetic process, and maintains metabolism. Plants then show the following diagnostic symptoms when nitrogen is limited:

- **Yellowing of Older Leaves:** Since nitrogen will be passed out of the senescent tissue to the actively growing ones, deficiency is first expressed in the older leaves.
- **Reduced Shoot Extension:** This is due to the deficiency in nitrogen levels, which stops extension, thus reducing the overall biomass.
- **Lower Fruit and Seed Yield:** Since nitrogen is used in the biosynthesis of nucleic acids and proteins, a nitrogen deficiency may threaten reproductive success and hence lower yield.

Nitrogen deficiency is common where the soil lacks organic matter or was leached due to too much rainfall, and as a result, it is one of the commonest forms of nutritional constraints in both the developed and developing areas [16, 22]. The nitrogen imbalances can be rectified through careful fertilizer application strategies, thus increasing the health and productivity of the plants [22].

6.3.2 Phosphorus Deficiency

Phosphorus is another macronutrient that is not easily dispensed with, and it helps in transferring and storing energy in the plant. As a constituent of adenosine triphosphate (ATP), it is central to cell division, root development, and flowering processes [16]. Plants exhibiting phosphorus deficiency commonly display the following symptoms:

- **Dark Green Leaves:** Darker leaves in plants turn reddish or purplish in color when they lack phosphorus, resulting in the formation of anthocyanin coloring.
- **Deformed Root Growth:** Phosphorus plays a major role in elongation as well as water absorption of roots, and its lack produces a stunted growth of roots.

6.3.3 Delayed Flowering and Poor Seed Formation

Phosphorus deficiency is common in soils that are acidic and in soils that have a high phosphate-fixing capacity, thus making the nutrient unavailable at large. The application of phosphatic fertilizers or the manipulation of soil pH by correct agronomic practice are some of the mitigation measures involved.

6.3.4 Potassium Deficiency

Potassium is an essential requirement in the activation of enzymes, protein synthesis, and in the maintenance of water balance in the plant [27]. Following a potassium deficiency, characteristic symptoms include

- **Yellowing and Necrosis of the Edges of the Leaves:** This happens when the plant cannot maintain balanced water, which then leads to a death-like reaction on the edges of the leaves.
- **Poor Stems:** Potassium helps to strengthen cell walls; without it, plants are more prone to disease and mechanical damage.
- **Weak Development of Fruits:** The plants lacking potential potassium often form fewer fruits or smaller fruits, thus decreasing the production of fruits and yield capacity.

Lack of potassium is common in sandy soils, where the nutrient is easily washed away, and the plants cannot get enough of the required amounts of potassium.

6.3.5 Iron Deficiency

Iron is essential to the formation of chlorophyll and in a variety of enzyme processes useful in the actual technique of photosynthesis. Plants whose leaves are iron-deficient exhibit interveinal chlorosis in which the veins of the young leaves are green and the area between them is yellow; this yellowing is unique compared to the normal appearance of green leaves

[28]. This is then followed by diminished photosynthesis since iron is needed in chlorophyll synthesis, and without it, the plant cannot produce energy in its tissues. The phenotypes of iron deficiency are also regular in alkaline soils or manure-deficient systems, and the severity is also aggravated by poor drainage or waterlogging in areas where the iron availability to the plants is disturbed.

6.3.6 How Deficiencies Impact Crop Yield and Quality

Deficiencies of nutrients undermine the external appearance and the yield and the quality of crops. Lack of essential elements prevents optimal physiological processes like photosynthesis, protein production, and cell division, which reduce biomass build-up and hamper fruit growth. In the current discussion, the impacts of nutrient deficiency on crop yield and quality are examined in detail.

6.3.7 Nitrogen Deficiency and Yield Reduction

Nitrogen is a very crucial component in the synthesis of amino acids and proteins, which are essential in cellular physiology and energy production. Nitrogen deficiency inhibits the production of proteins, thus restricting the growth of plants and causing stunting, chlorosis, and reduced biomass [27]. This makes the reproductive parts of the plant (flowers and seeds) continue to grow in an immature state, and this leads to decreased production of a crop. When it comes to cereal crops, the insufficient presence of nitrogen has been conventionally linked to the drop in the density of grain protein, hence altering the nutritional value of the whole harvest.

These changes in nitrogen levels have more or less the same effect on maize, wheat, and rice, which encompass outputs and poor quality. Additionally, crops that experience nitrogen shortage are easily affected by pests and diseases, thus enhancing the decline in production [20].

6.3.8 Phosphorus Deficiency and Reduced Yield

Phosphorus is important in root development, energy conversions, and reproductive success. Phosphorus deficiency can be identified by retarded root growth, improperly made flowers, and reduced seed set [18]. Plants with a deficiency in phosphorus also take a long time to achieve harvestable size, which in turn lowers the potential yield.

Nutritional investigations on the rice and soybean crops show that a lack of phosphorus may hamper the process of grain filling and degrade the quality of the seeds. Also, the purplish or reddish leaf coloration frequently accompanying phosphorus limitation is not a symptom of nutrient stress but, instead, comes as the result of poor utilization of nutrients, which lowers general growth and performance.

6.3.9 Potassium Deficiency and Crop Quality

Potassium cannot be dispensed with in osmotic adjustment, activation of enzymes, and photosynthesis. In a case where there is an insufficient amount of potassium in plants, the stem of the plant becomes weak, the initiation of the fruit is poor, and the character of the crop suffers. In addition, poor potassium affects resistance to diseases and drought as well as the quality and quantity of crops.

Deficiency of potassium in tomato, potato, and apple production systems causes smaller fruits, changes in pigmentation, and reduced sugar contents, hence, reducing the marketability [27]. Potassium deficiency reduces the plumpness of the grain and the quality of the whole grain in a grain crop, thus influencing yield and palatability.

6.4 Role of Computer Vision in Agriculture

Agriculture is a pillar of human civilization, which supplies the human race with its nutrition, as well as with an essential support of ecosystems. Throughout centuries, the agricultural practice has been developed under the influence of climate changes and advancements in science, as well as due to the introduction of new technologies. With the population of the world remaining on an upward trend and corresponding tensions such as those on climate change, declining arable land, and decreasing natural resources growing severe, the role of technology in contemporary agriculture takes center stage. CV is one of the most significant developments that has revolutionized agriculture today.

CV, a branch of AI, empowers machines to extract and interpret visual representations of the surrounding world. Its application in the farming setting has been successful in solving the old problems of crop production, pest infestations, disease detection, and nutrient deficiency. CV systems built around AI and ML algorithms can provide the tools that farmers need to analyze their crops and improve production, sustainability, and the health of their crops [12, 25]. As agriculture evolves to be more data-driven, the technology also enables effective resource use, highly specific interventions, and, eventually, a greener agricultural model.

The discussion that follows highlights how CV is transforming how agriculture is conducted, especially in the management of crops. The chapter takes into account its application in the diagnosis of nutrient deficiencies, crop monitoring, and optimization of farm management activities. CV systems, by applying advanced imaging modalities and complex ML algorithms, have the capacity to analyze crops and generate real-time actionable insights that enable farmers to make informed decisions, increase productivity, and sustain crop growth [26].

6.4.1 What Is Computer Vision?

Within the discipline of AI, CV addresses the problem of enabling machines to comprehend and interpret visual data. Similar to the human visual system, CV grants a digital agent the ability to see images, interpret their semantic information, and get insights into high-level concepts out of raw visual data [13]. Importantly, the computational engine most likely to succeed in CV can examine, categorize, and reason in chains of magnitude contrast to a human viewer at a similar moment, which ultimately dispenses with the time as well as cognitive limitations that interfere with humans.

Ways in which the CV systems actually work are by initially receiving the images or video streams, having them interpreted to reveal objects, visual information, pattern recognition, and implication of decision variables in the resultant visual projections. To give an example, where the human operator needs to visually identify a relevant object or a person, and to name it, a system based on the CV performs a thorough examination of the visual area, identifies the focal point, defines its route, and puts its activity in context something that would seem a primitive task to the human, but is performed by a computational device with greater precision and scale.

An arsenal of image-processing algorithms that aim to enhance, transform, and idealize the visual discourse is at the heart of such an autonomous operation. Edge detection, feature extraction, and pattern-recognition routines operate collectively to partition complex scenes into meaningful constituent parts, thereby rendering inferential tasks performed by algorithms such as convolutional neural networks (CNNs) more tractable.

The field is now applied to numerous commercial and scientific areas, with the most popular being autonomous vehicles, industrial automation, medical imaging, and modern agriculture. In the latter, CV-based solutions have become common in regular checking of crop health, identification of pests or diseases, or warning of nutritional deficiencies on-site; the ability to perform all those checks in real time and data-centered makes it possible to simplify a wide-ranging burden of growers needing to check it on-site all the time.

6.4.2 The Role of Computer Vision in Modern Agriculture

During the past few years, CV has emerged as a substitute for the traditional detection methods of crop surveillance. Positioned within AI, CV delivers automated analysis of imagery and video to appraise plant health and diagnose nutrient deficiencies. These systems obtain real-time imagery of crop fields by using their mounted cameras, drones, and satellite imagery, which are processed to identify visual evidence of nutrient stress [23]. After that, the ML algorithms classify and identify gaps, enabling prompt intervention.

The application of CV to agriculture has the potential to revolutionize crop observation and advance precision farming. Through real-time observations and early detection, farmers can make decisions on where to spread the fertilizer, how to schedule their irrigation time, and other crop management decisions. Such greater precision is expected to increase productivity and reduce waste.

6.4.3 The Role of Computer Vision in Monitoring Crop Health

CV has turned out to be a handy solution in monitoring and enhancing the health of crops in order to deal with the current issues of nutrient deficiencies and climate change. CV systems can be used to provide real-time information about the status of crops and enable farmers to detect any nutrient deficiency, pests, diseases, and other risks [12, 25] early enough through advanced imaging technologies and AI algorithms. This measure at this stage reduces any damage in the future and optimizes the crop yield.

One of the main benefits of CV in the field of agriculture is the possibility of detecting extremely early indicators of nutrient deficiency. An example is that a lack of phosphorus is likely to cause stunted growth, and a lack of nitrogen, which is an essential macroelement, but is also expressed through yellowing of the leaves. AI systems provide highly precise symptom recognition with hugely clear images taken by drones, satellites, or ground sensors. This early detection opens the way to specialized treatment, e.g., fertilization or irrigation, to fix the issue before it reaches the entire crop.

In addition to nutrient deficiencies, CV monitoring can also be used to observe other factors relating to the crop health systems, such as pest infestation, diseases, among others, as well as the general growth of crops. It is an all-inclusive outlook that gives the farmers insight into making intelligent decisions about managing resources, defending their crops, and maximizing their outputs.

6.4.4 How Does Computer Vision Work in Agriculture?

CV is currently in use in modern agronomical practice, which actively interrogates the high-resolution photographic or video coverage of crops, fields, or any other farming setting to distill informative signals. The workflow typically proceeds through multiple stages: (i) image acquisition, (ii) multispectral imaging, (iii) image processing, and (iv) data analysis. The description of each of these elements has been carried out in detail [12, 19].

6.5 Image Acquisition

Computer vision: Few people are aware that the key to any CV system is the ability to generate high-quality visual data. There are three main sources of the image or video streams in an agricultural environment:

- **Cameras:** We can use high-resolution cameras attached to any of the ground, tractor, or other agromachinery to obtain details of plants, soil, and crop status on the crop plane.
- **Drones:** Drones that are fitted with a high definition camera or multispectral sensors provide an aerial view, which allows large sections to be covered within a small timeframe, allowing crop growth, pest infestation, and pathology to be monitored in real-time without any risk of injury.
- **Satellites:** Satellites placed in the orbit of the earth provide medium and large-scale images, allowing a means to monitor the whole of the crop systems and especially in remote areas. Extensions of time series of satellite-derived data date back over many years and explain trend changes in crop dynamics and ecological stress patterns.

6.5.1 Multispectral Imaging

Multispectral imaging devices capture energy in different broad ranges of wavelength, both visible and nonvisible. Multiple spectral signatures simultaneously recorded by such systems allow sensing of phenomena that cannot be observed by the human eye, such as alterations in chlorophyll content or indices of water stress. The multispectral sensor data deliver quantitative data on the issues of physiological condition, nutritional availability, pathological causes, and abiotic stress. Multispectral and hyperspectral imaging detect hidden biochemical stress signals [29].

6.5.2 Data Analysis and Image Processing

Images are computationally manipulated and statistically analyzed postacquisition. Advanced ML architectures, most prominently CNNs, are trained on labeled datasets, wherein images of healthy vs. stressed plants are accompanied by annotations of their respective conditions. These models recognize deficiencies in nitrogen, phosphorus, or potassium through comparing visual appearance with trained correlations of crop health, nutrient concentrations, pests, or disease, and evidence of environmental stress generated by drought, over-watering, or pest pressure.

The results presented here provide an overview of the role played by CV in crop management, as visual information gathered at diverse spatial resolutions is analyzed. After incorporating multispectral Image capture, information-informed image processing, and statistical analysis, agronomists can obtain actionable information that can be used in influencing decisions made in different stages of production and agroecosystem conditions.

6.6 Detection of Plant Stress, Deficiencies, and Diseases

After processing the images, CV systems can identify specific issues affecting crops. Common applications in agriculture include:

6.6.1 Nutrient Deficiencies

Nutrient deficiencies frequently exhibit a suite of visually detectable symptoms, including chlorosis (yellowing of leaves), stunted growth, and anomalously shaped leaves. CV systems can detect these signs and, as such, warn farmers of possible deficiencies in the available nutrients. These systems also enable farmers to use fertilizers that are specific to an area of the field since the nutrient needs of certain areas can be determined. This reduces the amount of damage to the plants in general since the right nutrients will be supplied.

6.6.2 Pests and Diseases

CV systems have been demonstrated to be effective in pest, insect, and antipathogenic disease detection of crops, thus posing a threat to the farming of the crops. Early signs detection (holes in leaves or yellow discolored areas) enables farmers to implement interventions (use of pesticides or destruction of infected crops) early enough. This early detection reduces the spread of pests in the field, hence eliminating the resultant use of pesticides on a mass scale.

6.6.3 Environmental Stress

Physiological stressors that affect a plant are high environmental variables such as great changes in temperature, drought, and floods. With the help of CV techniques, farmers receive the possibility to monitor the state of crops in real time and also diagnose the distraction of exogenous factors on the performance of agronomic processes. As an example, foliar adherence to water shortage, which translates to obvious wilting or color shifts, is precisely captured by imaging systems.

6.6.4 Weed Detection

By competing actively with crops for water, nutrients, and sunlight, weeds pose a challenge to crop production as they continuously affect the crops. Current developments in CV have made it possible to have an accurate classification of plant species, and this has given an opportunity to farmers to use herbicides only on areas that are attacked by weeds. Such focused measures not only lower the usage of pesticides but also protect the crops.

6.6.5 Actionable Insights and Decision-Making

After the data has been analyzed and convincing trends have been derived, a CV system can provide beneficial information to the farmer. Such knowledge can include irrigation tips, fertilization, pest control, or harvesting tips as well. With such learning, farmers can use data-driven decision-making, which maximizes crop health and increases overall productivity.

As an example, when the nutrient deficiencies at the early stages are detected, the entire fertilizer may be applied to the desired location with the help of the system, and thus, the deficiency does not expand over the spatial area of the field. Secondly, the system could declare pest infestations or even prescribe specific pesticides or prescribe alternative ways of pest control, thus minimizing the use of chemical agents. Under environmental stress conditions, the system could recommend changes in irrigation rifts and covering of the crops with some protective covers.

The availability of real-time agricultural data would enable farmers to streamline their decision-making systems, which eventually would result in higher crop production, reduced costs, and the general introduction of greener farming choices.

6.6.6 Decision Support and Feedback

When the botanical deficiencies or stress factors in the environment are observed, with the help of the CV system, the data will be converted into constructive information that the farmers can use directly as input to their decision-making. The analysis that is created can bring about an automatic suggestion or an alert that provides instructions on what can be done. For example:

- **Recommendation on the Use of Fertilizer:** Diagnosis of nitrogen deficiency provokes recommendations on the specific nitrogen fertilization in the areas of its detection.
- **Pest Control Guidance:** after identifying the pests, suggestions on using specific pesticides, as well as nonchemical repellents in the infested areas, are given.
- **Modifications in Watering:** monitoring of unhealthy crops exposes water stress, which triggers programmed alterations to the currently existing crop irrigation programs.

With such insights about their fields and farmlands, through these comprehensive data, the system will allow the farmers and producers to employ data-driven decision-making tools, which will enhance the use of input-driven applications only in instances and areas where they are required.

6.7 Applications of Computer Vision in Agriculture

CV has a wide range of applications in agriculture. Below are some of the most significant ways in which it is used:

6.7.1 Detection of Nutrient Deficiencies

The application of CV to the area of crop management is a strong technological intervention [12]. Such systems can identify visual symptoms of nitrogen, phosphorus, and potassium deficiency, and visual symptoms of iron, magnesium, as well as zinc, through systematic capture and analysis of imagery collected by those crop canopies that are targeted by such systems [23]. These deficiencies can be noticed early in order to efficiently treat them with precision, whereby inefficiencies of applying excess fertilizers and wastage of resources are reduced.

6.8 Disease and Pest Monitoring

With CV, disease and pest infestation monitoring of plants has been a very crucial application. Using systematically analyzed plant images, the technology can identify all visible symptoms of fungal, bacterial, and viral infections, as well as the physical damage caused by insects, rodents, and other vermin. This results in early intervention as diseases can be identified within a short time, limiting the expansion of the diseases beyond control and reducing pesticide reliance. In addition, it reinforces integrated pest management (IPM) approaches that emphasize biological or ecological strategies for pest regulation [23, 25].

6.9 Weed Detection

Weed control is quickly becoming a CV tool. These systems allow the crops and weeds to be classified, which makes the application of herbicides more accurate and thus minimizes the amount of herbicides used and minimizes the effects on the environment that could be related to the use of such herbicides. At the level of precision agriculture, the developed technology of the detection of weeds can be integrated with robotic platforms to perform the procedure of weeding the fields without the manual involvement of the operator, which will be another addition to the efficiency of the farm operation [26, 30].

6.10 Yield Prediction

CV has become one of the most influential parts of the yield forecasting of all, allowing one to measure crop volume, shape, and colorimetric characteristics with high-resolution pictures [25]. Using the trend of plant growth, CV systems will be able to predict future harvests and, by extension, optimize planting times. These forecasts provide farmers with actual numbers of production expected, hence leading to better planning of the market and also the sales strategies.

6.11 Monitoring Plant Growth

CV not only detects the nutrient deficiencies but also monitors crop growth by comparing the changes in size, shape, and color of a plant from one period to another [19]. This kind of monitoring allows farmers to make informed decisions in monitoring their crop health and provides them with the opportunity to know the best time to harvest their crops. CV images in a time-lapse fashion have also helped to gain longitudinal tracking data at an individual plant level, as well as an individual field cohort, and by extension, provide improved data on farm-level productiveness.

6.12 Recent Advancements in AI Applications for Agriculture

Recent developments in AI have greatly improved agricultural monitoring and decision-making. Smart greenhouse systems today operate under CV and automated control to ensure optimal light and temperature, as well as nutrient levels to provide high-quality yield and resource-saving smartness [31]. Simultaneously, AI-driven biodiversity surveillance allows detecting environmental stress and ecosystem disequilibrium in the farmland environment early and allows farmers to preserve both crops and natural habitats [32]. These innovations demonstrate how AI can be improved to help in the management of controlled growing conditions and broader ecological monitoring to boost eco-friendly crop yields.

6.13 The Future of Computer Vision in Agriculture

The existing technological development trends suggest that CV will influence agricultural production to a considerable extent. AI, ML, and deep learning (DL) have now allowed researchers and practitioners to more readily identify crop diseases, track the health of plants, and determine the characteristics of different soils, as well as regulate the process of irrigation and pests [33]. It is anticipated that CV will play an important role in effective agronomic systems because many problems appear to be multiplying beneath it, such as climate change, population increase, and more sustainable agriculture. Higher productivity, efficiency, and environmental sustainability across a range of production operations are anticipated outcomes of current and upcoming technologies in this industry.

This section will examine the future of CV in agriculture, emphasizing the anticipated technological developments, integration with Internet of Things (IoT) devices, autonomous robotics, big data's role, and the growing significance of data-driven, real-time decision-making [34].

6.13.1 Advancements in Machine Learning and Deep Learning

The application of both ML and DL algorithms helps to implement CV, as the basic constructs of the CV system allow acquiring significant details belonging to large visual datasets. As DL frameworks, particularly CNNs, continue to evolve, they will provide progressively higher levels of accuracy and precision in the detection of plant health anomalies, ranging from nutrient deficiencies to pest infestations. The expected impact of such technological advancements is high reliability in monitoring crops, which will also help in early mitigation whenever there are some problems.

In thecoming time, more sophisticated DL algorithms will make it easier.

6.13.2 Detect Subtle Signs of Stress

Recent system theories have proven to be effective in detecting the gross evidence of plant stress, such as chlorosis on leaves and wilting of the vegetation. With the evolution of more advanced CV algorithms, less evident clues can now be identified, e.g., cases of nutrient depletion or experience with environmental stressful factors. This advanced detecting power can enable the agricultural practitioners to get involved much earlier with respect to the development of these stressors.

6.13.3 Improve Disease and Pest Identification

Blind proliferation of pathogens and crop pests is a lasting negative to agricultural crop production; its effects are far-reaching for plant health and production. Increased accuracy of CV systems distributed to farmers to screen and detect the presence of certain diseases or infestations of pests that are easily visible and predictable would also allow the farmers to intervene faster and more accurately, reducing the loss of crop and the use of pesticides, which may cause further damage.

6.13.4 Enhance Accuracy in Yield Prediction

With the introduction of DL in crop health samplings, it is projected to provide better clarification in quantitative estimation, consequently enabling enhanced prediction of agricultural output. This capability would enable the formation of harvest timetables, the fine-tuning of market-supply estimations, and the streamlining of crop management procedures elsewhere. Also, the technology will make it possible to efficiently distribute agricultural inputs, namely, fertilizer and water, using projected demand for the crop.

6.14 Integration of Internet of Things Devices

The synergistic deployment of IoT devices and CV systems promises to elevate the sophistication and precision of agricultural monitoring and management. The IoT devices are physical ones installed or fitted with some inbuilt sensors, software, and internet access, permitting them to gather and send data over networks. Combined with CV sensors, those devices offer a coherent platform for measuring crop conditions and environmental variables, thus allowing more subtle and targeted actions.

The key benefits of IoT integration into CV systems include:

6.14.1 Real-time Data Feedback

In contemporary agricultural practice, IoT devices such as soil moisture sensors, temperature sensors, and weather stations collect real-time environmental data. This becomes associated with a CV method to track crop well-being and also identify environmental stressors that affect crop growth. To take an example, in case a CV system identifies initial signs of nutrient deficiency, IoT sensors can establish the cause of the issue, ranging from soil nutrient contents, water stress, or other environmental factors.

6.14.2 Continuous Monitoring

IoT devices currently facilitate continuous monitoring of crop health, soil conditions, and environmental parameters. The subsequent streams of data can be used by CV systems to monitor crop health throughout the duration of the growing season to better analyze crop variability over time. Such real-time monitoring enables farmers to analyze the crop performance and recalculate the management approaches in real-time, maximizing growth and reducing the chance of unexpected crop failure.

6.14.3 Precision Farming

The tandem deployment of IoT and CV has dramatically enhanced precision farming capabilities. A good example is the use of IoT-powered irrigation systems, whereby sensors in spatially dispersed locations monitor soil moisture by sensing nodes and regulating the flow of water on-site based on instantaneous sensing. The CV system, at the same time, sends high-resolution images assessing the health of crops and identifying any anomaly, including water stress or improperly balanced nutrition, hence providing an opportunity to act in time. The combined use of these modalities leads to more effective use of inputs, hence preventing wastage of water, use of fertilizer, and dependence on pesticides.

6.14.4 Automated Decision-Making

In modern farming, sensor systems on IoT devices constantly track the progress of the crops and the state of the soil and thus provide a long-term stream of input to CV applications. These systems are programmed to provide real-time advice on matters related to fertilizing or altering irrigation. Such automation reduces the need to make manual interventions by repeatedly combining the sensor outputs with computational analysis; therefore, it eliminates the need to use conventional, non–data-driven labor-intensive processes to make decisions. The resulting efficiency levels translate into high crop harvests and low costs of operations.

6.15 Challenges with Traditional Detection Methods

For the past few decades, crop scientists have been using conventional methods of crop nutrient deficiency diagnosis that mostly consist of visual observations and soil analysis. In its turn, both methods have certain substantial drawbacks in terms of precision, time costs, and financial aspects. Below, the discussion on these flaws, as well as why they might not fit into modern and large-scale agricultural systems, is given.

6.15.1 Visual Inspection

The most common practice in the field is a visual diagnosis. Leaf color, leaf texture, and growth rate are always assumed to show nutrient imbalances. However, the limits in the effectiveness of this approach are rigidly determined by the subjective perception of a person, inter- and intra-plant variability, and the human ability to distinguish between subtle signs of physiological parameters [17]. As a result, the visual check method has been established to render error levels that are quite high compared to other methods of check that are quantitative.

6.15.2 Soil Testing

Soil assessment gives an indirect measurement and estimation of the nutrient adequacy and has a number of weaknesses. To start with, often the sampling procedures lack standardization and do not necessarily capture the in-field variability. Second, results derived using laboratory tests can take a long time, yet information is required within the shortest time so as to make decisions for management. Third, the values reported might be hard to interpret, as there was indeed a series of factors that affected the available nutrients in the soil. All this highlights the restricted practical value of soil testing as an independent diagnostic method [18].

6.16 Conclusion

In summary, the introduction of CV into modern farming is a significant step toward helping overcome the eternal challenges that food crop nutrition gaps and the mounting demands of international food output present. CV, as an inseparable part of AI, provides farmers with the real-time diagnostics they can work with to address the immediate issues with crop health, nutrient deficiency, pest invasion, and disease outbreaks.

The field-crop industry, which adds up to the main staples like wheat, maize, rice, and barley, lies at the forefront in ensuring the nourishment of the growing global population. Nevertheless, the growing stress resulting from the factors of climate change, soil degradation, and the declining nutrient availabilities requires new solutions. CV has thus become a disruptive tool that enables growers to become more aggressive in observing crop performance, as well as to remediate as quickly as possible at the first instances of stress or deficiency. Such precision agriculture ability allows fertilization, irrigation, and pest control to be applied selectively and to achieve optimal use of available resources and reduce waste.

CV systems use drones, satellite images, ground cameras, and sensors to obtain high-resolution crop images and then apply ML processes to that data to identify nutrient deficiencies, disease processes, and environmental stresses. These abilities not only increase crop production but also improve the nutritional value of foods, alleviating fears about the so-called hidden hunger based on low micronutrient levels in staple foods.

With the rising global demand for food, CV delivers the instrument to enhance agricultural systems in a way that they will meet the nutritional needs without a negative ecological impact. In combination with big-data systems, cloud services, and IoT equipment, CV provides farmers with a detailed understanding of the crop condition, which allows them to make decisions faster, automating the process of corrective measures and introducing precision farming. It is a typical technological convergence that makes farming operations data-driven, efficient, and adaptive with regard to climate variability.

Despite such improvements, there are still a number of setbacks. Large sums of capital, limited exposure to the necessary dataset, and insufficient teaching in complex systems perpetuate these barriers. However, continuous advancement in AI, the ubiquitous spreading of cheap drones, and the installation of cloud-based systems are the factors that would potentially increase access for smallholder farmers, thus driving the transformation toward smart farming.

The future of agriculture also depends on its smooth combination with AI, CV, and data analytics, all of which are the pillars of a stronger, more productive, and more sustainable agricultural and agri-food system. After such advances, not only does mankind have the capability of meeting the increasing worldwide food demand, but it will also be able to produce commodities that are healthier, more nutrient-dense, and are grown in a state of ecological protection. Now that CV technologies and complementary advancements are still developing, these are set to have a far-reaching and imperative effect on the destiny of global food security.

References

1 White, R.R. and Gleason, C.B. (2022). Global human-edible nutrient supplies, their sources, and correlations with agricultural environmental impact. *Scientific Reports* 12 (1): 16781.

2 Ray, D.K., Mueller, N.D., West, P.C., and Foley, J.A. (2013). Yield Trends Are Insufficient to Double Global Crop Production by 2050. *PLoS One* 8 (6): e66428.

3 FAO IUW and W (2022). The State of Food Security and Nutrition in the World 2022 [Internet]. The State of Food Security and Nutrition in the World 2022. FAO. Available from: http://www.fao.org/documents/card/en/c/cc0639en.

4 Neupane, D., Adhikari, P., Bhattarai, D. et al. (2022). Does climate change affect the yield of the top three cereals and food security in the world? *Earth* 3 (1): 45–71.

5 Godfray, H.C.J., Beddington, J.R., Crute, I.R. et al. (2010). Food security: the challenge of feeding 9 billion people. *Science* 327: 812–818.

6 FAOSTAT (2024). Food and Agriculture Organization of the United Nations. FAOSTAT Online Database.

7 Lele, U.J. and Goswami, S. (2021). Food for all: Setting the scene. In: Food for all: International Organizations and the Transformation of Agriculture (ed. U.J. Lele), 1–48. Oxford University Press https://doi.org/10.1093/oso/9780198854364.003.0001.
8 Bouis, H.E. and Saltzman, A. (2017). Improving nutrition through biofortification: a review of evidence from HarvestPlus, 2003 through 2016. *Global Food Security* 12: 49–58.
9 Barman, M., Mishra, R., and Trivedi, V. (2022). Advances in micronutrient fertilizer production and efficacy in plant nutrition. In: Soil Management for Sustainable Agriculture: New Paradigms and Advances (ed. N.C. Sarkar, S.K. Bohra, and R.S. Meena), 215–242. Apple Academic Press https://doi.org/10.1201/9781003184478-9.
10 Woldekidan, M.A., Arja, A., Worku, G. et al. (2024). The burden and trends of child and maternal malnutrition across the regions in Ethiopia, 1990–2019: the global burden of disease study 2019. *PLOS Global Public Health* 4 (7): e0002640.
11 Cakmak, I., McLaughlin, M.J., and White, P. (2017). Zinc for better crop production and human health. *Plant and Soil* 411: 1–2.
12 Bharman, P., Ahmad Saad, S., Khan, S. et al. (2022). Deep learning in agriculture: a review. *Asian Journal of Research in Computer Science* 13 (2): 28–47.
13 Kumari, K., Mirzakhani Nafchi, A., Mirzaee, S., and Abdalla, A. (2025). AI-driven future farming: achieving climate-smart and sustainable agriculture. *AgriEngineering* 7 (3): 89.
14 Shankar, S. and Shikha (2018). Impacts of climate change on agriculture and food security. In: Biotechnology for sustainable agriculture: Emerging approaches and strategies (ed. R.L. Singh and S. Mondal), 207–234. Elsevier https://doi.org/10.1016/B978-0-12-812160-3.00007-1.
15 United Nations (UN) (2022). World Population Prospects 2022: Summary of Results (UN DESA/POP/2022/NO.3). World Population Prospects 2022.
16 Srinivas, P., Veeraiah, V., Thejaswini, K.O. et al. (2025). The impact of electromagnetic and optical radiation on physical and biological systems. *Journal of Nano- and Electronic Physics* 17 (4): 04012. https://doi.org/10.21272/jnep.17(4).04012.
17 Singh, Y. and Singh, B. (2001). Efficient management of primary nutrients in the rice-wheat system. *Journal of Crop Production* 4 (1): 23–85.
18 Kihara, J., Bolo, P., Kinyua, M. et al. (2020). Micronutrient deficiencies in African soils and the human nutritional nexus: opportunities with staple crops. *Environmental Geochemistry and Health* 42 (9): 3015–3033.
19 Havrlentová, M., Kraic, J., Gregusová, V., and Kovácsová, B. (2021). Drought stress in cereals–a review. *Agriculture (Pol'nohospodárstvo)* 67 (2): 47–60.
20 Halford, N.G., Curtis, T.Y., Chen, Z., and Huang, J. (2015). Effects of abiotic stress and crop management on cereal grain composition: implications for food quality and safety. *Journal of Experimental Botany* 66 (5): 1145–1156.
21 Toreti, A., Deryng, D., Tubiello, F.N. et al. (2020). Narrowing uncertainties in the effects of elevated CO_2 on crops. *Nature Food* 1 (12): 775–782.
22 Jain, M., Bajwa, M.S., and Kumar, H. (2022). Agriculture assistant for crop prediction and farming selection using machine learning model with real-time data using imaging through UAV drone. In: Advances in Electrical and Computer Engineering, vol. 836 (ed. S. SenGupta, A.F. Zobaa, and K.S.S. Iyer), 45–59. Singapore: Springer Nature https://doi.org/10.1007/978-981-16-7365-8_4.

23 Pandey, J.K. (2025). Understanding the impact of electromagnetic and optical radiation on physical and biological systems. *Journal of Nano- and Electronic Physics* 17 (5): 05023. https://doi.org/10.21272/jnep.17(5).05023.

24 Venkatalakshmi, B. (2014). Decision support system for precision agriculture. *International Journal of Research in Engineering and Technology* 3 (19): 707–711. https://doi.org/10.15623/ijret.2014.0319130.

25 Ibrahim, S., Hasan, N., Sabri, N. et al. (2022). Palm leaf nutrient deficiency detection using a convolutional neural network (CNN). *International Journal of Nonlinear Analysis and Application* 13 (1): 1949–1956.

26 Beemkumar, N. et al. (2023). Activity recognition and IoT-based analysis using time series and CNN. In: Handbook of Research on Machine Learning-enabled IoT for Smart Applications Across Industries (ed. N. Goel and R.K. Yadav), 350–364. IGI Global https://doi.org/10.4018/978-1-6684-8785-3.ch018.

27 Al-Tamimi, N., Langan, P., Bernád, V. et al. (2022). Capturing crop adaptation to abiotic stress using image-based technologies. *Open Biology* 12 (2): 210353. https://doi.org/10.1098/rsob.210353.

28 Pandey, J.K., Kumar, S., Lamin, M. et al. (2022). A metaheuristic autoencoder deep learning model for an intrusion detector system. *Mathematical Problems in Engineering* 2022: 11, 3859155. https://doi.org/10.1155/2022/3859155.

29 Pandey, J.K., Yadav, A.K., Purushotham, M. et al. (2022). Colorimetric detection of deltamethrin pesticide from biological samples based on the peroxidase-mimic catalytic activity on MnO_2/SnO_2. *Key Engineering Materials* 928: 45–51. https://doi.org/10.4028/p-4ti1i0.

30 Sudhakar, M. and Swarna Priya, R.M. (2023). Computer vision-based machine learning and deep learning approaches for identification of nutrient deficiency in crops: a survey. *Nature, Environment and Pollution Technology* 22 (3): 1387–1399.

31 Hussain, D., Fauz, F., Almoustafa, T. et al. (2025). Smart greenhouse management: harnessing artificial intelligence for sustainable farming. *Scientific Reports* 15 (1): 402. https://doi.org/10.1038/s41598-024-51234-1.

32 Talukdar, V., Veeraiah, V., Roy, D.R. et al. (2023). An innovative approach: typicality-based collaborative filtering for recommender systems. *Proceedings of the 3rd International Conference on Advancement in Electronics & Communication Engineering (AECE)*, pp. 863–868. https://doi.org/10.1109/AECE59614.2023.10428576

33 Jawad, H.M., Nordin, R., Gharghan, S.K. et al. (2017). Energy-efficient wireless sensor networks for precision agriculture: A review. *Sensors* 17 (8): 1781. https://doi.org/10.3390/s17081781.

34 Bali, M.K. and Singh, M. (2024). Farming in the digital age: AI-infused digital twins for agriculture. *2024 3rd International Conference on Sentiment Analysis and Deep Learning (ICSADL)*, pp. 14–21. IEEE.

7

Soil Texture and Fertility Assessment Through Imaging

Sanket Sunil Kawade[1], *Panchakarla Sedyaaw*[2], *Huang Min*[3], *and Subrat Kumar Swain*[4]

[1] *Department of Aquatic Environment Management, College of Fisheries, Ratnagiri, Maharashtra, India*
[2] *Department of Fish Processing Technology, College of Fisheries, Ratnagiri, Maharashtra, India*
[3] *Department of Aquatic Biology, Faculty of Fisheries, Huazhong Agriculture University, China*
[4] *Department of Fisheries Resource Science, Graduate School of Agriculture, Forestry and Fisheries, Kagoshima University, Japan*

7.1 Introduction

It serves as the foundation for terrestrial ecosystems and agriculture, providing a medium for crop growth, nutrient cycling, and water regulation, while acting as a reservoir of biodiversity [1]. Soil texture, defined by the relative proportions of sand, silt, and clay, plays a crucial role in determining physical properties, water-holding capacity, aeration, and susceptibility to erosion and directly influences soil fertility and plant productivity [2]. Fertility, in turn, is a function of not only nutrient availability but also the physical and biological integrity of soil systems, a fact that underscores the interdependence between texture and fertility [3]. Traditionally, approaches to analyzing soil texture have used laboratory-based procedures such as the hydrometer, sieve, and pipette methods, while fertility has been assessed using chemical reagents to measure nitrogen, phosphorus, potassium, and organic matter [4]. While generally accurate, these methods are often time-consuming, expensive, and require skilled personnel and thus are poorly suited for site-specific soil management and extensive monitoring. Imaging approaches, in contrast, represent a non-destructive and rapid alternative to classical methods that can yield high-resolution information on soil properties [5].

The integration of imaging technologies with advanced computational models has revolutionized soil science, enabling researchers to capture high-resolution images of soil particles and extract relevant features for classification [6]. Recent studies employing digital image processing have successfully differentiated soil textures by analyzing particle color, shape, and distribution under controlled conditions, achieving accuracies as high as 99% with YOLOv8-based frameworks [7]. Similarly, hyperspectral imaging (HSI) systems have been deployed to detect soil nutrient status, organic carbon, and microbial activity, linking spectral reflectance patterns with fertility indicators [8]. Moreover, the application of artificial intelligence (AI), including machine learning (ML) and deep learning (DL) algorithms, has strengthened the predictive capacity of soil imaging, enabling automated classification and large-scale mapping [9]. Despite these advances, limitations such as heterogeneity in natural soils, dataset scarcity, and variability in imaging conditions continue to hinder wide-scale implementation. Thus, this review synthesizes available evidence on soil texture and fertility assessment through imaging, highlighting both technological breakthroughs and persisting challenges, while charting a course for future research and field applications [10].

7.2 Soil Texture and Fertility: Fundamentals

7.2.1 Soil Texture

Soil texture refers to the relative proportion of sand, silt, and clay particles in a soil sample, and it fundamentally determines the physical behavior of soils in relation to water retention, aeration, and root penetration. Sand particles are large, ranging from 0.05 to 2 mm, which enhances drainage but limits nutrient-holding capacity, whereas silt particles (0.002–0.05 mm) improve soil cohesiveness and water-holding ability [1]. Clay particles (<0.002 mm) contribute significantly to cation exchange capacity and chemical reactivity, but their dominance can reduce permeability and increase compaction risks [4]. According to the USDA soil texture triangle, twelve soil textural classes exist, each defined by unique proportions of sand, silt, and clay, and these classes influence land suitability for different crops [11]. Accurate determination of soil texture is vital for predicting crop productivity, managing irrigation schedules, and assessing susceptibility to erosion. Traditional approaches such as the hydrometer and pipette methods, though precise, are laborious, chemical-intensive, and unsuitable for large-scale, rapid assessments. This has fueled the development of imaging-based approaches capable of capturing structural and textural variations nondestructively [5] (Table 7.1 and Figure 7.1).

7.2.2 Soil Fertility

Soil fertility is the soil's capacity to supply essential nutrients in adequate amounts and proportions for crop growth, and it is shaped by interactions among physical, chemical, and biological properties. Fertility is influenced not only by macronutrients such as nitrogen (N), phosphorus (P), and potassium (K) but also by organic matter, pH, and microbial activity [1, 14]. For example, soils with higher organic carbon content and balanced texture often provide better nutrient retention and microbial habitat [3]. Conversely, sandy soils, though well-aerated, often exhibit poor fertility due to leaching, while clay-rich soils, though nutrient-dense, may restrict root growth due to poor aeration [2]. Conventional fertility testing relies heavily on chemical reagents and lab-based instruments, raising concerns of time, cost, and environmental safety [4]. Recent approaches highlight the potential of imaging technologies to detect fertility-related traits such as soil color, organic matter content, and microbial biomass signatures through spectral indices, thereby linking imaging directly to fertility prediction [6]. Thus, soil fertility assessment is increasingly

Table 7.1 Soil texture classes and particle distribution

Soil texture class	Sand (%)	Silt (%)	Clay (%)	Key characteristics	References
Sand	85–100	0–15	0–10	Coarse, rapid drainage, low nutrient holding	[12]
Loamy Sand	70–90	0–30	0–15	Moderately coarse, slightly higher fertility	[12]
Sandy Loam	50–70	20–50	0–20	Balanced texture, moderate fertility	[12]
Loam	40–60	30–50	10–25	Ideal agricultural soil, good water retention	[12]
Silt Loam	20–50	50–80	0–27	Smooth texture, retains water well	[12]
Clay Loam	20–45	15–35	27–40	Higher fertility, reduced aeration	[12]
Silty Clay Loam	0–20	40–60	27–40	Sticky when wet, good nutrient retention	[12]
Sandy Clay Loam	45–65	0–20	27–35	Moderately fine texture, firm soil structure	[12]
Clay	0–45	0–40	40–100	Very fine texture, high water, and nutrient capacity	[12]

moving toward noninvasive, imaging-driven approaches that combine physical texture recognition with nutrient mapping (Table 7.2).

7.3 Imaging Techniques for Soil Texture Analysis

7.3.1 Optical and Digital Imaging

Optical imaging, including digital photography, represents one of the simplest and most accessible methods for soil texture analysis, offering a low-cost and nondestructive approach. Soil images captured under controlled lighting conditions can be processed to extract particle features such as size, shape, and color distribution, which directly correlate with texture. Early studies demonstrated the utility of RGB imaging in differentiating sandy and clayey soils by analyzing color intensity values [19]. Smartphone-based photography has further democratized soil imaging, allowing on-field texture classification with reasonable accuracy, particularly when combined with preprocessing algorithms for noise reduction [18]. Advanced digital imaging techniques, when paired with convolutional neural networks (CNNs), achieve accuracies exceeding 95%, as networks automatically learn discriminative features of soil particles [7]. However, challenges remain in ensuring consistent results under uncontrolled natural conditions, where variations in illumination and soil moisture may introduce noise and reduce prediction reliability [9]. Thus, controlled imaging setups such as the Blackbox prototype

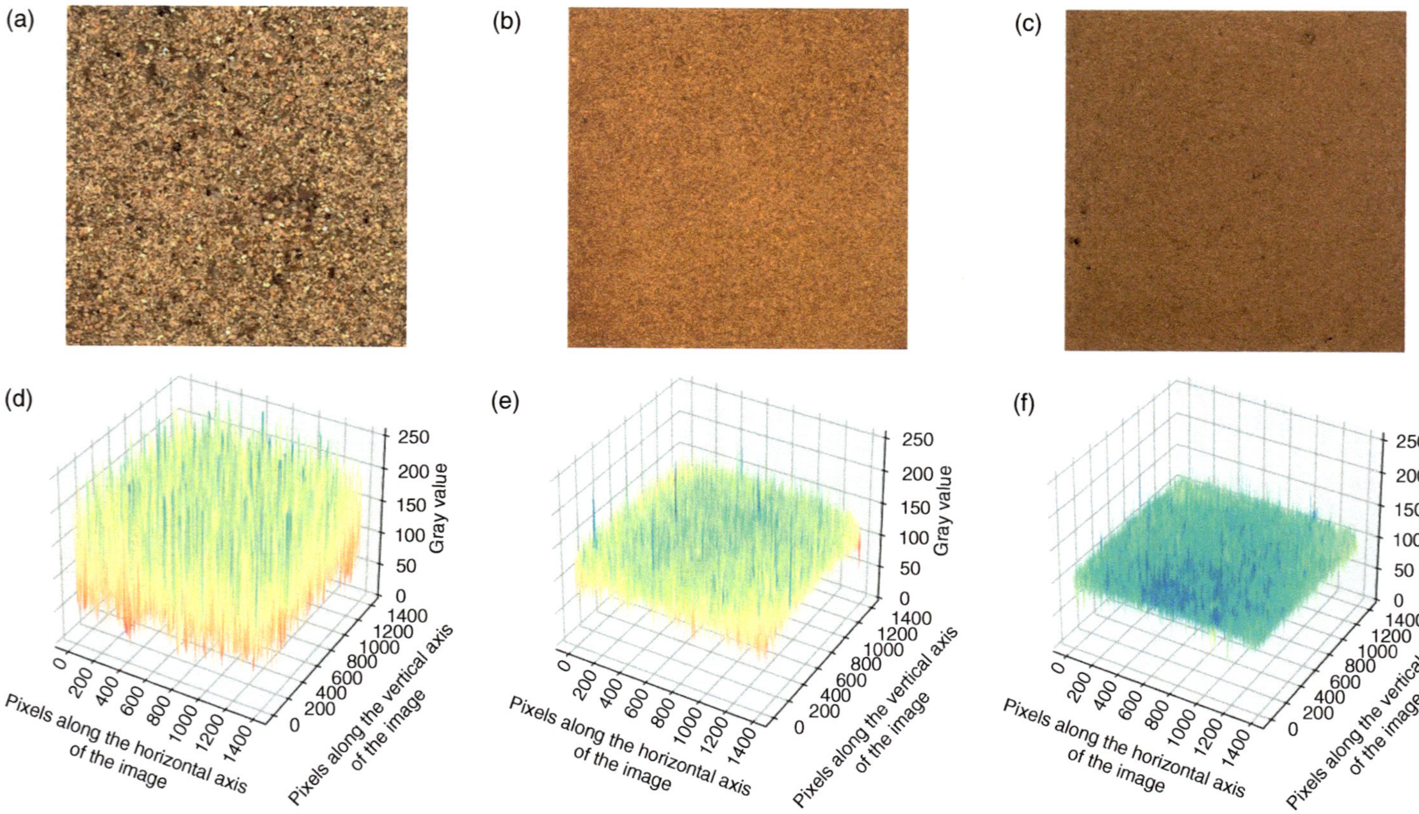

Figure 7.1 Soil surface image and grayscale 3D model diagram: (a) sand surface image; (b) silt surface image; (c) clay surface image; (d) sand grayscale image 3D model diagram; (e) silt grayscale image 3D model diagram; (f) clay image 3D model diagram. Note that (d–f) reflect the degree of change in grayscale values of three different types of soil; the higher the proportion of sand, the greater the change of amplitude it suggests. *Source:* Ma et al. [13]/ MDPI/CC BY 4.0.

Table 7.2 Comparison of ML models for soil classification.

Model	Strengths	Limitations	Example use	References
Random Forest (RF)	High accuracy, handles nonlinear data	Requires tuning, less interpretable	Soil fertility and organic matter prediction	[15]
Support Vector Machine (SVM)	Works well with small datasets, robust classification	Computationally expensive for large data	Soil texture classification using hyperspectral data	[16]
Convolutional Neural Network (CNN)	Excellent feature extraction, high accuracy	Requires large labeled data, GPU-intensive	Automated soil texture mapping from UAV imagery	[8]
Gradient Boosting (XGBoost)	High accuracy, handles missing data	Risk of overfitting	Soil nutrient and moisture prediction	[17]
Autoencoder + RF (Hybrid DL)	Dimensionality reduction, high predictive power	Complex training	Soil fertility estimation with hyperspectral data	[8]
Naïve Bayes	Fast, works on small data	Assumes independence, not suitable for correlated features	Soil type classification with limited inputs	[18]

have emerged, standardizing light intensity and camera position to enhance reproducibility and model performance [6] (Figure 7.2).

7.3.2 Hyperspectral and Multispectral Imaging

Hyperspectral imaging (HSI) extends conventional imaging by capturing reflectance information across hundreds of narrow spectral bands, making it a powerful tool for soil property characterization. Soil texture influences spectral reflectance due to differences in particle size, mineralogy, and surface roughness, allowing models to distinguish between sandy, silty, and clayey soils [20]. HSI systems have also been used to estimate soil organic carbon, moisture, and nutrient status, making them useful for linking texture and fertility in a single analysis [21]. Studies using portable hyperspectral sensors demonstrated rapid soil classification at field scales, reducing dependence on laboratory-based particle size analysis [22]. Similarly, multispectral imaging with fewer bands has proven effective for large-scale monitoring via drones and satellites, though with reduced resolution compared to hyperspectral systems [23]. Integration of HSI with ML techniques such as random forest and support vector machines has further enhanced predictive accuracy, often exceeding 90% for soil texture mapping [8]. Despite their potential, hyperspectral systems remain expensive and complex, requiring calibration protocols and robust data processing pipelines to achieve consistent results across diverse environments [24].

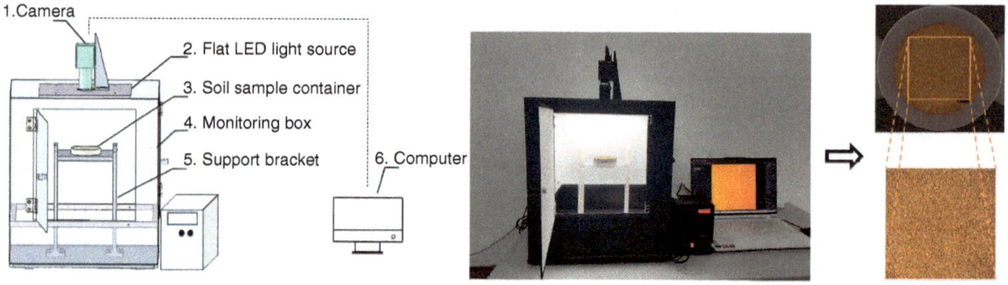

Figure 7.2 Image acquisition platform. From left to right: image acquisition platform, image acquisition setup, and regions of interest of the soil sample. *Source:* Ma et al. [13]/MDPI/CC BY 4.0.

7.3.3 Microscopic Imaging and Scanning Electron Microscopy (SEM)

Microscopic and scanning electron microscopy (SEM) techniques provide high-resolution images that reveal soil particle morphology and surface microstructure, enabling detailed examination of texture-related features. SEM has been applied to quantify particle aggregation, pore geometry, and fine-scale clay distributions, offering insights into soil–water interactions [25]. Unlike optical imaging, which provides macroscopic representations, SEM delivers microlevel precision, often used to validate models derived from image processing algorithms [26]. Advances in automated image segmentation allow SEM outputs to be coupled with particle-size distribution models, bridging laboratory precision with computational classification [27]. Additionally, digital light microscopy has been applied to soil thin sections, assisting in the visualization of aggregate stability and textural heterogeneity [28]. However, the cost, sample preparation requirements, and inability to scale SEM to large datasets limit its utility in routine soil assessments [29]. Current research trends emphasize combining SEM datasets with machine vision algorithms to automate classification, thereby reducing subjectivity in interpretation [30].

7.3.4 X-ray Computed Tomography (CT)

X-ray computed tomography (CT) has emerged as a nondestructive imaging technology capable of reconstructing three-dimensional models of soil structure, providing detailed insights into pore networks and particle distribution. CT scanning allows for volumetric quantification of soil textural properties without disturbing the natural arrangement of particles, making it especially useful for studying soil compaction and water movement [31]. High-resolution CT applications have been able to distinguish between different soil fractions and analyze root–soil interactions in situ [32]. Studies demonstrated that CT-derived texture parameters correlate strongly with hydraulic conductivity and bulk density, thereby linking imaging to soil functional properties [33]. The integration of CT imaging with ML has advanced soil classification by automating feature extraction from volumetric datasets [34]. While promising, CT systems are limited by high costs, radiation safety protocols, and the need for specialized expertise, restricting widespread adoption in agricultural contexts. Nonetheless, CT imaging represents a cutting-edge method for bridging microstructural soil analysis with computational modeling for texture assessment [35].

7.3.5 AI-Driven Imaging and Machine Vision Approaches

AI-based imaging approaches have revolutionized soil texture analysis by enabling automated feature extraction and classification from diverse image datasets. CNNs and advanced architectures such as YOLOv8 and Light-SoilNet have achieved classification accuracies exceeding 95%, even under heterogeneous conditions [36]. AI-driven models reduce the reliance on handcrafted features, as networks autonomously learn texture attributes, including particle distribution, color variations, and aggregate size [7]. The fusion of imaging with ML allows scalable soil mapping, with studies reporting strong correlations between predicted and laboratory-determined soil texture classes [8]. Additionally, hybrid models combining CNN with support vector machines or random forest algorithms have improved robustness against noisy or limited datasets [37]. Smartphone-assisted soil imaging has also benefited from lightweight AI frameworks, enabling rapid, field-level classification with real-time predictions [38]. Despite these successes, AI-driven models face challenges in dataset standardization, transferability across regions, and interpretability of deep-learning decisions, highlighting the need for explainable AI frameworks in soil science [9] (Table 7.3 and Figure 7.3).

Table 7.3 Imaging techniques vs. applications in soil studies.

Imaging technique	Principle	Applications in soil studies	Limitations	References
Multispectral Imaging	Captures reflectance in limited bands	Soil moisture estimation, classification	Lower resolution	[39]
RGB (Digital Photography)	Visible light spectrum	Soil color and surface texture	Limited depth analysis	[18]
Thermal Imaging	Infrared radiation mapping	Soil moisture, temperature profiling	Sensitive to the environment	[40]
Ground Penetrating Radar	Electromagnetic wave reflection	Subsurface soil structure and moisture	Expensive, requires expertise	[40]
LiDAR	Light detection and ranging	Topography, soil erosion analysis	Limited soil chemistry data	[17]
Drone-based Imaging	UAV-mounted multispectral/ hyperspectral	Precision agriculture mapping	Regulatory restrictions	[39]
Satellite Remote Sensing	Spaceborne multispectral/ hyperspectral	Regional/ continental soil surveys	Low temporal resolution	[16]

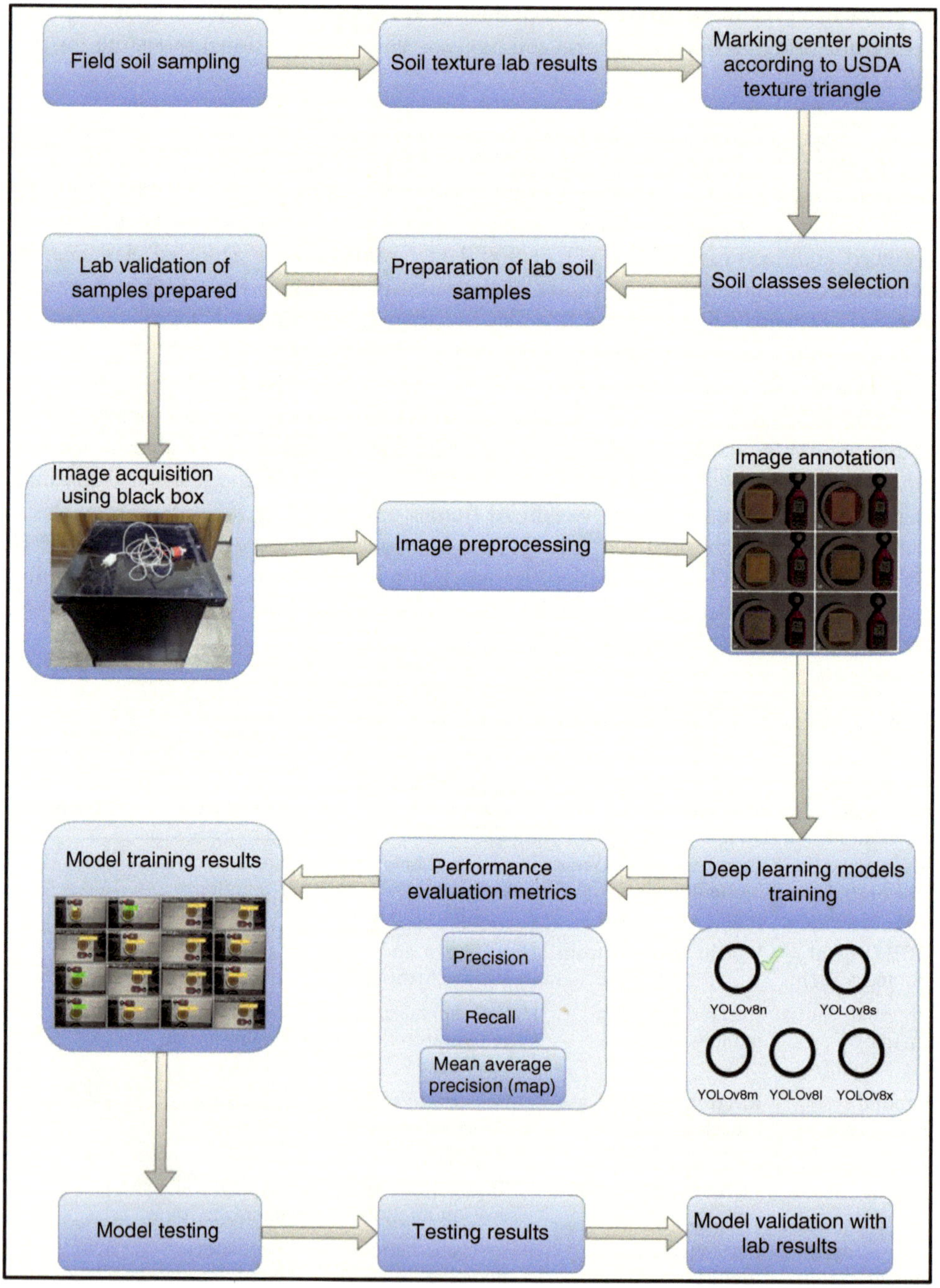

Figure 7.3 Methodology for soil texture analysis using controlled image processing. *Source:* Sattar et al. [6]/with permission of Elsevier.

7.4 Imaging Approaches for Soil Fertility Assessment

7.4.1 Visible and Near-Infrared (VNIR) Imaging

Visible and near-infrared (VNIR) imaging has become a widely used tool for assessing soil fertility because of its ability to capture spectral signatures associated with organic matter, mineralogy, and nutrient content. Soil organic carbon and nitrogen strongly absorb in the VNIR spectrum, enabling nondestructive prediction of fertility status [21]. VNIR systems, when integrated with regression models, have been shown to predict soil organic matter with high accuracy across diverse agricultural landscapes [41]. ML algorithms such as random forests and partial least squares regression enhance VNIR's predictive power by handling nonlinear relationships between spectra and soil fertility indicators [24]. Portable spectrometers now allow field-level applications, offering real-time measurements of nutrient status and reducing the need for laboratory testing [42]. Despite its potential, VNIR remains sensitive to soil moisture and surface conditions, necessitating calibration across regions and soil types for robust predictions [43].

7.4.2 Mid-Infrared (MIR) Imaging

Mid-infrared (MIR) imaging captures soil absorption features related to clay minerals, carbonates, and organic matter, making it highly effective for fertility assessments. Unlike VNIR, MIR spectra contain sharper and more distinct absorption bands, allowing more precise identification of fertility-related compounds. Studies have demonstrated that MIR is superior for predicting soil carbon and nitrogen due to its stronger molecular vibrations in these regions [44]. MIR spectroscopy has been applied to large soil libraries, providing reliable models for nutrient mapping at continental scales [45]. Integration of MIR with ML has further improved accuracy, with CNNs automating feature extraction from complex spectra [46]. However, MIR sensors are bulkier and more expensive than VNIR, limiting their direct field deployment, although advances in portable MIR instruments are bridging this gap [47].

7.4.3 Imaging of Soil Microbial Activity and Biota

Soil fertility is not only chemical but also biological, driven by microbial activity that regulates nutrient cycling. Imaging methods such as fluorescence microscopy and bioluminescence imaging have been applied to visualize microbial communities in soil aggregates [48]. Hyperspectral fluorescence imaging detects microbial metabolites and enzymes associated with carbon and nitrogen cycling [49]. Microbial biomass signatures captured through autofluorescence have been correlated with soil organic matter and fertility levels. Recent advances in confocal laser scanning microscopy allow three-dimensional imaging of microbial colonization in soil pores, linking biological activity to soil texture and fertility interactions [50]. While highly informative, these imaging techniques are mostly laboratory-based, requiring specialized equipment and limiting scalability to field contexts. Integration with spectral and thermal imaging may offer a pathway for field-level microbial fertility monitoring [51].

7.4.4 Fluorescence and Thermal Imaging

Fluorescence imaging provides noninvasive detection of organic compounds, chlorophyll-like pigments, and soil organic carbon that directly relate to fertility [49]. Thermal imaging complements this by detecting soil temperature variations influenced by moisture and organic matter, which are indirect indicators of fertility [52]. Together, these methods have been employed to assess soil respiration, microbial activity, and nutrient cycling under varying agricultural practices. Drone-based thermal sensors provide rapid fertility maps at farm scales, particularly useful for irrigation scheduling and water-nutrient balance monitoring [53]. ML enhances the interpretation of fluorescence and thermal datasets, linking patterns to fertility indices and crop yield predictions. Although these approaches remain supplementary, their integration with hyperspectral and MIR imaging strengthens holistic fertility assessments [54].

7.4.5 Remote Sensing Indices for Fertility Mapping

Remote sensing has enabled large-scale fertility mapping using vegetation indices such as the Normalized Difference Vegetation Index (NDVI), Soil-Adjusted Vegetation Index (SAVI), and Enhanced Vegetation Index (EVI). These indices serve as indirect proxies of soil fertility by monitoring vegetation vigor, canopy greenness, and chlorophyll content [55]. High-resolution satellite data from Sentinel and Landsat missions provide valuable inputs for regional fertility assessments [56]. Fertility indices derived from NDVI have been correlated with nitrogen availability and organic matter status, making them essential for precision nutrient management. Drone-based multispectral platforms further refine field-level fertility mapping, enabling variable-rate fertilization strategies. Integration of remote sensing indices with soil imaging and GIS allows multiscale assessments, bridging field and regional fertility predictions. However, vegetation-based proxies must be complemented with soil-level imaging to capture subsurface fertility dynamics [57].

7.5 Integration of Imaging with Advanced Tools

7.5.1 Artificial Intelligence and Machine Learning in Soil Imaging

AI and ML have become integral in processing soil imaging datasets due to their ability to handle complex, nonlinear relationships between soil spectral or visual features and fertility or texture classes. CNNs have achieved remarkable success by learning features directly from soil images, eliminating the need for handcrafted feature extraction [7]. DL frameworks such as YOLOv8 and autoencoder-based CNNs have reported classification accuracies above 95%, even under heterogeneous conditions [8]. Hybrid approaches that integrate CNN with Random Forest or Support Vector Machine classifiers have improved robustness in predicting soil fertility and texture under variable field conditions [37]. Additionally, lightweight models optimized for smartphone-based imaging have enabled real-time field classification, making AI-driven soil assessment more accessible for farmers [38]. Despite these advances, challenges remain in standardizing soil image datasets, ensuring cross-regional applicability, and addressing the “black box” problem of interpretability in deep models [9].

7.5.2 GIS-Based Soil Fertility Mapping

Geographic Information Systems (GIS) have revolutionized soil fertility mapping by integrating imaging data with spatial analysis to generate site-specific soil fertility maps. Imaging data from hyperspectral sensors or drones can be georeferenced and incorporated into GIS platforms, enabling high-resolution mapping of nutrient availability and soil texture variability [58]. GIS has been widely used for precision agriculture by overlaying soil fertility maps with crop yield data, supporting site-specific management practices that optimize fertilizer use [57]. Remote sensing indices such as NDVI and SAVI, when combined with GIS, provide insights into spatial variations in fertility across fields and regions [55]. Integration with GPS-based sampling enhances the accuracy of spatial interpolation models such as kriging, resulting in precise fertility maps at the farm scale [56]. Challenges in GIS integration include ensuring consistent calibration of imaging datasets across diverse soil types and the cost of developing GIS infrastructure for smallholder farmers [59].

7.5.3 Integration with Proximal and In Situ Sensors

The fusion of imaging with proximal and in situ sensors has opened new pathways for real-time soil fertility monitoring. Proximal sensors such as portable spectrometers, electrical conductivity meters, and pH probes, when combined with imaging data, provide complementary information for fertility assessment [60]. In situ sensor networks integrated with IoT systems allow continuous monitoring of soil moisture, temperature, and nutrient dynamics, which can be correlated with imaging data to improve predictive models [3]. Proximal soil sensing integrated with imaging has also enabled multiscale assessments, from field plots to regional landscapes, thereby bridging the gap between lab-based accuracy and field-scale application [47]. Coupling these datasets with ML further enhances predictions by reducing noise and improving model generalizability [46]. However, barriers such as sensor calibration, maintenance, and data integration complexity must be overcome to achieve practical scalability [14].

7.5.4 Big Data Analytics and Predictive Modeling

Soil imaging generates massive volumes of data, especially when derived from hyperspectral sensors, drones, and satellite systems, necessitating the use of big data analytics. Cloud-based platforms now allow real-time storage and analysis of soil imaging datasets, facilitating global-scale fertility monitoring [61]. Predictive models powered by big data approaches combine imaging with auxiliary datasets such as climate, topography, and management practices, improving fertility predictions across diverse agro-ecological zones. ML pipelines deployed on big data frameworks such as Hadoop and Spark have been applied to process terabytes of hyperspectral soil images efficiently [62]. Moreover, predictive modeling frameworks integrating big data with AI are increasingly used for forecasting crop yields, optimizing fertilizer use, and monitoring soil degradation. Challenges remain in ensuring data quality, harmonization of soil imaging protocols, and addressing privacy concerns in cloud-based agricultural platforms [63].

7.6 Applications in Agriculture and Environmental Management

7.6.1 Precision Agriculture and Site-Specific Management

Imaging technologies have played a crucial role in advancing precision agriculture by enabling detailed monitoring of soil texture and fertility variability within fields. High-resolution imaging from drones, satellites, and proximal sensors allows for site-specific nutrient management, reducing fertilizer waste while improving crop yields [64]. VNIR and HSI, integrated with GIS, support variable-rate fertilizer applications tailored to soil fertility patterns, increasing resource-use efficiency [42]. ML models trained on soil images provide automated recommendations for irrigation scheduling and nutrient allocation, particularly in water-scarce regions. Drone-based imaging systems further allow rapid field surveys that can detect fertility hotspots and texture-related water retention variability, guiding farmers in real-time decision-making [23]. These advances align with global goals for sustainable agriculture by reducing input costs, minimizing environmental footprints, and enhancing food security [54].

7.6.2 Soil Degradation and Erosion Monitoring

Soil degradation, including erosion, compaction, and nutrient depletion, poses significant threats to agricultural sustainability and environmental health. Imaging systems, especially remote sensing and hyperspectral tools, have been employed to detect erosion-prone areas and monitor topsoil loss [65]. Thermal imaging has been used to identify compacted soils with reduced water infiltration capacity, linking physical degradation to fertility decline [52]. Satellite imagery coupled with vegetation indices such as NDVI and SAVI can map soil erosion indirectly through vegetation cover decline [55]. X-ray CT imaging has enabled detailed quantification of soil structural changes under erosion stress, offering new insights into microstructural impacts of degradation [31]. By integrating imaging with predictive models, soil degradation risk maps can be generated, helping policymakers and land managers prioritize conservation practices [35].

7.6.3 Fertility Mapping for Crop Yield Prediction

Fertility mapping through imaging and remote sensing has become indispensable for predicting crop yields at both farm and regional scales. HSI provides detailed information on soil nutrient availability, particularly nitrogen and organic carbon, which strongly correlate with yield outcomes [21]. Fertility maps generated from VNIR and MIR imaging combined with GIS can identify spatial variability in soil nutrients, enabling more accurate yield forecasting [41]. ML algorithms enhance yield prediction models by linking fertility indices derived from imaging with historical crop performance data [62]. Remote sensing indices like NDVI and EVI serve as effective proxies for soil fertility, enabling near-real-time monitoring of vegetation vigor and growth [59]. Fertility imaging integrated with crop simulation models has also been applied to assess the impacts of nutrient management strategies on long-term yield sustainability [61].

7.6.4 Climate Change Impacts on Soil Quality: Imaging Perspectives

Climate change introduces multiple stressors, such as increased drought, flooding, and temperature variability, that alter soil fertility and texture dynamics. Imaging provides tools to monitor these impacts, offering early-warning indicators of soil quality decline [66]. Remote sensing datasets have been used to track climate-induced shifts in vegetation cover, indirectly reflecting soil fertility changes [57]. Thermal imaging identifies temperature-driven changes in soil respiration, while hyperspectral systems detect organic matter depletion under drought stress [52]. AI-integrated imaging platforms can model long-term fertility shifts under climate scenarios, providing essential data for adaptation strategies. Moreover, imaging-based soil monitoring supports climate-smart agriculture initiatives by linking soil resilience metrics with management practices [14]. By integrating imaging into climate adaptation frameworks, soil management strategies can be tailored to mitigate fertility losses and ensure sustainable agricultural productivity [63].

7.7 Current Trends and Future Prospects in Soil Texture and Fertility Assessment Through Imaging

The current trends in soil texture and fertility assessment emphasize the integration of advanced imaging technologies such as hyperspectral, multispectral, and X-ray CT, combined with AI and ML algorithms. HSI enables detection of fine-scale variations in soil organic matter and mineral composition, while CT provides three-dimensional visualization of pore structure and compaction patterns [8]. The shift toward AI-driven analysis, including CNNs and YOLOv8 models, has automated feature extraction, improving classification accuracy of soil textures beyond 95% [7]. Integration with GIS and Internet of Things (IoT) networks facilitates real-time data visualization, empowering precision agriculture through site-specific nutrient management. Moreover, cloud-based big data platforms now process massive imaging datasets for predictive modeling, supporting sustainable land management [58].

Prospects envision multimodal imaging combining hyperspectral, thermal, and fluorescence sensors to provide holistic assessments of soil health. Portable smartphone-based imaging tools are expected to democratize soil assessment among smallholder farmers [9]. Additionally, integration with climate modeling and 3D soil reconstruction may transform imaging systems into central tools for global soil health monitoring, ensuring sustainability, resilience, and food security.

7.8 Conclusion

The comprehensive exploration of this chapter underscores the paradigm shift in soil science from traditional laboratory-based techniques toward advanced, nondestructive, and scalable imaging methods. The integration of optical, hyperspectral, microscopic, and computed tomography imaging has enabled soil scientists to characterize texture and fertility with unprecedented precision. These techniques provide detailed insights into soil composition, particle size distribution, and nutrient variability, thereby supporting the growing demand for precision agriculture and environmental monitoring. The synergy between imaging technologies and AI, particularly through CNNs, YOLO architectures, and hybrid ML models, has revolutionized data analysis, allowing automated classification and fertility prediction with exceptional accuracy. Furthermore, imaging-based soil assessment facilitates the simultaneous evaluation of physical and chemical properties, bridging the gap between texture and fertility. VNIR and MIR imaging allow rapid estimation of organic carbon, nitrogen, and microbial activity, while fluorescence and thermal imaging contribute to assessing soil biological health. Future research should emphasize multimodal imaging integration and the fusion of imaging with in situ sensors and IoT systems for real-time monitoring. Ultimately, imaging-driven soil texture and fertility assessment represent a transformative frontier for sustainable land management, offering an intelligent, precise, and environmentally conscious pathway to safeguard global food security.

References

1 Dexter, A.R. (2004). Soil physical quality: part I. Theory, effects of soil texture, density, and organic matter, and effects on root growth. *Geoderma* 120 (3–4): 201–214. https://doi.org/10.1016/j.geoderma.2003.09.004.

2 Oliveira Morais, P.A., de Souza, D.M., de Melo Carvalho, M.T. et al. (2019). Predicting soil texture using image analysis. *Microchemical Journal* 146: 455–463. https://doi.org/10.1016/j.microc.2019.01.009.

3 Mohapatra, H. and Rath, A.K. (2022). IoE-based framework for smart agriculture: networking among all agricultural attributes. *Journal of Ambient Intelligence and Humanized Computing* 13 (1): 407–424. https://doi.org/10.1007/s12652-021-02908-4.

4 Faé, G.S., Montes, F., Bazilevskaya, E. et al. (2019). Making soil particle size analysis by laser diffraction compatible with standard soil texture determination methods. *Soil Science Society of America Journal* 83 (4): 1244–1252. https://doi.org/10.2136/sssaj2018.10.0385.

5 Beemkumar, N., Gupta, S., Bhardwaj, S. et al. (2023). Activity recognition and IoT-based analysis using time series and CNN. In: Handbook of Research on Machine Learning-enabled IoT for Smart Applications Across Industries (ed. N. Goel and R.K. Yadav), 350–364. IGI Global https://doi.org/10.4018/978-1-6684-8785-3.ch018.

6 Sattar, K., Maqsood, U., Hussain, Q. et al. (2024). Soil texture analysis using controlled image processing. *Smart Agricultural Technology* 9: 100588. https://doi.org/10.1016/j.atech.2024.100588.

7 Lanjewar, M. G., & Gurav, O. L. (2022). Convolutional neural networks-based classifications of soil images. *Multimedia Tools and Applications*, 81(7), 10313–10336. https://doi.org/10.1007/s11042-022-12200-y

8 Zhao, Z., Feng, W., Xiao, J. et al. (2022). Rapid and accurate prediction of soil texture using an image-based deep learning autoencoder convolutional neural network random forest (DLAC-CNN-RF) algorithm. *Agronomy* 12 (12): 3063. https://doi.org/10.3390/agronomy12123063.

9 Babalola, E.O., Asad, M.H., and Bais, A. (2023). Soil surface texture classification using RGB images acquired under uncontrolled field conditions. *IEEE Access*. https://doi.org/10.1109/ACCESS.2023.3290907 11: 67140–67155.

10 Abishek, J., Kannan, P., Nirmala Devi, M., Prabhaharan, J., Sampathkumar, T., & Kalpana, M. (2023). Soil texture prediction using a machine learning approach for sustainable soil health management. *International Journal of Plant & Soil Science*, 35(19), 1416–1426. https://doi.org/10.9734/IJPSS/2023/v35i193685

11 USDA Natural Resources Conservation Service (NRCS) (2023). Soil Texture Calculator. Retrieved from https://www.nrcs.usda.gov/resources/education-and-teaching-materials/soil-texture-calculator.

12 USDA (2017). Soil texture triangle and classification system. United States Department of Agriculture, Natural Resources Conservation Service. Retrieved from https://www.nrcs.usda.gov.

13 Ma, R., Jiang, J., Ouyang, L. et al. (2024). Toward flexible soil texture detection by exploiting deep spectrum and texture coding. *Agronomy* 14: 2074. https://doi.org/10.3390/agronomy14092074.

14 Ahmed, R.A., Hemdan, E.E.D., El-Shafai, W. et al. (2022). Climate-smart agriculture using intelligent techniques, blockchain, and internet of things: concepts, challenges, and opportunities. *Transactions on Emerging Telecommunications Technologies* 33 (11): e4607. https://doi.org/10.1002/ett.4607.

15 Mirzaeitalarposhti, R., Shafizadeh-Moghadam, H., Taghizadeh-Mehrjardi, R., and Demyan, M.S. (2022). Digital soil texture mapping and spatial transferability of machine learning models using Sentinel-1, Sentinel-2, and terrain-derived covariates. *Remote Sensing* 14 (21): 5909. https://doi.org/10.3390/rs14215909.

16 Liu, L., Ji, M., and Buchroithner, M. (2022). Transfer learning for soil spectroscopy based on convolutional neural networks and its application in soil clay content mapping using hyperspectral imagery. *Sensors* 22 (9): 3169. https://doi.org/10.3390/s22093169.

17 Gao, J., Li, Y., Zhang, X., and Xu, C. (2022). Artificial intelligence for soil and water conservation: recent advances and applications. *Computers and Electronics in Agriculture* 193: 106702. https://doi.org/10.1016/j.compag.2022.106702.

18 Swetha, R.K., Bende, P., Singh, K. et al. (2020). Predicting soil texture from smartphone-captured digital images and an application. *Geoderma* 376: 114562. https://doi.org/10.1016/j.geoderma.2020.114562.

19 Chung, S.O., Cho, K.H., Kong, J.W. et al. (2010). Soil texture classification algorithm using RGB characteristics of soil images. *IFAC Proceedings Volumes* 43 (1): 71–76. https://doi.org/10.3182/20101206-3-JP-3009.00005.

20 Talukdar, V., Veeraiah, V., Roy, D. R., Pandey, J. K., Gupta, A., & Dhabliya, D. (2023). An innovative approach: typicality-based collaborative filtering for recommender systems. *Proceedings of the 3rd International Conference on Advancement in Electronics & Communication Engineering (AECE)*, pp. 863–868. https://doi.org/10.1109/AECE59614.2023.10428576

21 Pandey, J.K., Yadav, A.K., Purushotham, M. et al. (2022). Colorimetric detection of deltamethrin pesticide from biological samples based on the peroxidase-mimic catalytic activity on MnO_2/SnO_2. *Key Engineering Materials* 928: 45–51. https://doi.org/10.4028/p-4ti1i0.

22 Ji, W., Viscarra Rossel, R.A., and Shi, Z. (2015). Accounting for soil condition in spectral soil prediction models with factor analysis. *European Journal of Soil Science* 66 (3): 712–724. https://doi.org/10.1111/ejss.12241.

23 Peng, Y., Zhu, Q., Chen, Y., and Ma, J. (2021). Estimating soil organic carbon using UAV-based hyperspectral imaging and random forest. *Remote Sensing* 13 (6): 1150. https://doi.org/10.3390/rs13061150.

24 Vašát, R., Kodešová, R., Borůvka, L. et al. (2020). Predicting soil properties by NIR spectroscopy: considering soil moisture as an explanatory variable. *Soil and Tillage Research* 196: 104436. https://doi.org/10.1016/j.still.2019.104436.

25 Mermut, A.R. and Cano, A.F. (2001). SEM and soil studies: applications and limitations. *Advances in Agronomy* 73: 93–132. https://doi.org/10.1016/S0065-2113(01)73005-5.

26 Jien, S.H. and Chen, T.H. (2013). Effects of biochar on soil properties and erosion potential in a highly weathered soil. *Catena* 110: 225–233. https://doi.org/10.1016/j.catena.2013.06.021.

27 Basaran, M., Yilmaz, K., and Apaydin, A. (2017). Use of SEM and image analysis for the determination of soil structure. *Catena* 150: 135–145. https://doi.org/10.1016/j.catena.2016.11.006.

28 Stoops, G. (2003). Guidelines for analysis and description of soil and regolith thin sections. *Soil Science Society of America Journal.* ISBN (Hardback): 0-89118-842-8, (xvi + 184 pp.).

29 Cardoso, E.J.B.N., Vasconcellos, R.L.F., Bini, D. et al. (2021). Microscopic and molecular approaches to soil microbial community analysis. *Applied Soil Ecology* 157: 103736. https://doi.org/10.1016/j.apsoil.2020.103736.

30 Pandey, J. K., Kumar, S., Lamin, M., Gupta, S., Dubey, R.K. and Sammy, F. (2022). A metaheuristic autoencoder deep learning model for an intrusion detector system. *Mathematical Problems in Engineering,* 2022, Article 3859155, 11. https://doi.org/10.1155/2022/3859155

31 Taina, I.A., Heck, R.J., and Elliot, T.R. (2008). Application of X-ray computed tomography to soil science: a literature review. *Canadian Journal of Soil Science* 88 (1): 1–20. https://doi.org/10.4141/CJSS06010.

32 Tracy, S.R., Black, C.R., Roberts, J.A., and Mooney, S.J. (2015). Exploring the interacting effect of soil texture and bulk density on root system development in tomato. *Environmental and Experimental Botany* 114: 19–31. https://doi.org/10.1016/j.envexpbot.2015.02.002.

33 Katuwal, S., Moldrup, P., Lamandé, M. et al. (2015). Linking air and water transport in intact soils to macropore characteristics inferred from X-ray computed tomography. *Soil Science Society of America Journal* 79 (2): 559–571. https://doi.org/10.2136/sssaj2014.08.0326.

34 Houston, A.N., Otten, W., Baveye, P.C., and Hapca, S. (2017). Quantifying and visualizing the effect of soil structure on hydraulic conductivity using X-ray computed tomography and image-based modeling. *Water Resources Research* 53 (4): 2586–2602. https://doi.org/10.1002/2016WR019727.

35 Rabot, E., Wiesmeier, M., Schlüter, S., and Vogel, H.J. (2018). Soil structure as an indicator of soil functions: a review. *Geoderma* 314: 122–137. https://doi.org/10.1016/j.geoderma.2017.11.009.

36 Pandiri, D.K., Murugan, R., and Goel, T. (2024). Smart soil image classification system using a lightweight convolutional neural network. *Expert Systems with Applications* 238: 122185. https://doi.org/10.1016/j.eswa.2023.122185.

37 Abeje, B.T., Salau, A.O., Gela, B.M., and Mengistu, A.D. (2024). Soil type identification model using a hybrid computer vision and machine learning approach. *Multimedia Tools and Applications* 83 (1): 575–589. https://doi.org/10.1007/s11042-023-15692-4.

38 Barman, U., Choudhury, R.D., Sarma, S.K. et al. (2022). Smartphone-assisting convolutional neural networks for soil texture classification in dry and wet humid conditions in West Guwahati, Assam. *Journal of Applied and Natural Science* 14 (4): 1351–1359. https://doi.org/10.31018/jans.v14i4.3966.

39 Song, Q., Gao, X., Li, Q. et al. (2023). Estimation and mapping of soil texture content based on unmanned aerial vehicle hyperspectral imaging. *Scientific Reports* 13: 14097. https://doi.org/10.1038/s41598-023-40384-2.

40 Karimi, Y., Mahmoudi, S., and Sadeghi, S.H.R. (2020). Ground-penetrating radar application for soil structure and moisture studies. *Journal of Applied Geophysics* 179: 104083. https://doi.org/10.1016/j.jappgeo.2020.104083.

41 Pandey, J.K. (2025). Understanding the impact of electromagnetic and optical radiation on physical and biological systems. *Journal of Nano- and Electronic Physics* 17 (5): 05023. https://doi.org/10.21272/jnep.17(5).05023.

42 Rossel, R.A.V. and Behrens, T. (2010). Using data mining to model and interpret soil diffuse reflectance spectra. *Geoderma* 158 (1–2): 46–54. https://doi.org/10.1016/j.geoderma.2009.12.025.

43 Srinivas, P., Veeraiah, V., Thejaswini, K.O. et al. (2025). The impact of electromagnetic and optical radiation on physical and biological systems. *Journal of Nano And Electronic Physics* 17 (4): 04012. https://doi.org/10.21272/jnep.17(4).04012.

44 Janik, L.J., Merry, R.H., and Skjemstad, J.O. (2007). Can mid-infrared diffuse reflectance analysis replace soil extractions? *Australian Journal of Experimental Agriculture* 47 (5): 523–529. https://doi.org/10.1071/EA05299.

45 Shepherd, K.D. and Walsh, M.G. (2007). Infrared spectroscopy—enabling an evidence-based diagnostic surveillance approach to agricultural and environmental management in developing countries. *Journal of Near Infrared Spectroscopy* 15 (1): 1–19. https://doi.org/10.1255/jnirs.716.

46 Ng, W., Minasny, B., and McBratney, A.B. (2019). Convolutional neural network for soil mid-infrared spectra. *The Soil* 5 (1): 79–90. https://doi.org/10.5194/soil-5-79-2019.

47 Terra, F.S., Demattê, J.A.M., and Viscarra Rossel, R.A. (2018). Proximal soil sensing for monitoring soil fertility in tropical soils. *Geoderma Regional* 13: e00189. https://doi.org/10.1016/j.geodrs.2018.e00189.

48 Nunan, N., Wu, K., Young, I.M. et al. (2001). Spatial distribution of bacterial communities and their relationships with the micro-architecture of soil. *FEMS Microbiology Ecology* 44 (2): 203–215. https://doi.org/10.1111/j.1574-6941.2003.tb01094.x.

49 Schulz, H., Baranska, M., and Baranski, R. (2013). Potential of fluorescence and Raman spectroscopy for analysis of plant cell wall polysaccharides. *Journal of Agricultural and Food Chemistry* 61 (42): 10241–10249. https://doi.org/10.1021/jf402323m.

50 Vos, M., Wolf, A.B., Jennings, S.J., and Kowalchuk, G.A. (2013). Micro-scale determinants of bacterial diversity in soil. *FEMS Microbiology Reviews* 37 (6): 936–954. https://doi.org/10.1111/1574-6976.12023.

51 Kravchenko, A.N., Guber, A.K., Razavi, B.S. et al. (2019). Microbial spatial footprint as a driver of soil carbon stabilization. *Nature Communications* 10 (1): 3121. https://doi.org/10.1038/s41467-019-11057-4.

52 Lu, Y., Lu, X., and Chen, F. (2019). Application of thermal imaging in soil and plant research: a review. *Plant Methods* 15: 109. https://doi.org/10.1186/s13007-019-0480-0.

53 Hoffmann, H., Nieto, H., Jensen, R. et al. (2016). Estimating evaporation with thermal UAV data and two-source energy balance models. *Hydrology and Earth System Sciences* 20 (2): 697–713. https://doi.org/10.5194/hess-20-697-2016.

54 Shen, Z., Li, Z., Ma, Y. et al. (2021). Soil fertility monitoring using UAV thermal and multispectral imaging integrated with machine learning. *Remote Sensing* 13 (17): 3429. https://doi.org/10.3390/rs13173429.

55 Huete, A.R. (1988). A soil-adjusted vegetation index (SAVI). *Remote Sensing of Environment* 25 (3): 295–309. https://doi.org/10.1016/0034-4257(88)90106-X.

56 Gao, F., Anderson, M., Zhang, X. et al. (2020). Toward mapping crop progress at field scales through fusion of Landsat and Sentinel-2 data. *Remote Sensing of Environment* 238: 111181. https://doi.org/10.1016/j.rse.2019.111181.

57 Liu, Y., Yue, H., Guo, W., and Wang, J. (2018). Integration of multi-source remote sensing data for regional soil fertility evaluation. *Computers and Electronics in Agriculture* 155: 165–173. https://doi.org/10.1016/j.compag.2018.10.015.

58 Gupta, R., Singh, K., and Sharma, R. (2019). GIS-based soil fertility mapping and management. *International Journal of Plant & Soil Science* 30 (4): 1–10. https://doi.org/10.9734/ijpss/2019/v30i430181.

59 Shiratsuchi, L.S., Ferguson, R.B., Shanahan, J.F. et al. (2014). Remote sensing for mapping soil organic matter. *Agronomy Journal* 106 (5): 1970–1979. https://doi.org/10.2134/agronj14.0005.

60 Viscarra Rossel, R.A., Behrens, T., Ben-Dor, E. et al. (2015). A global spectral library to characterize the world's soil. *Earth-Science Reviews* 155: 198–230. https://doi.org/10.1016/j.earscirev.2015.01.008.

61 Huang, J., Gómez-Dans, J., Huang, H. et al. (2020). Assimilation of remote sensing into crop growth models for better agricultural monitoring. *Science of the Total Environment* 728: 138502. https://doi.org/10.1016/j.scitotenv.2020.138502.

62 Bhatia, N., Singh, A., and Kaur, G. (2021). Big data analytics in precision agriculture: a survey. *Journal of King Saud University, Computer and Information Sciences* 33 (8): 101010. https://doi.org/10.1016/j.jksuci.2019.06.010.

63 Sun, J., Wang, S., Huang, W. et al. (2022). Privacy-preserving big data analytics in agriculture. *Computers and Electronics in Agriculture* 193: 106653. https://doi.org/10.1016/j.compag.2021.106653.

64 Mulla, D.J. (2013). Twenty-five years of remote sensing in precision agriculture: key advances and remaining knowledge gaps. *Biosystems Engineering* 114 (4): 358–371. https://doi.org/10.1016/j.biosystemseng.2012.08.009.

65 Vrieling, A. (2006). Satellite remote sensing for water erosion assessment: a review. *Catena* 65 (1): 2–18. https://doi.org/10.1016/j.catena.2005.10.005.

66 IPCC (2021). Climate change 2021: the physical science basis. Contribution of Working Group I to the Sixth Assessment Report of the Intergovernmental Panel on Climate Change. Cambridge University Press. https://doi.org/10.1017/9781009157896

8

Vision-Assisted Edge Framework for Real-Time Food and Nutrition Analysis

Hansa Rajput

Department of Computer Science and Applications, Unique College, Parasia, Madhya Pradesh, India

8.1 Introduction

Nutritional monitoring plays a critical role across all age groups in maintaining public health, preventing lifestyle-related diseases, and informing both clinical practice and policy decisions. Reliable dietary assessment is essential not only for managing individual health but also for designing effective nutrition programs at the population level. The global rise in cardiovascular disorders, diabetes, obesity, and micronutrient deficiencies highlights the pressing need for continuous and accurate dietary tracking. Traditional approaches such as self-reporting, food diaries, and periodic clinical consultations are often limited by recall bias, inaccurate portion estimation, and inconsistent adherence, particularly when applied to large or geographically dispersed populations.

To overcome these challenges, there is a growing demand for intelligent, automated, and real-time systems capable of capturing dietary behaviors with greater precision and minimal user effort. Recent advances in artificial intelligence (AI), combined with the widespread adoption of Internet of Things (IoT) devices, have opened new avenues for nutritional monitoring. In particular, the integration of Edge artificial intelligence (Edge AI) with smart IoT-enabled cameras enables local data processing directly on compact, energy-efficient devices. This approach reduces reliance on cloud servers, minimizes latency, improves data privacy, and conserves bandwidth, making it highly suitable for clinical, household, and low-resource environments.

This chapter presents the design and application of a comprehensive Edge AI–IoT framework for real-time nutritional monitoring. It details how deep learning algorithms, intelligent sensors, and image-processing techniques can work together to recognize food items, estimate portion sizes, and map visual data to nutritional information. Furthermore, it demonstrates how this integration can shift dietary assessment from a static and reactive process to one that is dynamic, personalized, and predictive, thereby contributing to improved individual well-being and public health outcomes.

8.2 Background and Motivation

8.2.1 Current Landscape of Nutritional Monitoring Systems

Nutritional monitoring plays a crucial role in clinical nutrition, public health, and preventive medicine. Traditionally, dietary assessments have depended on self-reported methods such as food frequency questionnaires, 24-hour recalls, and food diaries. Although these techniques are widely applied in both research and clinical settings, they often underestimate actual nutrient and calorie intake due to recall bias and irregular reporting [1–3]. The advent of digital technologies, including wearable sensors and smartphone-based applications, has enhanced automation and continuity in dietary tracking. Wearable devices monitor physiological indicators such as heart rate and glucose levels to estimate energy expenditure, while mobile applications allow users to record meals and retrieve nutritional information through barcode scanning [1, 4]. Despite these advancements, challenges persist regarding measurement accuracy, user compliance, and contextual adaptation across diverse dietary habits. Table 8.1 provides a comparative overview of conventional and emerging nutritional monitoring approaches.

8.2.2 Challenges in Data Collection, Accuracy, and User Compliance

Despite notable technological progress, achieving accuracy in nutritional monitoring remains challenging due to issues of data reliability, collection techniques, and user compliance. Self-reported dietary tools are susceptible to recall bias, human error, and underreporting. Studies suggest adults may underestimate calorie intake by nearly 20% [1, 5]. Although mobile applications enhance convenience, they demand sustained user interaction, often leading to fatigue and disengagement. Similarly, wearable devices provide continuous physiological tracking but offer limited insights into food composition and portion estimation, restricting their effectiveness for precise nutritional evaluation [4, 6, 7]. Figure 8.1 illustrates the major sources of error in dietary monitoring systems, including user misreporting, portion misestimating, and variability in food presentation.

Table 8.1 Comparative overview of nutritional monitoring methods.

Method	Advantages	Limitations	Use case
24-hour recall	Low cost, easy administration	Memory bias, not real-time	Clinical nutrition
Food diaries	Detailed, customizable	High user burden, low compliance	Research studies
Mobile apps	Automated logging, notifications	Requires a smartphone, occasional user error	Personalized nutrition
Wearable sensors	Continuous monitoring	Cannot identify food types/portions	Health tracking, research
IoT + Edge AI (emerging)	Real-time, automated, privacy-preserving	Technical complexity, model training required	Preventive healthcare, remote monitoring

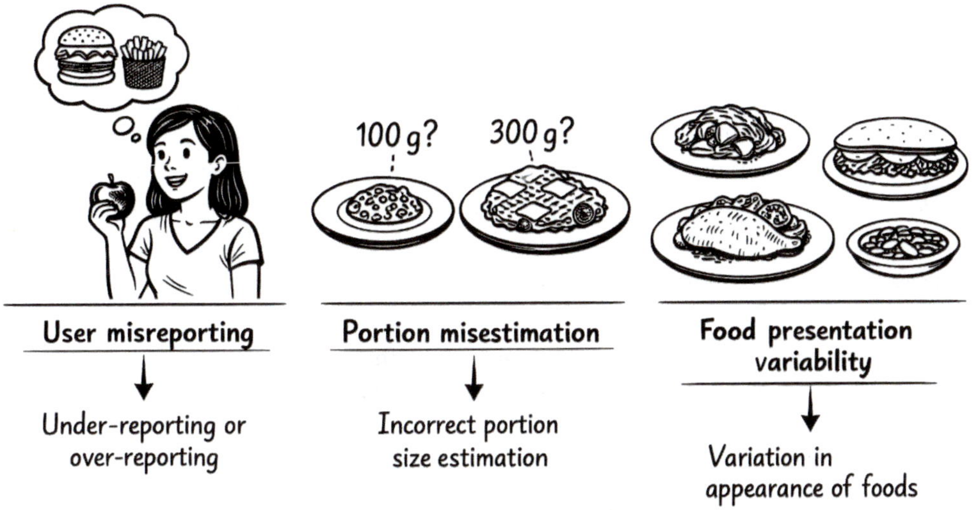

Figure 8.1 Sources of error in nutritional monitoring.

8.2.3 Role of AI and IoT in Healthcare and Nutrition

The convergence of AI and the IoT has facilitated intelligent, automated, and scalable nutritional monitoring systems [8–10]. Convolutional neural networks (CNNs) can accurately recognize food items, estimate portion sizes, and associate them with nutritional databases, while IoT-enabled sensors support continuous data acquisition with minimal user involvement. Edge AI further strengthens operational efficiency and privacy by performing computations locally on devices, thereby safeguarding personal data, minimizing latency, and reducing bandwidth consumption [9, 11, 12]. Figure 8.2 depicts a typical Edge AI-enabled

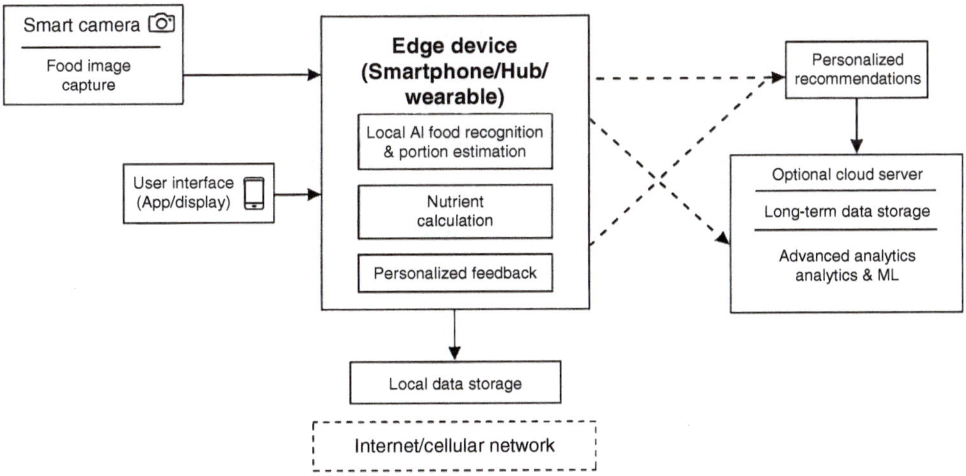

Figure 8.2 Edge AI-enabled IoT nutritional monitoring architecture.

Table 8.2 Edge vs. cloud-based nutritional monitoring systems.

Feature	Edge AI	Cloud AI
Latency	Low, real-time feedback	Higher, network-dependent
Bandwidth	Minimal data transfer	High, requires continuous connection
Privacy	Local processing preserves data	Centralized storage, higher risk
Scalability	Distributed deployment	Limited by server capacity
Energy efficiency	Optimized for low-power devices	High consumption

IoT nutritional monitoring framework, highlighting the interaction between smart cameras, edge devices, and optional cloud servers for analytics.

8.2.4 Motivation for Integrating Edge Computing with IoT Devices

Integrating edge computing with IoT devices effectively addresses the latency, privacy, and bandwidth constraints associated with cloud-based systems. Unlike conventional cloud processing, which depends on continuous connectivity and introduces transmission delays, edge computing performs local data analysis to deliver faster responses and enhanced data confidentiality [11, 12]. Moreover, federated learning (FL) strengthens both model accuracy and security by enabling distributed training across multiple edge devices without transferring sensitive information [11, 13].

8.2.5 Key Benefits: Low Latency, Energy Efficiency, Privacy, and Scalability

Edge AI-based nutritional monitoring provides significant advantages over both traditional and cloud-dependent approaches. Its low-latency processing enables instant feedback and real-time dietary recommendations, facilitating quicker behavioral adjustments. Energy-efficient edge devices ensure sustainable operation in households, clinics, and rural environments while maintaining optimal performance and power efficiency. By processing data locally, these systems enhance privacy through minimal external transmission, and distributed edge architectures enable scalable population-wide monitoring without burdening centralized infrastructure [11, 12, 14]. Table 8.2 summarizes the comparative benefits of edge-enabled versus cloud-based nutritional monitoring systems.

8.3 Core Architecture of Edge AI-based Nutritional Monitoring

8.3.1 System Overview

A paradigm shift from conventional dietary assessment techniques to an automated, intelligent, and real-time framework is represented by the proposed Edge AI-based nutritional monitoring system. To offer a comprehensive platform for dietary evaluation, the system combines a number of technological elements, such as edge devices, cloud infrastructure, IoT-enabled cameras, and mobile applications. The idea of local data processing lies at the heart of this architecture, guaranteeing quick inference and reducing reliance on centralized servers [9].

8.3.1.1 Edge Devices, IoT Cameras, Cloud Integration, and Mobile Applications

Edge devices execute on-device image recognition and nutrient estimation, utilizing IoT-enabled cameras installed in dining or healthcare environments to capture food images. The cloud layer supports FL, large-scale data storage, and analytical processing, while mobile applications deliver real-time feedback and personalized dietary recommendations to users. Figure 8.3 illustrates the multi-tiered architecture of the system.

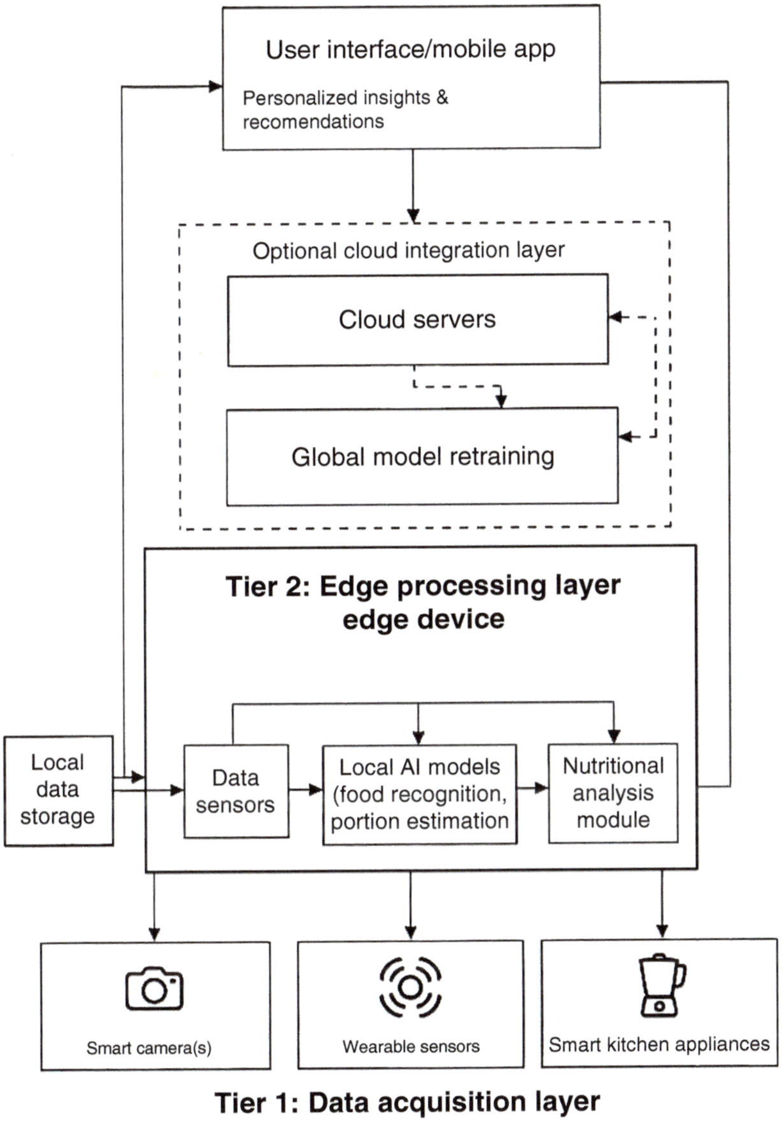

Figure 8.3 Multitier Edge AI nutritional monitoring architecture.

Table 8.3 Data flow in Edge AI nutritional monitoring system.

Stage	Description	Key function
Image capture	An IoT camera acquires food images	Initial data acquisition
Preprocessing	Background subtraction, lighting normalization, segmentation	Prepare data for accurate recognition
Food recognition	CNN-based classification and portion estimation	Identify food items and quantify intake
Nutritional mapping	Linking identified food to nutrient databases	Calculate calories, macros, micronutrients
User feedback	Mobile app or cloud dashboard	Real-time personalized recommendations

8.3.1.2 Data Flow: Image Capture → Processing → Nutritional Assessment

IoT cameras capture food images and transmit them to nearby edge devices for processing. Preprocessing algorithms isolate food items, remove background noise, and adjust lighting. Deep learning models such as CNNs then perform food classification and portion estimation [1, 5]. Nutrient values are derived by linking identified items with nutritional databases [7], and the processed data are relayed to users in real time via cloud dashboards or mobile apps. Table 8.3 summarizes the data flow components and their functions.

8.3.2 Edge AI Processing

By processing complicated computational tasks locally, Edge AI reduces the need for cloud connectivity. This is especially important in situations where bandwidth, latency, or data privacy are major issues [15].

8.3.2.1 Local Image Processing vs. Cloud Processing

Compared to traditional cloud-based techniques, local image processing on edge devices has several benefits. Through on-device visual data analysis, the system lowers response time and network dependency by transmitting less raw data [9]. Cloud processing adds latency and uses more energy and bandwidth, which makes it less appropriate for real-time dietary interventions, even though it is good for large-scale analytics and historical data aggregation. Figure 8.4 contrasts the performance of local versus cloud processing in terms of latency and data transfer.

8.3.2.2 Advantages: Latency Reduction, Bandwidth Optimization, Data Privacy

The edge-based framework supports real-time behavioral interventions by providing dietary insights quickly. Transmitting only processed results rather than full-resolution images optimizes bandwidth, which is crucial in remote or low-bandwidth settings. Additionally, by keeping sensitive dietary data on-device, local computation significantly improves user privacy and complies with legal requirements like GDPR and HIPAA [11]. Table 8.4 highlights the comparative benefits of edge versus cloud processing for nutritional monitoring.

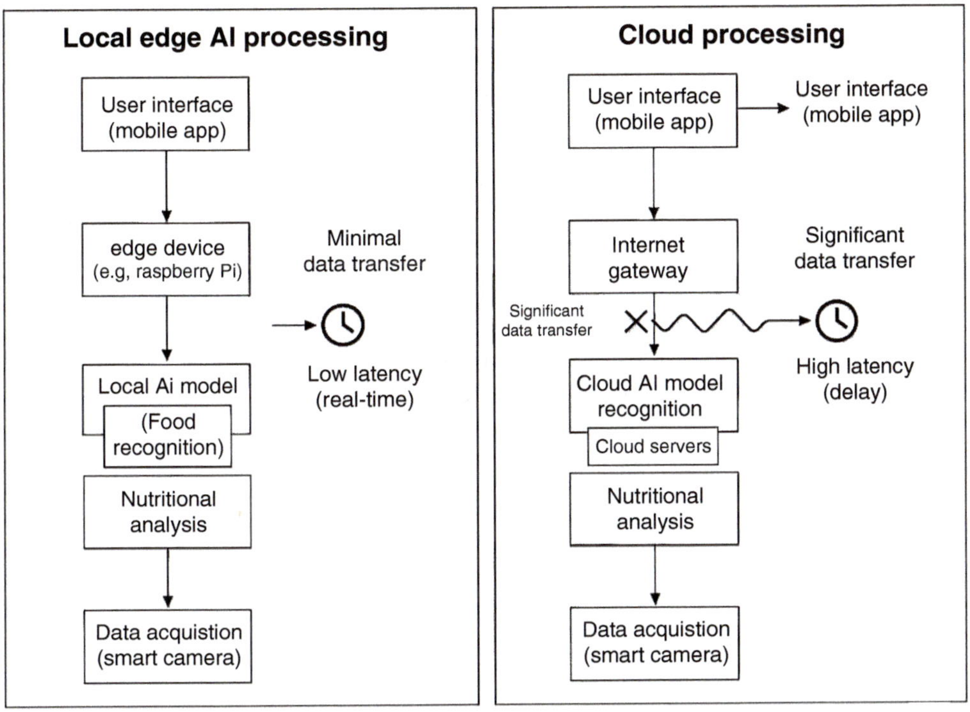

Figure 8.4 Local Edge AI processing vs. cloud processing.

Table 8.4 Benefits of Edge AI processing.

Feature	Edge AI	Cloud AI
Latency	Low, real-time feedback	High, dependent on the network
Bandwidth	Minimal, only processed data	High, requires raw image transfer
Privacy	Local data storage, secure	Centralized storage, higher risk
Reliability	Operates offline	Dependent on continuous connectivity
Energy efficiency	Optimized for low-power devices	Higher energy consumption

8.3.3 IoT Camera Integration

The system's sensory backbone is made up of IoT cameras, which record visual information in natural environments. Their implementation and design have a significant impact on the monitoring system's precision, dependability, and usability.

8.3.3.1 Smart Sensors, Connectivity, and Adaptive Configurations

High-resolution imaging sensors, motion detection, and adaptive exposure control are features that modern IoT cameras have to deal with changing lighting conditions. Wi-Fi, LTE, and LoRaWAN are connectivity options that enable flexible deployment in homes,

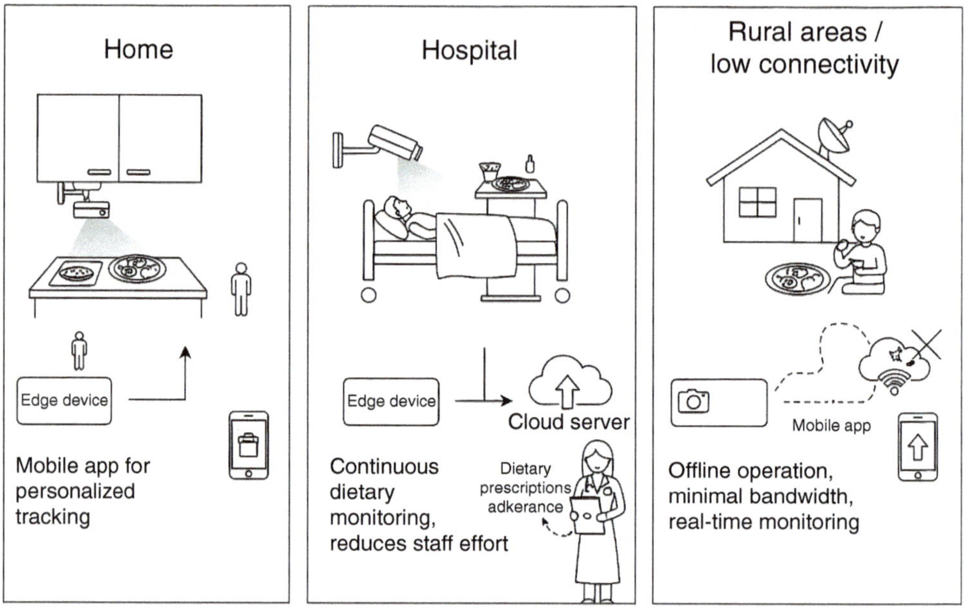

Figure 8.5 Deployment scenarios for Edge AI-based nutritional monitoring.

hospitals, and remote locations. While cameras constantly adjust to changes in the environment and meal presentation to ensure accurate food capture, edge devices receive and process images locally [13].

8.3.3.2 Deployment Scenarios: Home, Hospital, Rural Areas

The architecture supports diverse deployment scenarios. In homes, cameras discreetly capture meals for personalized tracking, while in hospitals, continuous dietary monitoring reduces staff effort and ensures adherence to prescriptions. In rural or low-connectivity regions, edge devices operate offline with minimal bandwidth, extending real-time nutritional monitoring to resource-limited settings [8]. Figure 8.5 illustrates the various deployment scenarios.

Edge AI-based nutritional monitoring integrates IoT cameras, edge devices, cloud systems, and mobile apps into a unified framework. Through smart sensors and local image processing, it enables accurate, real-time, and privacy-preserving dietary assessment. This architecture forms the foundation for scalable and adaptable next-generation nutrition monitoring platforms.

8.4 Food Recognition and Nutritional Analysis

The success of intelligent nutritional monitoring depends on its ability to recognize foods, estimate portions, and compute nutrient composition in real time. This process integrates computer vision, deep learning, and structured nutritional databases within Edge AI systems, transforming unstructured visual data from IoT or mobile devices into actionable dietary metrics for personalized nutrition and preventive healthcare [14].

8.4.1 Image Processing Techniques and Feature Extraction

Images captured in real environments often suffer from illumination changes, occlusion, and background clutter, reducing model accuracy. Preprocessing enhances both efficiency and reliability [6]. Techniques such as Retinex or CLAHE normalization correct lighting, while HSV or LAB segmentation removes background noise. Affine or homography transformations restore geometric alignment, ensuring consistent scale interpretation in varied environments like hospital canteens or rural kitchens [7]. Table 8.5 presents the relative contribution of each preprocessing stage to the overall food recognition accuracy.

After normalization, deep convolutional encoders extract key visual features such as color, texture, and shape. Object detection models like YOLOv5, EfficientDet, and Faster R-CNN accurately localize multiple food items, surpassing traditional methods [5]. For edge or low-power devices, lightweight variants such as Tiny-YOLOv4 and MobileNet-SSD balance speed and accuracy, achieving sub-25 ms inference with mAP above 0.85.

8.4.2 Deep Learning for Food Identification and Portion Estimation

Deep learning, particularly CNNs, forms the analytical foundation of visual food recognition by automatically learning hierarchical spatial features. State-of-the-art architectures such as ResNet-50, EfficientNet-B3, and Vision Transformers (ViT-B/16) have demonstrated classification accuracies between 90% and 95% on benchmark datasets including Food-101, UECFood-256, and IndianFoodDB [16]. Figure 8.6 illustrates the CNN architecture used for food identification, including convolutional, pooling, and fully connected layers deployed on an edge device.

To enable real-time deployment on lightweight hardware, model compression techniques including pruning, quantization, and knowledge distillation are applied, reducing the memory footprint by up to 70% with minimal accuracy degradation [17]. For portion estimation, geometric modeling or depth-aware imaging infers food volume and converts it to mass using density parameters. Calories are then calculated as:

$$\text{Calories} = \sum_{i=1}^{n} (\text{Weight}_i \times \text{EnergyDensity}_i) \tag{8.1}$$

where Weight represents estimated portion mass, and EnergyDensity denotes kilocalories per gram obtained from the linked nutrient dataset [3]. Table 8.6 presents representative

Table 8.5 Preprocessing stages and their contribution to overall food recognition accuracy.

Preprocessing step	Primary algorithm	Functional purpose	Performance impact
Lighting normalization	CLAHE/Retinex	Corrects uneven brightness	+6% accuracy gain
Background subtraction	HSV segmentation	Removes nonfood elements	+8% accuracy gain
Perspective correction	Affine/Homography transform	Rectifies camera angle distortion	+5% accuracy gain

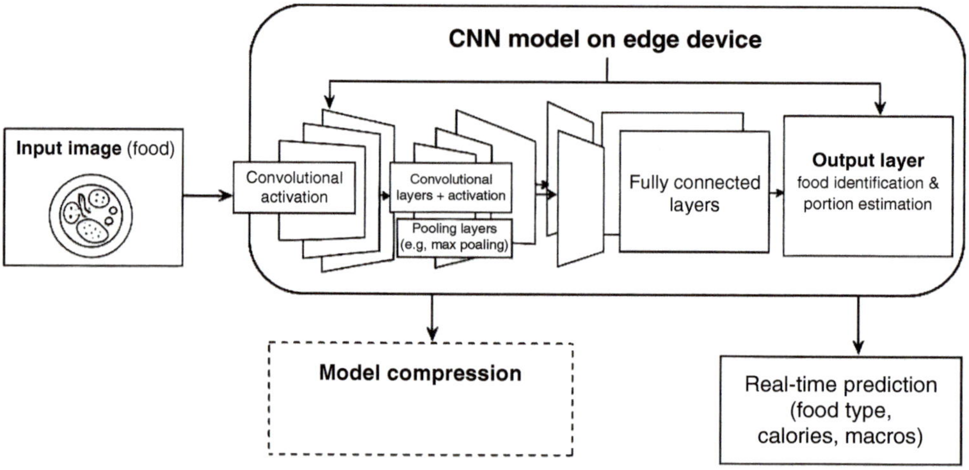

Figure 8.6 Architecture for food identification on an edge device.

Table 8.6 Comparative analysis of portion estimation results across sample food types.

Food item	Actual weight (g)	Predicted weight (g)	Error (%)
Rice (cooked)	150	138	8.0
Grilled chicken	120	126	5.0
Lentil curry	100	93	7.0
Salad (mixed)	90	96	6.7

results from experimental trials comparing predicted and actual portion sizes, with an average estimation error below 8%.

8.4.3 Nutrient Composition Mapping and Cultural Adaptation

After food recognition and portion estimation, the system maps detected items to structured nutritional databases such as IFCT and USDA FoodData Central [18]. Ontology-based and fuzzy logic methods resolve mismatches between informal names and standardized categories, ensuring accurate nutrient mapping [17]. Each food item is linked to detailed macro- and micronutrient profiles, with multilingual descriptors supporting diverse cuisines. Complex meals like Indian thalis are segmented using models such as Mask R-CNN or attention-based U-Net, trained on datasets like FoodX-251 and IndianFoodDB [14].

8.4.4 Analytical Significance and Future Implications

By combining preprocessing, deep learning, and nutrient mapping, Edge AI systems provide a continuous, objective, and real-time framework for dietary monitoring. Unlike manual food logging, these systems can correlate nutrient intake with biometric data from wearables, enabling the early detection and correction of dietary imbalances [16]. The integration of multimodal data, visual and sensor-based, marks a transition from reactive to preventive healthcare, facilitating proactive nutrition management. Future research should prioritize hardware-level optimization for micro-edge devices, cross-cuisine adaptability, and FL to allow personalized recognition without data sharing. Ultimately, AI-enabled food recognition and nutritional analysis extend beyond computation to promote health equity and sustainable well-being, linking technological innovation with public health advancement.

8.5 Enhancing Model Accuracy on Edge Devices

Deploying AI-based nutritional monitoring on edge devices demands balancing accuracy and efficiency under strict hardware constraints. Unlike cloud-based systems, edge computing enables privacy and real-time response but requires highly optimized ML models capable of running on low-power devices. This section outlines the major challenges, optimization strategies, and FL mechanisms that sustain reliable performance in such environments.

8.5.1 Hardware-Constrained Challenges

8.5.1.1 Computational Limits and Memory Restrictions

Edge devices such as IoT cameras and wearables often rely on low-power GPUs or ARM microcontrollers that cannot support large deep learning workloads. Running convolutional models for real-time food recognition may cause latency and overheating. Devices like the NVIDIA Jetson Nano or Google Coral TPU must sustain real-time inference despite limited computational power. Table 8.7 presents a comparative analysis of commonly used architectures, highlighting their inference latency and accuracy performance on Jetson Nano hardware.

Table 8.7 Comparative performance of deep learning architectures on edge hardware.

Model architecture	Parameters (millions)	FLOPs (billions)	Inference latency (ms) on Jetson Nano	Accuracy (Top-1, %)
ResNet-50	25.6	4.1	210	76.2
VGG-16	138	15.5	340	71.5
MobileNetV2	3.4	0.3	45	72.0
SqueezeNet	1.2	0.25	38	70.1
EfficientNet-B0	5.3	0.39	52	76.3

8.5.1.2 Energy Efficiency Considerations

Battery-driven devices must operate for long durations while capturing images and computing nutrient estimates. Event-triggered computation and dynamic voltage and frequency scaling (DVFS) are widely adopted to reduce active power use. Specialized neural accelerators further enhance performance per watt [15].

8.5.2 Optimization Techniques

Achieving high model accuracy within the resource limitations of edge devices depends on extensive optimization. Efficient architectures, reduced parameter redundancy, and model compression enable real-time inference without substantially sacrificing predictive performance [13].

8.5.2.1 Model Pruning, Quantization, and Knowledge Distillation

Pruning removes redundant neurons or filters, cutting model size by 60–80% while preserving up to 95% accuracy [13]. Quantization lowers precision from 32-bit to 8-bit integers, reducing memory and computation cost, particularly on ARM-based accelerators [15]. Knowledge distillation trains smaller "student" networks from larger "teacher" models, improving interpretability and efficiency. Table 8.8 summarizes the comparative outcomes of these optimization techniques.

8.5.3 Federated Learning

8.5.3.1 Distributed Training Across Edge Devices

Federated learning (FL) enables devices to train local models on private data while sharing only model updates with a central server. This collaborative process enhances the global model without exposing raw user data and reduces dependency on cloud resources.

8.5.3.2 Privacy Preservation and Improved Model Generalization

Since food images and medical data never leave the device, FL naturally aligns with GDPR and HIPAA standards [11]. Further protection is provided by differential privacy and secure aggregation methods. Learning from heterogeneous devices improves model generalization across different cuisines and regions.

Table 8.8 Effect of optimization techniques on model size and accuracy.

Optimization technique	Compression ratio	Accuracy retention (%)	Inference speed improvement (%)	Remarks
Structured pruning	4×	95	60	Effective for CNNs with redundant filters
Quantization (INT8)	3×	93	70	Ideal for ARM-based AI accelerators
Knowledge distillation	2×	97	40	Best for teacher–student model frameworks

8.5.4 Integration Outlook

Edge AI for nutritional monitoring is advancing toward a hybrid ecosystem of privacy-aware, hardware-efficient, and federated models. With neural accelerators and better energy management, edge devices will soon handle complex AI workloads independently. These integrated systems promise accurate, ethical, and sustainable AI-driven nutrition monitoring at both individual and population levels.

8.6 Multimodal Data Integration

Real-time nutritional monitoring has advanced significantly with the integration of visual data with physiological and behavioral data from wearable sensors. Although image-based dietary assessment offers important information about what people eat, it is unable to explain how their bodies react to that consumption. By combining continuous physiological tracking from gadgets like glucose monitors, heart rate sensors, and fitness trackers with food image analysis, multimodal data integration fills this gap. The system can produce context-aware, customized nutritional insights that can dynamically adjust to a person's metabolic state, lifestyle, and health objectives thanks to this convergence.

8.6.1 Combining Visual Data with Wearable Sensors

Edge AI-based IoT cameras capture meal images to estimate food type, portion size, and nutritional composition. When this data is combined with streams from wearable sensors such as fitness trackers and continuous glucose monitors (CGMs), it gains physiological relevance [18]. Table 8.9 summarizes how each sensor complements food recognition data,

Table 8.9 Types of wearable sensors and their contribution to nutritional monitoring.

Sensor type	Data captured	Nutritional insight enabled	Integration with Edge AI system
Continuous glucose monitor (CGM)	Blood glucose trends	Postmeal glucose response, glycemic load	Correlates food intake with real-time glucose levels
Heart rate monitor	Heart rate, HRV	Stress-related eating, metabolism indicators	Affects portion-size recommendations and meal timing
Fitness tracker	Steps, activity, calories burned	Energy balance estimation	Adjusts calorie recommendations dynamically
Bioimpedance sensor	Body composition, hydration	Fat–muscle ratio and hydration tracking	Supports personalized diet plans
Smart thermometer	Body temperature changes	Detects illness or metabolic rate shifts	Modulates dietary suggestions during health anomalies

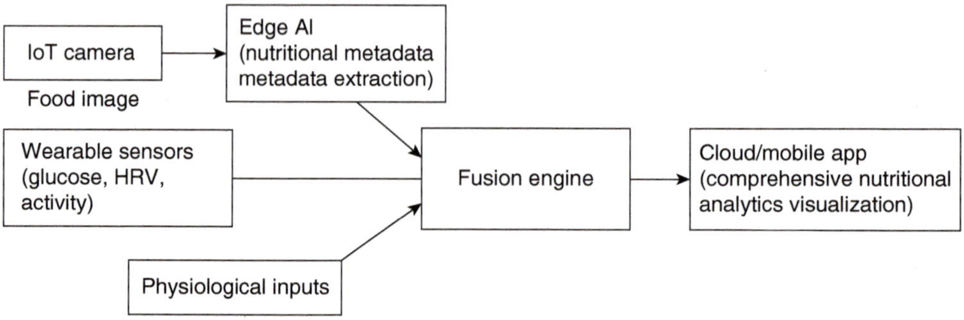

Figure 8.7 Multimodal fusion of visual and wearable data streams.

enriching static dietary logs with metabolic context and transforming them into actionable intelligence.

At the edge or fog computing level, the fusion layer combines and aligns time-stamped food photos with streams of multi-sensor data. This fusion process is depicted in Figure 8.7, which also shows how synchronized data integration improves the breadth and interpretability of dietary monitoring.

8.6.2 Contextualized Nutritional Insights for Personalized Healthcare

Once fused, the system generates contextualized insights reflecting each user's physiology, lifestyle, and health status. Rather than static calorie counts, it interprets real-time responses, for instance, recommending low-glycemic foods if glucose levels rise after a carbohydrate-heavy meal. By correlating glucose recovery, postmeal activity, and sleep quality, the system supports Precision Nutrition, continuously adapting through machine learning [18]. Table 8.10 demonstrates how multi-source inputs yield real-time, context-rich feedback, empowering informed dietary decisions.

Table 8.10 Example of contextualized feedback generated by multimodal integration.

Data source	Observation	AI-generated insight	Recommended action
Food image (rice + curry)	High carbohydrate meal detected	Anticipated glucose spike	Suggest pairing with a fiber-rich salad
Glucose monitor	Glucose level 160 mg/dl postmeal	Rapid spike detected	Recommend a smaller portion for the next meal
Fitness tracker	Low physical activity for 3 h postmeal	Poor glucose utilization	Suggest light walking after lunch
Heart rate sensor	Elevated HR and stress indicators	Possible stress-related eating	Advise relaxation or hydration
Sleep tracker	Reduced REM sleep post-heavy dinner	Late meal interference	Recommend earlier dinner timing

Beyond immediate insights, multimodal data enables predictive intelligence that foresees potential metabolic risks. Models such as temporal graph neural networks (TGNNs) and long short-term memory (LSTM) networks learn sequential dependencies between meal patterns and physiological reactions, detecting early signs of disorders like insulin resistance or obesity [19]. Emerging trends include federated multimodal learning for adaptive personalization, lightweight transformer models for efficient on-device processing, and explainable AI (XAI) frameworks to enhance clinical transparency [17]. These advances will ensure that nutritional monitoring evolves into a scalable, interpretable, and ethically responsible ecosystem for global health management.

8.7 Applications and Use Cases

The convergence of Edge AI and the IoT has transitioned nutritional monitoring from conceptual research into diverse real-world deployments. These systems enable automated, continuous, and scalable assessment of food intake, thereby connecting individualized dietary management with broader public health strategies.

8.7.1 Clinical Nutrition: Real-Time Monitoring for Obesity, Diabetes, and Cardiovascular Risk

Edge AI facilitates real-time dietary tracking for conditions such as obesity, diabetes mellitus, and cardiovascular disorders. Unlike conventional self-reporting methods, often affected by recall bias [12], IoT-enabled cameras and wearables collect meal images and biometric data instantaneously. Local edge processors evaluate nutritional composition and deliver predictive recommendations such as portion control or low-glycemic alternatives supporting precision-based dietary therapy. Integration with Electronic Health Records (EHR) enables continuous supervision and early anomaly detection, improving adherence and glycemic stability by more than 40% [12]. Figure 8.8 illustrates this ecosystem, where edge-based inference and IoT connectivity enable dynamic clinician dashboards for personalized nutrition management.

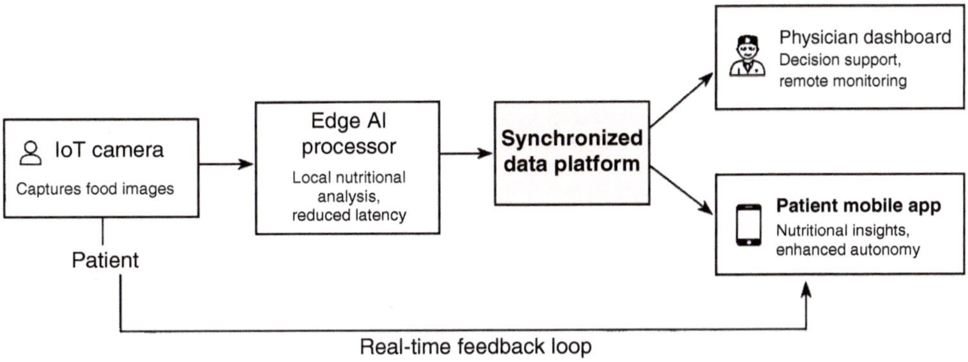

Figure 8.8 Real-time clinical nutrition monitoring using Edge AI and IoT cameras.

Table 8.11 Clinical outcomes improved by real-time Edge AI-based nutritional monitoring.

Clinical condition	Traditional approach	AI-enabled monitoring advantage	Expected outcome
Type-2 diabetes	Manual food logging	Glucose–meal correlation with predictive feedback	Improved glycemic control
Obesity	Calorie estimation via recall	Automated portion recognition and energy balance analysis	Sustainable weight reduction
Cardiovascular risk	General diet counseling	Real-time fat/sodium detection with adaptive recommendations	Lower LDL levels and improved compliance

Empirical studies reveal that AI-driven nutrition systems can enhance dietary reporting accuracy by approximately 60% and improve compliance by more than 40% [12]. Through real-time linkage of dietary and physiological data, these systems advance precision treatment outcomes. Table 8.11 summarizes clinical improvements achieved through AI-enabled dietary surveillance.

8.7.2 Pediatric and Geriatric Care: Tailored Dietary Assessments and Interventions

Children and older adults are particularly vulnerable to nutritional imbalance. Edge AI–based systems enable age-specific, noninvasive dietary monitoring through IoT cameras and wearables in homes or care facilities [19]. Models trained on regional food data detect issues like protein or iron deficiency in children, while sensors identify dehydration or swallowing difficulties in the elderly, enabling early and sustained nutritional management.

8.7.3 Preventive Healthcare: Behavioral Nudges, Real-Time Feedback, and Personalized Recommendations

In preventive healthcare, which focuses on early detection and the development of healthy habits rather than disease treatment, Edge AI-based nutritional systems play a pivotal role. These systems foster self-regulation and dietary discipline through real-time feedback and behavioral analytics. For example, context-aware nudges such as:

- "Your lunch today contained high sodium; consider balancing with low-salt dinner options."
- "Walk for 10 minutes to improve metabolism; post-meal glucose is slightly elevated."

When provided instantly and contextually, these micro-interventions enhance user engagement and promote long-term behavioral improvement. Predictive analytics further supports proactive health management by identifying emerging risk trends. Table 8.12 highlights how adaptive AI feedback aligns dietary behavior with physiological well-being, forming a continuous data-driven health loop.

Table 8.12 Examples of real-time behavioral feedback mechanisms in preventive care.

Data type	Detected event	AI recommendation	Behavioral outcome
Food image	High sugar dessert detected	Suggest a fruit alternative	Reduced daily sugar intake
Glucose reading	Slow glucose recovery	Alert to monitor future meals	Improved glycemic control
Fitness data	Low postmeal activity	Suggest mild exercise	Increased calorie burn
Sleep pattern	Late-night heavy meal	Recommend an earlier dinner	Improved sleep quality

8.7.4 Remote and Resource-Constrained Deployments: Rural Healthcare Applications and Low-Bandwidth Environments

Edge AI-based nutrition systems are crucial for rural and low-connectivity regions where cloud models face bandwidth and reliability challenges. Local edge processing minimizes latency, safeguards privacy, and functions without constant internet access [9]. As shown in Figure 8.9, solar-powered IoT cameras analyze food images on-site and send only summarized results. Using affordable devices like Raspberry Pi and Jetson Nano, deployments in rural India and Sub-Saharan Africa have improved data accuracy by about 60% and supported public health planning through mHealth integration [19].

8.7.5 Broader Implications

Edge AI and IoT-based nutritional systems are transforming the interaction between food, health, and data across clinical, preventive, and rural domains. Beyond individual care, they

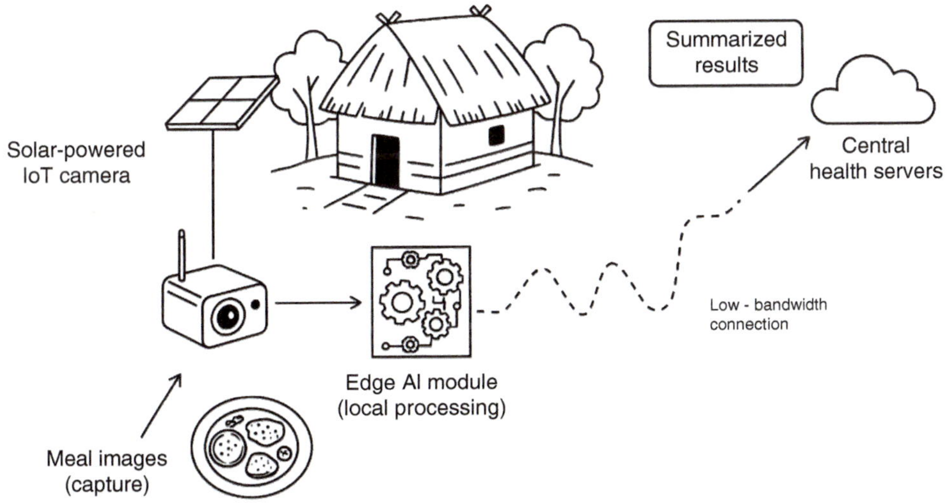

Figure 8.9 Edge AI-based nutritional monitoring in rural environments.

enable population-level insights that inform health policies, agriculture, and public nutrition planning. By combining localized adaptability with real-time intelligence, these technologies promote a decentralized and human-centered model of nutrition monitoring aligned with the vision of sustainable digital health [17].

8.8 Regulatory, Ethical, and Social Implications

IoT-based nutritional monitoring and Edge AI technologies hold substantial promise for advancing public health outcomes. Yet, their widespread adoption brings with it several ethical, social, and regulatory concerns that require careful attention. Proactively addressing these issues is vital for maintaining accountability, transparency, and user trust in real-time dietary assessment and monitoring systems.

8.8.1 Ensuring Responsible Deployment of AI and IoT in Nutrition

Strong governance frameworks must be followed when integrating AI-driven nutrition monitoring into healthcare ecosystems to avoid sensitive data being misused. To ensure responsible deployment, AI systems must be transparent, auditable, and explicable, particularly when automated nutritional recommendations have the potential to directly influence long-term dietary patterns or medical decisions [17]. The three main phases of the responsible AI lifecycle for nutritional systems design, deployment, and continuous evaluation are depicted in Figure 8.10.

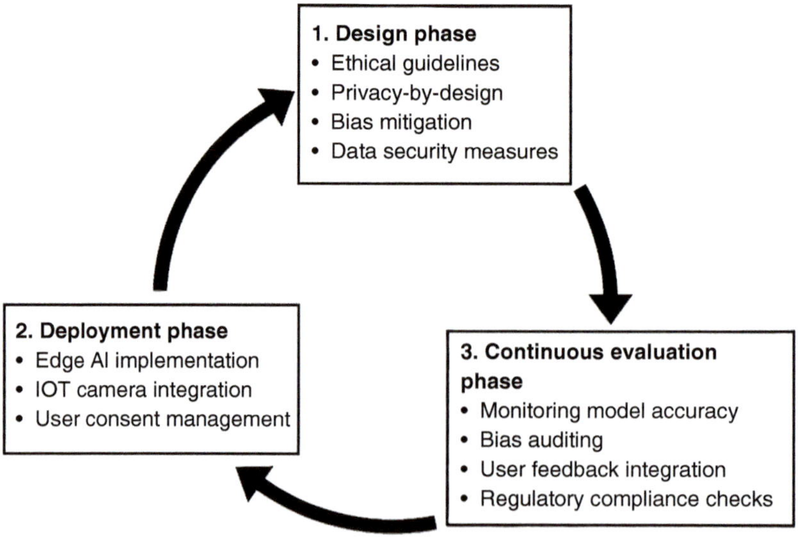

Figure 8.10 Responsible AI lifecycle for Edge AI-based nutritional monitoring systems.

8.8.2 User Consent and Data Protection Frameworks

User consent becomes the cornerstone of ethical deployment since wearable technology and IoT cameras continuously collect sensitive visual and biometric data [11]. Transparent consent procedures that let users specify usage limits, opt out of data sharing, and revoke participation at any moment must be included in the system. Table 8.13 provides a comparative analysis of the main international data protection laws and their applicability to AI-driven nutritional systems, including the US Health Insurance Portability and Accountability Act (HIPAA), India's Digital Personal Data Protection Act (DPDPA), and the EU's General Data Protection Regulation (GDPR).

8.8.3 Addressing Bias, Fairness, and Inclusivity in AI Models

Biases in culture and demographics present in training data are frequently reflected in AI-driven food recognition systems. For instance, when it comes to recognizing foods specific to a given region, such as East Asian mixed dishes, African stews, or Indian thalis, models that were trained primarily on Western dietary datasets may perform poorly. This results in differences in accuracy, which may exacerbate inequalities in medical outcomes [19].

8.8.4 Compliance with Healthcare Standards and IoT Regulations

AI- and IoT-enabled nutrition systems must operate in alignment with existing healthcare and IoT regulations to uphold safety, interoperability, and user protection standards [14]. Frameworks such as ISO/IEC 27001, IEEE 2413, and ISO/TS 82304-2 provide essential guidance for building secure and trustworthy architectures, while medical-grade devices are further required to comply with MDR or FDA SaMD regulatory standards. However, true ethical governance extends beyond basic compliance – it demands transparency, procedural fairness, and deliberate action to narrow the digital divide. Adopting a "Responsible AI by Design" philosophy ensures that these nutrition technologies remain human-centric, privacy-conscious, and socially responsible as they are integrated into broader digital healthcare ecosystems.

Table 8.13 Comparative overview of data protection regulations relevant to AI-based nutrition monitoring systems.

Regulation	Region	Key provisions	Applicability to Edge AI + IoT systems
GDPR	European Union	Data minimization, explicit consent, right to be forgotten	Ensures transparent data handling and limited cross-border transfer of user data
HIPAA	United States	Medical data confidentiality, audit trails, and access control	Regulates nutritional data when integrated with medical records
DPDPA	India	User rights, purpose limitation, and data fiduciary accountability	Strengthens privacy for AI-based health applications within the Indian jurisdiction

8.9 Future Directions and Research Opportunities

Edge AI and IoT-based nutritional monitoring systems have already shown strong potential across clinical and preventive contexts. Even though the field is still developing, it is becoming clear that sustained progress will depend on advances in hardware design, algorithmic efficiency, and user-centered implementation. Broader acceptance and integration will ultimately determine how far these systems can extend their impact in real-world health management.

8.9.1 Scaling and Integration of Edge AI Systems for Public Health and Smart Cities

Most Edge AI-based dietary monitoring systems currently operate at community levels. Scaling them globally could revolutionize nutrition surveillance and policymaking by detecting population dietary trends and guiding targeted interventions [19]. Efficient processing of visual and biometric data through distributed edge frameworks reduces bandwidth use and latency. Ensuring interoperability across sensors and data formats maintains system reliability. Integrated within smart cities, such networks can link personal nutrition tracking with community health, adapting recommendations to local factors like food access and air quality [20].

8.9.2 Multimodal AI: Vision, IoT, and Biometrics for Holistic Health

A promising direction for future research lies in combining multiple data modalities to achieve a holistic nutritional understanding. Integrating visual inputs from IoT cameras with biometric data such as glucose monitoring, heart rate, and physical activity creates a richer, more individualized health profile. Environmental and behavioral data can further enhance this personalization. Such an approach can:

- Detect metabolic irregularities before clinical symptoms appear
- Deliver tailored, real-time dietary recommendations based on physiological signals
- Refine guidance by incorporating lifestyle metrics such as sleep and exercise

This integration relies on cross-modal learning, where AI models identify relationships between visual, biometric, and contextual inputs. As these systems mature, nutritional monitoring could evolve from simple observation into predictive and preventive health management [21].

8.9.3 Advanced Predictive Analytics for Dietary Trends and Health Outcomes

Predictive analytics represents a key opportunity for improving both individual and population-level health forecasting. By combining biometric indicators, IoT sensor data, and prior consumption patterns, AI models can anticipate dietary risks and nutrient deficiencies. They may also:

- Identify individuals vulnerable to metabolic disorders or obesity
- Simulate possible outcomes under alternative dietary choices
- Generate long-term predictions to support preventive healthcare

Integrating behavioral insights into predictive modeling can further improve system effectiveness. Real-time nudges, such as recommending smaller portions or healthier substitutes, can enhance user engagement and promote sustainable eating behaviors [22]. At a policy level, these insights can guide public awareness campaigns and inform food system planning for healthier and more sustainable diets.

8.9.4 Research Opportunities

Ongoing research in intelligent nutrition systems may explore several priority areas:

a) **Energy-Efficient and Federated Edge AI Models**: Develop computationally efficient AI methods suitable for edge devices, especially in low-resource areas. FL can allow collaborative model updates while keeping user data private
b) **Cross-Cultural** Food Recognition: Construct inclusive food datasets that account for diverse cuisines, preparation methods, and cultural variations to minimize recognition bias
c) **Integration with Genomic, Microbiome, and Long-Term Health Data:** Pursue precision nutrition approaches by linking genetic, microbial, and dietary data to design truly personalized dietary interventions
d) **Ethical and Regulatory Frameworks:** Create comprehensive guidelines addressing privacy, data ownership, and algorithmic transparency to ensure trust in AI-based nutrition technologies
e) **Behavioral and Social Analytics:** Investigate how sustained interaction with AI systems influences long-term diet patterns, motivation, and adherence across demographics

By pursuing these pathways, researchers can design nutritional monitoring systems that are technically sound, socially responsible, and globally scalable. Such solutions have the potential to strengthen preventive care, empower individuals, and inform public policy with reliable, real-time nutrition intelligence [23].

8.10 Conclusion

The revolutionary potential of combining Edge AI with IoT-enabled cameras for real-time nutritional monitoring has been emphasized in this chapter. The suggested framework allows for automated, intelligent, and ongoing monitoring of food intake, portion sizes, and nutrient composition by going beyond conventional dietary assessment techniques like self-reporting, manual logs, and intermittent clinical observations. CNNs for precise food identification, FL for model training that protects privacy, and multimodal data integration that combines visual, biometric, and environmental inputs to produce contextualized nutritional insights are some of the major innovations covered. Together, these developments transform dietary monitoring from a static, reactive procedure into a dynamic, proactive system that can provide early health warnings, behavioral cues, and tailored recommendations. Such a system has significant ramifications, especially for high-risk groups like the elderly, children, and people with diabetes, obesity, and cardiovascular diseases. Edge AI-based nutritional monitoring can help prevent the onset of lifestyle-related diseases, improve clinical nutrition outcomes, and increase adherence to healthy eating patterns by providing real-time feedback and personalized dietary interventions. Furthermore, incorporating these systems into community-level monitoring platforms and digital health ecosystems can promote population-level preventive healthcare by providing useful information to legislators and healthcare professionals. In the end, this chapter outlines a vision for an ecosystem of intelligent, noninvasive, real-time nutritional monitoring that seamlessly combines wearable, IoT, and AI technologies. Through ongoing, data-driven, and individualized nutrition management, such a system not only enables people to make educated food choices but also advances a larger public health goal by promoting healthier communities. This method represents a paradigm shift in the way nutrition is tracked, evaluated, and optimized in both individual and societal contexts by tackling issues with accuracy, scalability, privacy, and fairness.

References

1 Wang, H., Tian, H., Ju, R. et al. (2024). Nutritional composition analysis in food images: an innovative swin transformer approach. *Frontiers in Nutrition* 11: 1454466. https://doi.org/10.3389/fnut 2024.1454466.

2 Rewane, R. and Chouragade, P.M. (2019). Food nutritional detection, visualization and recommendation for health monitoring using image processing. *Proceedings of the 3rd International Conference on Trends in Electronics and Informatics (ICOEI)* (April 2019). https://doi.org/10.1109/ICOEI.2019.8862550.

3 Zhu, F., Bosch, M., Boushey, C.J., and Delp, E.J. (2010). An image analysis system for dietary assessment and evaluation. *Proceedings of the International Conference on Image Processing (ICIP)*, pp. 1853–1856, https://doi.org/10.1109/ICIP.2010.5650848.

4 Meenu, M., Kurade, C., Neelapu, B.C., and Kalra, S. (2021). A concise review on food quality assessment using digital image processing. *Trends in Food Science & Technology* 118 (4): 106–124. https://doi.org/10.1016/j.tifs.2021.09.014.

5 Karabay, A., Varol, H.A., and Chan, M.Y. (2025). Improved food image recognition by leveraging deep learning and data-driven methods with an application to Central Asian food scene. *Scientific Reports* 15: 14043. https://doi.org/10.1038/s41598-025-95770-9.

6 Pandey, J.K., Kumar, S., Lamin, M. et al. (2022). A metaheuristic autoencoder deep learning model for an intrusion detector system. *Mathematical Problems in Engineering* 2022: 11, 3859155. https://doi.org/10.1155/2022/3859155.

7 Bossard, L., Guillaumin, M., and Van Gool, L. (2014). Food-101 – Mining discriminative components with random forests. In: Fleet, D., Pajdla, T., Schiele, B., and Tuytelaars, T. (Eds.), Computer Vision – ECCV 2014, pp. 446–461. Springer. https://doi.org/10.1007/978-3-319-10599-4_29

8 Agrawal, K., Goktas, P., Kumar, N., and Leung, M.-F. (2025). Artificial intelligence in personalized nutrition and food manufacturing: a comprehensive review of methods, applications, and future directions. *Frontiers in Nutrition* 12: 1636980. https://doi.org/10.3389/fnut 2025.1636980.

9 Beemkumar, N., Gupta, S., Bhardwaj, S. et al. (2023). Activity recognition and IoT-based analysis using time series and CNN. In: *Handbook of Research on Machine Learning-enabled*

IoT for Smart Applications Across Industries (ed. N. Goel and R.K. Yadav), 350–364. IGI Global https://doi.org/10.4018/978-1-6684-8785-3.ch018.

10 Kutyauripo, I., Rushambwa, M., and Chiwazi, L. (2023). Artificial intelligence applications in the agrifood sectors. *Journal of Agriculture and Food Research* 11: 100502. https://doi.org/10.1016/j.jafr.2023.100502.

11 Dedeoglu, V., Malik, S., Ramachandran, G.S., and Pal, S. (2023). Blockchain meets edge-AI for food supply chain traceability and provenance. In: Comprehensive Analytical Chemistry, Vol. 101, pp. 251–275. Elsevier. https://doi.org/10.1016/bs.coac.2022.12.001.

12 Kasera, R.K., Gour, S., and Acharjee, T. (2024). A comprehensive survey on IoT and AI-based applications in different pre-harvest, during-harvest, and post-harvest activities of smart agriculture. *Computers and Electronics in Agriculture* 216 (3): Art. no. 108522, https://doi.org/10.1016/j.compag.2023.108522.

13 Javaid, M., Haleem, A., Singh, R.P., and Suman, R. (2022). Artificial intelligence applications for Industry 4.0: a literature-based study. *Journal of Industrial Integration and Management* 7 (1): 83–111. https://doi.org/10.1142/S2424862221300040.

14 Revelou, P.-K., Tsakali, E., Batrinou, A., and Strati, I.F. (2025). Applications of machine learning in food safety and HACCP monitoring of animal-source foods. *Food* 14 (6): 922. https://doi.org/10.3390/foods14060922.

15 Reddy, K.V., Reddy, B.S.K., Veerapu, G., and Mahesh, M. (2024). Edge AI in sustainable farming: deep learning-driven IoT framework to safeguard crops from wildlife threats. *IEEE Access* 12: 77707–77723. https://doi.org/10.1109/ACCESS.2024.3406585.

16 Aijaz, N., Yaqub, M., Lan, H., and Raza, T. (2025). Artificial intelligence in agriculture: advancing crop productivity and sustainability. *Journal of Agriculture and Food Research* 20 (6): 101762. https://doi.org/10.1016/j.jafr.2025.101762.

17 Pandey, J.K. (2025). Understanding the impact of electromagnetic and optical radiation on physical and biological systems. *Journal of Nano- and Electronic Physics* 17 (5): 05023. https://doi.org/10.21272/jnep.17(5).05023.

18 Omia, E., Bae, H., Park, E. et al. (2023). Remote sensing in field crop monitoring: a comprehensive review of sensor systems, data analyses and recent advances. *Remote Sensing* 15 (2): 354. https://doi.org/10.3390/rs15020354.

19 Allcock, H.R. (2008). *Introduction to Materials Chemistry*. Hoboken, NJ: Wiley.

20 Nyakuri, J.P., Nkundineza, C., Gatera, O., and Nkurikiyeyezu, K. (2025). AI and IoT-powered edge device optimized for crop pest and disease detection. *Scientific Reports* 15 (1): https://doi.org/10.1038/s41598-025-06452-5.

21 Talukdar, V., Veeraiah, V., Roy, D.R. et al. (2023). An innovative approach: typicality-based collaborative filtering for recommender systems. *Proceedings of the 3rd International Conference on Advancement in Electronics & Communication Engineering (AECE)*, pp. 863–868. https://doi.org/10.1109/AECE59614.2023.10428576

22 Pandey, J.K., Yadav, A.K., Purushotham, M. et al. (2022). Colorimetric detection of deltamethrin pesticide from biological samples based on the peroxidase-mimic catalytic activity on MnO_2/SnO_2. *Key Engineering Materials* 928: 45–51. https://doi.org/10.4028/p-4ti1i0.

23 Srinivas, P., Veeraiah, V., Thejaswini, K.O. et al. (2025). The impact of electromagnetic and optical radiation on physical and biological systems. *Journal of Nano- and Electronic Physics* 17 (4): 04012. https://doi.org/10.21272/jnep.17(4).04012.

9

Deep Learning Approaches for Nutritional Profiling of Food Plates

Hamadou Mamoudou[1,2,3]

[1] *Department of Biological Sciences, Biochemistry, Bioinformatics, and Therapeutic Innovations, Research Unit, Faculty of Science, University of Maroua, Maroua, Cameroon*
[2] *Department of Chemistry, COMSATS University, Islamabad, Pakistan*
[3] *Department of Basic Science Education, National Advanced School of Engineering, University of Maroua, Maroua, Cameroon*

9.1 Introduction

Nutritional profiling uses pictures of food to guess how many macronutrients and micronutrients it has, how many calories it has, and what kinds of foods people eat. This feature helps solve global health problems by making it possible to use data to fight diet-related diseases like obesity and diabetes [1, 2]. A precise evaluation of food habits facilitates individualized nutrition, clinical oversight, and public health initiatives [2, 3]. Self-reported food records, dietitian assessments, and chemical analysis are among the oldest ways to keep track of what you eat, but they have some big problems [4]. Self-reporting is susceptible to recall bias and mistakes due to limited user knowledge [5], whereas chemical analysis, despite its accuracy, is time-consuming and prohibitively expensive for large-scale use [5]. These limitations highlight the necessity for automated, scalable, and accurate nutritional profiling systems that cater to varied and intricate eating habits.

Deep learning (DL) techniques currently provide automated nutritional analysis with great precision by directly extracting features from food photos [6–8]. DL models create hierarchical representations on their own from raw pixel data, unlike classical machine learning, which uses handcrafted features [8, 9]. Convolutional neural networks (CNNs), like ResNet and EfficientNet, are great at recognizing food and figuring out how much of it is in a picture because they can see spatial hierarchies in images [7, 10, 11]. In more recent times, vision transformers (ViTs) have used self-attention processes that model global contextual linkages to increase performance in complicated settings with many food items or messy backdrops [7, 10, 11]. These technological improvements help with mobile health, clinical nutrition, and large-scale nutritional monitoring.

The use of DL in nutritional profiling signifies a synthesis of computer vision, nutritional research, and health informatics [6, 12–14]. Automated systems enable consumers to make informed dietary decisions, support physicians in creating tailored therapies, and facilitate public health researchers in examining population-level dietary trends [15, 16]. However, there are still problems, such as the dataset being biased toward certain cuisines, the model not being able to be used in many different situations, and the fact that it can't be used in real time. To get around these problems, we need to make strong models, collect datasets that represent a wide range of cultures, and create algorithms that work well in the actual world.

This chapter looks at DL methods for figuring out what foods are good for you by looking at pictures of them. We assess model efficacy, scrutinize methodological constraints, and suggest avenues for future research. Our synthesis includes CNN- and transformer-based architectures as well as generative models to give a full picture of what is possible now and what is coming up. We want to encourage collaboration among different fields and move the area forward by bringing together ideas from computer vision, nutrition, and health informatics. Our goals are to critically evaluate current approaches, point out useful applications, and find new ways to solve problems that keep coming up.

9.2 Literature Review

9.2.1 Early Approaches: Feature Engineering and Classical Classifiers

Initial attempts to automate nutritional profiling relied on classical machine learning coupled with manual feature engineering. Research conducted in the early 2010s predominantly utilized classifiers such as support vector machines (SVMs) and random forests [17–19]. These systems depended on handcrafted descriptors, specifically color histograms, texture features like the Scale-Invariant Feature Transform (SIFT), and shape profiles, to identify food items [7, 14, 15, 20, 21].

However, these methods demonstrated limited robustness when applied to complex food images characterized by diverse textures, occlusion, and variable lighting conditions. Multi-food recognition accuracy rarely exceeded 70%, as the reliance on manual feature selection created a bottleneck for scalability [17–19]. Furthermore, these early architectures lacked the capacity to estimate portion sizes or infer volumetric data, necessitating labor-intensive postprocessing to derive nutritional content. These limitations highlighted the need for architectures capable of learning hierarchical features directly from raw data.

9.2.2 Evolution of Deep Learning Architectures

The transition to DL addressed the limitations of manual feature extraction by enabling end-to-end feature learning. CNNs, including ResNet, Inception, and EfficientNet, established new baselines for performance. Liu [22] reported a top-1 accuracy exceeding 80% on the Food-101 dataset utilizing a fine-tuned Inception-V3 model. These architectures are particularly effective at capturing spatial hierarchies, which aids in distinguishing both single and multiple food plates.

Later improvements were made to granularity and context. Semantic segmentation techniques, such as Mask R-CNN, made it easier to guess the number of portions by separating individual food items in messy scenes (pixel-level classification). ViTs have been used more recently to simulate global contextual interactions using attention mechanisms. A 2023 assessment revealed that a Swin Transformer surpassed the F1-score of a ResNet-101 backbone by 5% on the UE-Food dataset [22]. Generative adversarial networks (GANs) are also being used more and more to add data by making fake food pictures to fix class imbalance and make models more robust when there isn't much data.

9.2.3 Benchmark Datasets and Bias

The creation of strong nutritional profile models relies heavily on the existence of open-source benchmark datasets. Food-101, UE-Food, and Nutrition5k are the best-known examples. Food-101, which has 101,000 photos in 101 classes, is still the best way to test food categorization tasks. However, it is very biased toward Western cuisines, which makes it less useful for worldwide dietary assessment [8, 10, 22]. The UE-Food dataset tries to cover multiple meal situations; however, its limited sample size makes it harder for DL models to make generalizations. Nutrition5k, on the other hand, offers unique value with its precise macro- and micronutrient annotations. However, researchers have pointed out that its micronutrient labeling isn't always consistent and that some data is missing [8, 10]. These current datasets all have problems, such as not having enough cultural variety, having annotation mistakes, and not having enough variation in imaging circumstances. People are looking at ways to fix these problems, such as crowdsourcing and making fake data, but the lack of a large, culturally diverse, and well-annotated dataset is still the biggest problem that keeps these models from being used in the real world.

9.2.4 Performance Metrics

Evaluation techniques in computational nutrition are specifically designed for particular purposes. In image-based food categorization, top-1 and top-5 accuracies are standard measures. The best architectures now achieve top-1 accuracy rates of 85–90% on the Food-101 benchmark dataset [22, 23]. The mean absolute error (MAE) measure is what nutritional estimation is based on. It is used to guess the amount of calories and nutrients in food. Recent studies show that the MAE values for single-item plates in the Nutrition5k dataset are between 10 and 15 kcal [7, 24, 25]. Transformer-based models have also been able to cut this error by about 20% relative to earlier CNN baselines [7, 24, 25]. The F1-score is the best way to find several food items since it strikes a good mix between precision and recall, which is important when items are likely to overlap.

Many obstacles still exist in spite of these developments. Estimating micronutrients is still prone to large errors, mostly because there isn't enough dense training data. Also, when models developed using curated lab datasets like Food-101 are tested on more chaotic "in-the-wild" datasets like UE-Food, their performance often drops significantly. This shows a big generalization gap that makes it hard to use them in the actual world [7, 22, 24, 25].

9.3 Deep Learning Methods for Food Profiling

9.3.1 Food Recognition and Classification

The main goal of automated profiling is to effectively classify food items from unstructured images. CNNs have been the go-to framework for this task, with models like VGG, Inception, and ResNet setting the early performance benchmarks [22, 24]. VGG architectures showed that deep, layered structures can effectively extract textural features, but their high number of parameters leads to heavy computational demands, making them less ideal for mobile use [22, 24]. On the other hand, inception models use multi-scale convolutional filters, enabling the network to handle food items of different sizes and shapes effectively. Some versions of this architecture have reached top-1 accuracies of 80–85% on the Food-101 benchmark [22, 26]. ResNet tackles the degradation issue in deep networks with residual connections, making it easier to train much deeper models and achieving over 85% accuracy in single-food classification [22, 26]. CNNs have seen success, but they do have limitations due to their local receptive fields. Recent studies have shifted focus to ViT and Swin Transformers [22, 27]. These architectures use self-attention mechanisms to capture global context and long-range relationships in images, unlike CNNs. This feature is especially helpful for identifying food in messy settings where background noise can resemble food textures.

Table 9.1 shows a comparison of key public datasets often used for food recognition and nutrition profiling. Food-101 is a popular benchmark with a dataset of 101,000 labeled images across 101 food categories [10, 22, 23]. Even with its wide reach and popularity, it faces limitations due to a Western-focused food classification system and missing portion size labels, which are crucial for accurate nutritional evaluation. UE-Food may have around 10,000 images, but it offers multi-food plate representations that better reflect actual meal compositions [10, 22, 23]. However, it often faces visual blockages and poor annotation quality, which can hinder model generalization.

The Nutrition5k dataset is important for estimating nutrients because it provides accurate macronutrient details from real meals [24, 25]. The nutritional details make it useful for training and assessing diet tracking models, but it lacks thorough micronutrient information and has limitations in geographic and dietary variety. Recipe1M stands out because it combines recipe text, food images, and nutrient labels uniquely. The dataset has over a million samples, but its reliance on recipe-based nutrient estimates and varying image quality can create noise, so careful preprocessing is needed for effective machine learning use

Table 9.1 Overview of key public datasets for nutritional profiling and food recognition.

Dataset	Approximate size	Content type	Available nutritional labels	Primary limitations
Food-101	101,000 images	Single-food images	None	Geographic bias (Western-centric); lacks portion information
UE-Food	~10,000 images	Multi-food plates	Limited	Small-scale dataset; frequent occlusions in food items
Nutrition5k	~5000 images	Real-world meals	Macronutrients	Sparse micronutrient coverage; limited diversity
Recipe1M	~1,000,000 recipes	Multimodal (text + images)	Inferred from recipes	High variance in image quality; indirect nutrient estimation

[28, 29]. These datasets show a balance between scale, realism, and detail in annotations that needs to be considered when creating or assessing AI models for food and nutrition informatics.

9.3.2 Estimating Portion Sizes and Volumes

Getting portion sizes and volumes right is essential for turning food recognition into useful nutritional information. Techniques like U-Net and DeepLab are great at picking out individual food items in complicated plate images [30, 31]. U-Net's encoder-decoder design, featuring skip connections, is great for pixel-wise classification, reaching IoU scores of 0.85–0.90 for food segmentation in datasets like Nutrition5k [24, 25]. DeepLab uses atrous convolutions to improve segmentation by capturing context at multiple scales, especially for oddly shaped foods. Depth estimation and 3D reconstruction help calculate volume, which is key for accurate calorie and nutrient estimation. Techniques such as monocular depth estimation, which use CNNs trained on depth-annotated datasets, help estimate food height and shape. However, the accuracy relies on the quality of the image and camera calibration [22, 26]. Multi-view imaging and stereoscopy provide great ways to create 3D food models from different angles, with research showing volume estimation errors under 10% for standardized food items. Yet, challenges remain in real-world situations, as inconsistent lighting and missing depth references make estimation tricky. Using multi-view data and advanced depth sensors can improve accuracy, leading to better portion size analysis in various settings.

9.3.3 Approximation of Nutritional Content

Estimating the nutritional content in food images means predicting macronutrients like carbs, proteins, and fats, as well as micronutrients, based on food recognition and portion size. Regression models often use CNNs to connect visual features with nutritional values, relying on datasets like Nutrition5k [24–26]. A refined ResNet-50 model can predict the caloric content of single food plates with an average error of 10–15 kcal. End-to-end models that combine food recognition, segmentation, and nutritional mapping have made this process easier [24]. Transfer learning and fine-tuning help tailor models to different cuisines or diets since nutritional profiles differ widely between cultures. Fine-tuning with local datasets can boost accuracy by 10–15% for non-Western cuisines. Estimating micronutrients is tough because of limited data and differences in how food is prepared [3, 22, 32]. To improve prediction accuracy, we need to integrate external nutritional databases with contextual data like cooking techniques.

9.3.4 Multimodal Methods

Multimodal methods improve nutritional profiling by combining image data with other sources like text (recipes and ingredient lists) or sensor data from wearables [8, 33, 34]. Multimodal transformers use cross-modal attention to blend visual and text features, enhancing recognition and estimation accuracy. Using a transformer-based model for food images alongside recipe texts improved macronutrient prediction by 7% over models that relied only on images [8, 33, 34]. Wearable sensors like glucose monitors and activity trackers offer valuable insights into how diet affects us, helping to personalize nutrition advice. Cross-modal learning connects features from different modalities, helping to overcome the drawbacks of single-modality models [33, 35]. Challenges include bringing together different data sources and making sure everything runs efficiently. New multimodal architectures, like CLIP-based models, are set to improve accuracy by learning together.

9.3.5 Data Augmentation and Synthetic Data

Data augmentation and synthetic data can help improve nutritional profiling by tackling dataset limitations and biases. Rotation, flipping, and color changes help mimic different lighting and presentation styles, making models more robust. Generative models like GANs and diffusion models create synthetic food images to improve databases [31, 36]. A 2023 study showed that photos created by GANs helped reduce overfitting and improved UE-Food categorization by 8%. Diffusion models boost dataset variety, but they need a lot of computing power. One big challenge is making sure synthetic data truly reflects nutritional content and cultural diversity. Using domain knowledge to create processes can improve synthetic datasets [31, 36].

It shows how DL can change nutritional profiles. These methods improve food image analysis using advanced architectures, segmentation techniques, multimodal integration, and data augmentation. We need to keep researching multi-food recognition, volume estimation, and dataset limitations to create effective and scalable solutions for real-world dietary assessment.

9.4 Datasets and Metrics

9.4.1 Datasets

DL models for nutritional profiling depend on strong, labeled datasets [7, 10]. Datasets like Food-101, UE-Food, Nutrition5k, and Recipe1M have their own strengths, but they also come with challenges. Food-101 is extensive but leans toward Western cuisines, whereas Nutrition5k offers detailed nutritional info but has a smaller scale [14, 24, 25]. UE-Food specializes in analyzing multiple food plates, though its sample size is quite limited. These datasets show the importance of having culturally diverse and globally representative data with accurate nutritional details. Generating synthetic data with GANs or diffusion models has become a great way to enhance datasets [14, 24, 25, 37].

9.4.2 Assessment Metrics

Assessing DL models necessitates a range of metrics that are specifically designed for particular tasks. In the context of food recognition, the metrics commonly utilized include accuracy, precision, recall, and F1-score. The concept of top-k accuracy plays a significant role in the recognition of multiple food items [24, 33]. In the context of segmentation, the Intersection over Union (IoU) and the Dice coefficient serve as important metrics for measuring the degree of overlap between the predicted regions and the actual ground-truth regions. Estimating portions involves the use of MAE and root mean square error

Table 9.2 Comparative model performance across datasets.

Model architecture	Dataset	Top-1 accuracy (%)	F1-score	Mean absolute error (MAE)
ResNet-50	Food-101	85–88%	0.80–0.85	15–20 kcal
Swin Transformer	Food-101	90–92%	0.85–0.90	12–15 kcal
Mask R-CNN	Nutrition5k	Not reported	0.85 (IoU)	12–18% (nutrient error rate)
End-to-End Pipeline	Nutrition5k	Not reported	Not reported	10–15% (nutrient error rate)

(RMSE) as key metrics [30, 31, 38, 39]. The metrics used for nutritional estimation primarily concentrate on the inaccuracies found in predicting calories and macronutrients [38, 39]. It is important to develop standardized metrics that take into consideration the rich tapestry of cultural and contextual diversity.

Table 9.2 provides an overview of the performance metrics associated with advanced DL models that have been assessed across different food datasets. In the context of Food-101, both ResNet-50 and the Swin Transformer architectures demonstrate strong classification capabilities. Notably, the Swin Transformer achieves a higher Top-1 accuracy ranging from 90% to 92% and F1-scores reaching up to 0.90 [10, 22, 24, 29]. The findings highlight the advantages of transformer-based models in handling complex food images, which can be attributed to their enhanced capacity to understand spatial relationships and contextual connections.

On the other hand, we usually evaluate model performance on Nutrition5k by looking at how accurately it estimates nutrients instead of focusing on classification. Mask R-CNN achieves an F1-score of 0.85 for food segmentation, helping with nutrient estimation [24, 25, 40]. The MAE for caloric estimation in both Mask R-CNN and an end-to-end model pipeline ranges from 10% to 18%, showing good accuracy for real-world diet analysis [26, 38, 40]. Shifting from classification accuracy in Food-101 to nutrient prediction error in Nutrition5k shows the need for benchmarks tailored to specific tasks when evaluating AI models in food applications.

These performance comparisons show how important it is to choose models that fit the specific needs of applications like visual food classification, portion estimation, or nutrient analysis. We need to keep working on standardizing evaluation methods for different datasets and integrating multi-task learning approaches to tackle the complexities of nutritional data.

9.4.3 Benchmarking

Comparisons of DL models on standard datasets indicate nutritional profiling progress and limitations. On Food-101, CNN models such as ResNet-50 achieve top-1 accuracies of 85–88%, whereas ViT such as Swin Transformer attain greater rates, reaching 90–92%. This demonstrates superior efficacy in capturing intricate image details [22, 24]. Nutrition5k benchmarks indicate that end-to-end models achieve an MAE of 12–15% in macronutrient prediction, but transformer-based approaches reduce errors by 3–5% relative to CNNs. The performance on UE-Food is suboptimal, with F1-scores averaging between 75% and 80% due to occlusion and background noise [8, 24, 25]. Cross-dataset generalization is a substantial challenge. Models trained on Food-101 frequently exhibit a decline in accuracy of 10–15% when applied to UE-Food or Nutrition5k, attributable to variations in image composition and annotation methodologies [25, 27].

The dependence on datasets characterized by insufficient diversity and inconsistent annotations, along with challenges in generalizing across datasets, underscores the necessity for improved, more inclusive data and standardized evaluation methodologies to enhance the applicability of nutritional profiling in real-world contexts.

9.5 Applications and Impact

9.5.1 Personalized Nutrition

DL-driven nutritional profiling creates customized diets based on health needs and tastes, improving personalized nutrition. Models like CNNs and ViT can analyze food plate images to identify food items, estimate portion sizes, and predict nutritional content, giving users valuable insights into their diet [10, 23]. Liu [22] showed that a model combining ResNet-50 with regression heads reached an MAE of 12 kcal for estimating calories, allowing for accurate tracking of diet patterns. This precision helps people tailor their diets to meet health goals, like managing weight or dealing with conditions like diabetes. Connecting with wearable devices like fitness trackers and glucose monitors improves this ability by linking nutrition data with health metrics. Combining image data with health information from users helps make personalization even better. Using multimodal transformers for food image analysis boosted dietary recommendation accuracy by 8% over image-only models [8, 28]. These advancements provide personalized dietary advice, encouraging healthier eating habits and supporting individual preventive healthcare.

9.5.2 Health and Commercial Applications

DL improves nutritional profiling in public health, business, and clinical settings. Automated analysis of food images in public health helps monitor dietary trends on a large scale, guiding efforts to tackle obesity and related diseases [10]. A 2022 study used a Mask R-CNN model to analyze food images collected by the community, revealing trends in high-calorie consumption with an F1-score of 0.87 [40]. Nutritional profiling in epidemiological research helps link diet to disease, allowing researchers to connect eating habits with health outcomes.

In the commercial world, DL for nutritional profiling improves customer experiences and makes operations smoother. Food delivery apps now use food recognition models to provide nutritional info for menu items. In clinical settings, dietitians use these tools to track patients' diets more accurately than traditional food logs [27, 41, 42]. A 2023 clinical trial showed that a mobile app using a Swin Transformer for food analysis cut dietary assessment errors by 15% in patients with obesity [43].

DL in nutritional profiling provides scalable, precise, and actionable dietary information for personalized nutrition, public health, and commercial-clinical sectors. These applications aim to change how people, communities, and industries think about nutrition by using new technologies and tackling implementation challenges, leading to healthier societies.

9.6 Challenges and Limits

9.6.1 Limitations of the Dataset

DL models for nutritional profiling rely on the quality and diversity of training datasets; unfortunately, existing datasets are biased and impede progress. Datasets such as Food-101 and Nutrition5k are significant; nevertheless, they predominantly emphasize Western cuisines and largely exclude African, South Asian, or indigenous meals. Food-101's 101 categories predominantly emphasize popular dishes such as pizza and burgers, neglecting other culinary traditions. This leads to models that exhibit a 10–20% decline in accuracy when processing non-Western food imagery. This cultural bias constrains the application of nutritional profiling in global health, where dietary practices vary significantly. The quality of annotations presents a significant challenge. Nutrition5k delineates macronutrients; nonetheless, laboratory-based nutritional analysis is intricate, resulting in often inadequate or inconsistent micronutrient annotations [24, 25, 43, 44]. Crowdsourcing annotations may result in inaccuracies, as nonexpert annotators could misclassify portion amounts or food categories. Scalability presents a hurdle; datasets such as UE-Food, containing fewer than 10,000 photos, are insufficient for training robust models capable of handling real-world variations. GANs have been employed to augment datasets; nevertheless, achieving nutritional accuracy in synthetic images poses challenges [24, 25, 43, 44]. These constraints underscore the necessity for varied, well-annotated datasets to facilitate comprehensive and precise nutritional assessment.

The application of DL models for nutritional assessment in practical environments encounters significant computational and logistical challenges. Advanced models require substantial computational resources, rendering them challenging to implement on mobile devices [15, 35]. Model pruning and quantization reduce computational requirements; however, they may occasionally compromise accuracy. Privacy concerns arise as nutritional profiling utilizes food photographs from users, potentially containing personal information [10]. Federated learning demonstrates potential; nonetheless, it remains in its nascent phase and faces several hurdles [45, 46].

9.6.2 Generalization and Resilience

DL nutritional profiling algorithms seek to generalize across various food types, lighting conditions, and cultural contexts. Models trained on datasets such as Food-101 perform effectively in controlled settings but encounter significant accuracy declines of 15–25% in real-world scenarios due to variations in lighting, angles, or cluttered backgrounds [27, 37, 43, 44]. A model trained on distinct, single-item photos struggles to recognize foods on plates containing many things, particularly in the presence of shadows or impediments. The cultural context complicates generalization; models trained on Western datasets frequently misidentify items from underrepresented cuisines, such as Ethiopian injera or Indian thali, because of their distinctive textures and presentations [27, 37, 43, 44]. Overfitting is a prevalent problem, particularly with limited datasets such as Nutrition5k, where models often memorize training samples rather than identifying overarching trends. Methods such as transfer learning and adversarial training have demonstrated efficacy in improving resilience, with research indicating a 5–10% enhancement in accuracy when models are adapted to new cuisines [13, 24–27, 37, 43, 44]. These approaches require additional labeled data from target domains, which is often difficult to obtain. The development of models that perform effectively across diverse real-world scenarios is an important topic for enhancement.

9.6.3 Real-world Deployment

Deploying DL models for nutritional profiling in real-world applications is hindered by significant computational and practical challenges. State-of-the-art models are computationally intensive, requiring high-end GPUs that are impractical for mobile deployment. Techniques like model pruning and quantization have reduced computational requirements, but often at the cost of accuracy. Privacy concerns also arise, as nutritional profiling relies on user-generated food images that may contain sensitive information. Federated learning offers a potential solution, but its implementation is nascent and challenging.

9.6.4 Dietary Precision

Assessing nutritional content, particularly micronutrients, and accounting for cooking methods are notably difficult [47–49]. We have become proficient in forecasting macronutrients; nevertheless, calculating micronutrients remains challenging due to insufficient data and variability in food composition [8, 43, 50, 51]. Cooking techniques such as frying and steaming can alter nutritional composition; however, models frequently neglect these variations. The absence of standardized nutritional databases exacerbates this issue. Integrating diverse data sources may enhance accuracy [28, 33]; nevertheless, synchronizing them with visual data remains a significant technical challenge.

Utilizing DL for nutritional profiling presents difficulties stemming from dataset biases, generalization challenges, deployment constraints, and mistakes in nutritional information. Addressing these difficulties necessitates interdisciplinary collaboration, integrating advancements in dataset management, model enhancement, and nutritional science to develop effective and equitable solutions for dietary evaluation.

9.7 Prospects

9.7.1 Enhanced Datasets

Diverse, culturally inclusive, and well-annotated datasets that reflect worldwide food trends are essential to DL for nutritional profiling. The datasets Food-101 and Nutrition5k are significantly skewed toward Western cuisines and underrepresent African, South Asian, and indigenous foods, resulting in models that underperform in varied contexts by 10–20% [10, 52]. International partnerships are needed to organize statistics on a variety of culinary traditions, including Ethiopian injera and Southeast Asian street cuisine. Crowdsourcing systems like Amazon Mechanical Turk provide scalable annotation solutions, but quality control techniques like expert validation and inter-annotator agreement measures are needed to prevent mistakes, which can lower annotation accuracy by 15% [24, 30, 52].

Synthetic data with nutritional metadata like macronutrient profiles or cooking process specifics may be more useful. To provide robust and fair models across global populations, future initiatives should target open-access, standardized datasets with extensive macro- and micronutrient annotations.

9.7.2 Advanced Architectures

Advances in DL architectures promise to improve nutritional profiling. Hybrid models that combine CNN spatial feature extraction with vision transformer global contextual knowledge may improve multi-food recognition and nutritional estimation. CNN-transformer hybrids showed 5% lower caloric estimate MAE than solo ResNet models on Nutrition5k [24, 43, 44]. Nutritional datasets generally lack labeled data; therefore, self-supervised learning (SSL) is another option [53]. SSL methods like contrastive learning let computers build strong feature representations from unlabeled food photos [53]. Sparse transformers and graph neural networks may improve performance in complicated scenarios with overlapping edibles. These advances require careful testing against existing models to ensure scalability and computational efficiency, especially in resource-constrained contexts.

9.7.3 Edge and Real-time Computing

Computational restrictions make real-time nutritional monitoring on mobile and embedded devices difficult [42]. Modeling, pruning, and knowledge distillation provide lightweight, accurate models with low computational needs. Federated learning trains models using decentralized user data to protect privacy [26, 45]. Optimizing lightweight edge device models and standardizing federated learning protocols should be a future study.

9.7.4 Interdisciplinary Integration

Computer scientists, nutritionists, and public health professionals must collaborate to improve nutritional profiling. Public health professionals can use profiling technologies for large-scale food surveillance, while nutritionists can generate datasets. Genomic and metabolomics integration enables a hyper-personalized diet. Joint research consortia and open-source platforms can expedite advancement.

9.8 Conclusion

DL makes it possible to automatically and better estimate dietary profiles using food plate images. Using CNNs, ViT, and multimodal models, researchers have improved food recognition, portion size estimation, and nutritional content prediction. Models now estimate calories on Nutrition5k with MAEs of 10–15 kcal and top-1 accuracies of 85–92% on

Food-101. These advances enable individualized nutrition, public health surveillance, and therapeutic applications. This empowers consumers, healthcare providers, and policymakers to address global issues like obesity and chronic diseases. However, major issues remain. These include dataset biases that make it hard to include people from different cultures, problems applying models to different foods and real-life situations, limited mobile computing power, and inaccurate micronutrient estimates due to few annotations and different cooking methods. Different datasets, hybrid designs, lightweight models, and multimodal data may solve these issues. Its combination of computer vision, nutrition science, and public health reveals its potential to change the world. Computer scientists, dietitians, and health specialists must collaborate to better analyze data, refine models, and ensure ethical use. In the future, combining artificial intelligence with new areas like genomics and edge computing could completely change nutritional science, making it possible for highly customized dietary interventions and health solutions that can be used by many people. DL can make nutritional profiling a key part of precision health, making people all over the world better by fixing the problems that are holding it back now and encouraging new ideas.

Funding

No funding.

Declaration of Competing Interest

The author declares that he has no known competing financial interests or personal relationships that could have appeared to influence the work reported in this chapter.

References

1 Sosa-Holwerda, A., Park, O.-H., Albracht-Schulte, K. et al. (2024). The role of artificial intelligence in nutrition research: a scoping review. *Nutrients* 16: 2066. https://doi.org/10.3390/nu16132066.

2 Nganso Ditchou, Y.O., Leutcha, P.B., Miaffo, D. et al. (2024). In vitro and in silico assessment of antidiabetic and antioxidant potencies of secondary metabolites from Gymnema sylvestre. *Biomedicine & Pharmacotherapy* 177: 117043. https://doi.org/10.1016/j.biopha.2024.117043.

3 Hamadou, M. (2025). Bioactive peptides and metabolic health: a mechanistic review of the impact on insulin sensitivity, lipid profiles, and inflammation. *Applied Food Research* 5: 101056. https://doi.org/10.1016/j.afres.2025.101056.

4 Aleixandre, A. and Miguel, M. (2011). Food peptides as antihypertensive agents. In: Bioactive Food Proteins and Peptides, 1e (ed. B. Hernández-Ledesma and C.C. Hsieh), 131–180. Boca Raton: CRC Press https://doi.org/10.1201/b11217-10.

5 Munteanu, I.G. and Apetrei, C. (2021). Analytical methods used in determining antioxidant activity: a review. *International Journal of Molecular Sciences* 22: 3380. https://doi.org/10.3390/ijms22073380.

6 Srinivas, P., Veeraiah, V., Thejaswini, K.O. et al. (2025). The impact of electromagnetic and optical radiation on physical and biological systems. *Journal of Nano- and Electronic Physics* 17 (4): 04012. https://doi.org/10.21272/jnep.17(4).04012.

7 Theodore Armand, T.P., Nfor, K.A., Kim, J.-I., and Kim, H.-C. (2024). https://doi.org/10.3390/nu16071073). Applications of artificial intelligence, machine learning, and deep learning in nutrition: a systematic review. *Nutrients* 16: 1–24.

8 Cheng, S.-T., Lyu, Y.-J., and Teng, C. (2025). Image-based nutritional advisory system: employing multimodal deep learning for food classification and nutritional analysis. *Applied Sciences* 15: 4911. https://doi.org/10.3390/app15094911.

9 Li, Y., Li, X., Liu, Y. et al. (2022). MPMABP: a CNN and Bi-LSTM-based method for predicting multi-activities of bioactive peptides. *Pharmaceuticals* 15: 707. https://doi.org/10.3390/ph15060707.

10 Liu, D., Zuo, E., Wang, D. et al. (2025). Deep learning in food image recognition: a comprehensive review. *Applied Sciences* 15: 7626. https://doi.org/10.3390/app15147626.

11 An, X., Zhang, W., Rong, C., and Liu, S. (2023). Understanding Ramachandran plot for dipeptide: a density functional theory and i information-theoretic approach study. *Journal of the Chinese Chemical Society* 70: 243–252. https://doi.org/10.1002/jccs.202200444.

12 Lai, L., Liu, Y., Song, B. et al. (2025). Deep generative models for therapeutic peptide discovery: a comprehensive review. *ACM Computing Surveys* 57: 1–29. https://doi.org/10.1145/3714455.

13 Lei, Y., Wang, X., Fang, M. et al. (2024). PepGB: facilitating peptide drug discovery via graph neural networks. http://arxiv.org/abs/2401.14665.

14 Kaushal, S., Tammineni, D.K., Rana, P. et al. (2024). Computer vision and deep learning-based approaches for detection of food nutrients/nutrition: new insights and advances. *Trends in Food Science and Technology* 146: 104408. https://doi.org/10.1016/j.tifs.2024.104408.

15 Esteva, A., Chou, K., Yeung, S. et al. (2021). Deep learning-enabled medical computer vision. *npj Digital Medicine* 4: 5. https://doi.org/10.1038/s41746-020-00376-2.

16 Pandey, J.K. (2025). Understanding the impact of electromagnetic and optical radiation on physical and biological systems. *Journal of Nano- and Electronic Physics* 17 (5): 05023. https://doi.org/10.21272/jnep.17(5).05023.

17 Qiu, X., Gao, J., Yang, J. et al. (2020). A comparison study of machine learning (random survival forest) and classic statistic (cox proportional hazards) for predicting progression in high-grade glioma after proton and carbon ion radiotherapy. *Frontiers in Oncology* 10: https://doi.org/10.3389/fonc.2020.551420.

18 Tripathi, A., Goswami, T., Trivedi, S.K., and Sharma, R.D. (2021). A multi-class random forest (MCRF) model for classification of small plant peptides. *International Journal of Information Management Data Insights* 1: 100029. https://doi.org/10.1016/j.jjimei.2021.100029.

19 Sanders, W.S., Johnston, C.I., Bridges, S.M. et al. (2011). Prediction of cell penetrating peptides by support vector machines. *PLoS Computational Biology* 7: e1002101. https://doi.org/10.1371/journal.pcbi.1002101.

20 Thieme, A.H., Zheng, Y., Machiraju, G. et al. (2023). A deep-learning algorithm to classify skin lesions from mpox virus infection. *Nature Medicine* 29: 738–747. https://doi.org/10.1038/s41591-023-02225-7.

21 Romero-Tapiador, S., Tolosana, R., Morales, A. et al. (2024). Leveraging automatic personalised nutrition: food image recognition benchmark and dataset based on nutrition taxonomy. *Multimedia Tools and Applications* 84: 1945–1966. https://doi.org/10.1007/s11042-024-19161-4.

22 Pandey, J.K., Kumar, S., Lamin, M. et al. (2022). A metaheuristic autoencoder deep learning model for an intrusion detector system. *Mathematical Problems in Engineering* 2022: 11, 3859155. https://doi.org/10.1155/2022/3859155.

23 Rosaline, T.M.L., Suciawan, C.A., Ng, W. et al. (2024). AI-powered mobile application for image-based food ingredient detection and recipe generation. *2024 International Conference on Information Management and Technology (ICIMTech)*, pp. 123–128. IEEE. https://doi.org/10.1109/ICIMTech63123.2024.10780810.

24 Bianco, R., Coluccia, S., Marinoni, M. et al. (2025). 2D prediction of the nutritional composition of dishes from food images: deep learning algorithm selection and data curation beyond the Nutrition5k project. *Nutrients* 17: 2196. https://doi.org/10.3390/nu17132196.

25 Yanuartanti, I., Suciati, N., and Fatichah, C. (2025). Deep learning model comparison for food nutrient prediction using the Nutrition5k dataset. *2025 International Conference on Smart Computing, IoT and Machine Learning (SIML)*, pp. 1–6. IEEE. https://doi.org/10.1109/SIML65326.2025.11081021.

26 Ding, H., Hou, H., Wang, L. et al. (2025). Application of convolutional neural networks and recurrent neural networks in food safety. *Food* 14: 247. https://doi.org/10.3390/foods14020247.

27 Dhal, S.B. and Kar, D. (2025). Leveraging artificial intelligence and advanced food processing techniques for enhanced food safety, quality, and security: a comprehensive review. *Discover Applied Sciences* 7: 75. https://doi.org/10.1007/s42452-025-06472-w.

28 Marin, J., Biswas, A., Ofli, F. et al. (2019). Recipe1m+: a dataset for learning cross-modal embeddings for cooking recipes and food images. http://arxiv.org/abs/1810.06553.

29 Vyshnavi, K.S., Thaniya, S., and Fancy, C. (2025). A deep learning framework for recipe generation from multi-input food data. *2025 6th International Conference on Inventive Research in Computing Applications (ICIRCA)*, pp. 1–6. https://doi.org/10.1109/ICIRCA65293.2025.11089828.

30 Goyzueta, C.A.R., De la Cruz, J.E.C., Machaca, W.A.M. (2021). Integration of U-Net, ResU-Net and DeepLab architectures with intersection over union metric for cells nuclei image segmentation. *2021 IEEE Engineering International Research Conference (EIRCON)*, pp. 1–4. IEEE. https://doi.org/10.1109/EIRCON52903.2021.9613150.

31 Miao, Y., Wang, R., Jing, Z. et al. (2024). CT image segmentation of foxtail millet seeds based on semantic segmentation model VGG16-UNet. *Plant Methods* 20: 169. https://doi.org/10.1186/s13007-024-01288-y.

32 Mamoudou, H. and Mune Mune, M.A. (2024). Investigating Bambara bean (*Vigna subterranea* [Verdc.] L.) protein and hydrolysates: a comprehensive analysis of biological and biochemical properties. *Applied Food Research* 4: 100489. https://doi.org/10.1016/j.afres.2024.100489.

33 Lun, Z., Wu, X., Dong, J., and Wu, B. (2025). Deep learning-enhanced spectroscopic technologies for food quality assessment: convergence and emerging frontiers. *Food* 14: 2350. https://doi.org/10.3390/foods14132350.

34 Ruan, W., Lyu, Y., Zhang, J. et al. (2025). Large language models for bioinformatics. https://doi.org/10.48550/arXiv.2501.06271.

35 Pandey, J.K., Yadav, A.K., Purushotham, M. et al. (2022). Colorimetric detection of deltamethrin pesticide from biological samples based on the peroxidase-mimic catalytic activity on MnO_2/SnO_2. *Key Engineering Materials* 928: 45–51. https://doi.org/10.4028/p-4ti1i0.

36 Abbas, S., Ahmed, F., Khan, W.A. et al. (2025). Intelligent skin disease prediction system using transfer learning and explainable artificial intelligence. *Scientific Reports* 15: 1746. https://doi.org/10.1038/s41598-024-83966-4.

37 Lin, E., Lin, C.-H., and Lane, H.-Y. (2022). De novo peptide and protein design using generative adversarial networks: an update. *Journal of Chemical Information and Modeling* 62: 761–774. https://doi.org/10.1021/acs.jcim.1c01361.

38 Dikmen, F., Demir, A., Özkaya, B. et al. (2025). AI-driven wastewater management through comparative analysis of feature selection techniques and predictive models. *Scientific Reports* 15: 25347. https://doi.org/10.1038/s41598-025-07124-0.

39 Zhao, N., Wei, J., Long, Z. et al. (2023). An integrated method for tunnel health monitoring data analysis and early warning: Savitzky–Golay smoothing and wavelet transform denoising processing. *Sensors* 23: 7460. https://doi.org/10.3390/s23177460.

40 Talukdar, V., Veeraiah, V., Roy, D.R. et al. (2023). An innovative approach: typicality-based collaborative filtering for recommender systems. *Proceedings of the 3rd International Conference on Advancement in Electronics & Communication Engineering (AECE)*, pp. 863–868. https://doi.org/10.1109/AECE59614.2023.10428576

41 Rossi, S., Gemma, S., Borghini, F. et al. (2025). Agri-food traceability today: advancing innovation towards efficiency, sustainability, ethical sourcing, and safety in food supply chains. *Trends in Food Science and Technology* 163: 105154. https://doi.org/10.1016/j.tifs.2025.105154.

42 Revelou, P.-K., Tsakali, E., Batrinou, A., and Strati, I.F. (2025). Applications of machine learning in food safety and HACCP monitoring of animal-source foods. *Food* 14: 922. https://doi.org/10.3390/foods14060922.

43 Wang, H., Tian, H., Ju, R. et al. (2024). Nutritional composition analysis in food images: an innovative Swin Transformer approach. *Frontiers in Nutrition* 11: https://doi.org/10.3389/fnut.2024.1454466.

44 Bianco, R., Marinoni, M., Coluccia, S. et al. (2024). Tailoring the nutritional composition of Italian foods to the US Nutrition5k dataset for food image recognition: challenges and a comparative analysis. *Nutrients* 16: 3339. https://doi.org/10.3390/nu16193339.

45 Beemkumar, N., Gupta, S., Bhardwaj, S. et al. (2023). Activity recognition and IoT-based analysis using time series and CNN. In: Handbook of Research on Machine Learning-enabled IoT for Smart Applications Across Industries (ed. N. Goel and R.K. Yadav), 350–364. IGI Global https://doi.org/10.4018/978-1-6684-8785-3.ch018.

46 Gavai, A., Bouzembrak, Y., Mu, W. et al. (2023). Applying federated learning to combat food fraud in food supply chains. *npj Science of Food* 7: 46. https://doi.org/10.1038/s41538-023-00220-3.

47 Mamoudou, H., Obadias, F.V., Samuel René, M., and Martin Alain, M.M. (2024). Physical characteristics, chemical composition, and antioxidant properties of defatted grain legumes cultivated in Diamare division (Far North Region, Cameroon). *Applied Food Research* 4: 100498. https://doi.org/10.1016/j.afres.2024.100498.

48 Mamoudou, H., Başaran, B., Mune, M.A.M. et al. (2024). Bioactive peptides derived from the enzymatic hydrolysis of cowhide collagen for the potential treatment of atherosclerosis: a computational approach. *Intelligent Pharmacy* 2: 456–466. https://doi.org/10.1016/j.ipha.2024.05.004.

49 Mamoudou, H., Abdoulaye, A.H., Ditchou, N.Y.O. et al. (2025). Computational investigation of Plectranthus neochilus essential oil phytochemicals interaction with dipeptidyl peptidase 4: a potential avenue for antidiabetic drug discovery. *Current Pharmaceutical Analysis* 21: 169–178. https://doi.org/10.1016/j.cpan.2025.03.002.

50 Bayang, J.P., Touwang, C., Mamoudou, H. et al. (2025). Variation of nutrients and bioactive compounds of five wild edible leafy vegetables from the far north region of Cameroon. *Food Chemistry Advances* 6: 100849. https://doi.org/10.1016/j.focha.2024.100849.

51 Hamadou, M., Martin Alain, M.M., Obadias, F.V. et al. (2022). Consumption of underutilised grain legumes and the prevention of type II diabetes and cardiometabolic diseases: evidence from field investigation and physicochemical analyses. *Environmental Challenges* 9: 100621. https://doi.org/10.1016/j.envc.2022.100621.

52 Batuwatta-Gamage, C.P., Jeong, H., Karunasena, H. et al. (2025). Physics-informed machine learning for microscale drying of plant-based foods: a systematic review of computational models and experimental insights. https://doi.org/10.48550/arXiv.2501.09034.

53 Peng, A., He, J., Zhu, F. (2023). Self-supervised visual representation learning on food images. http://arxiv.org/abs/2303.09046.

10

Consumer Preferences and Market Drivers for Varieties of Millet Products: A Study Among Urban and Semi-Urban Consumers

P. Sasikumar, T.M. Hemalatha, and V. Priyanka

School of Business and Commerce, Rathinam Global Deemed to be University, Coimbatore, Tamil Nadu, India

10.1 Introduction

10.1.1 Background of Millets: History, Nutritional Value, and Market Relevance

Millets are among the first domesticated crops, having been grown for thousands of years as a collection of small-seeded cereal grains [1]. Because of their adaptability to dry and semi-arid conditions, millet was a staple grain for rural and tribal societies in Asia and Africa as early as 5000 BCE, according to archaeological finds [2]. In difficult climes, these grains provide food security since they grow well on poor soils and need little water [3]. Even if rice and wheat were preferred in contemporary agriculture, millets are still important to culture and provide nutrients in many traditional cuisines.

Because of their high content of complex carbohydrates, dietary fiber, and vital micronutrients including iron, magnesium, phosphorus, potassium, and B vitamins, millets are referred to as "nutri-cereals" in the nutrition industry [4]. They are advantageous for people with gluten intolerance because they are naturally gluten-free [5]. Millets also include antioxidants, including flavonoids and phenolic compounds, which have been linked to a lower risk of chronic illnesses like heart disease and some types of cancer [6]. While pearl millet is a good source of iron and helps with typical nutritional inadequacies, finger millet is particularly well-known for its high calcium content [7].

Concerns about sustainability and growing health consciousness are reviving interest in millets in the modern food market [8]. Compared to rice and wheat, they are more environmentally friendly due to their flexibility and reduced water footprint [9]. In response, the food industry has promoted the functional food qualities of millet-based goods by creating new flours, snacks, cereals, and noodles [10]. Government programs encourage the production and consumption of millet, especially in India, to boost rural economies and nutritional security [11]. Notwithstanding these initiatives, issues with accessibility, taste preferences, and consumer awareness still exist in urban areas [12].

10.1.2 Increasing Global and Local Interest in Millet Products

Due to rising consumer awareness of environmental sustainability, nutrition, and health, millets have seen a comeback in popularity recently, both domestically and internationally.

Ancient grains are becoming more popular worldwide as customers look for alternatives to highly processed and traditional cereals. Since they are gluten-free and high in nutrients, millets have gained popularity as "superfoods" with a host of health advantages, such as better glycemic control, better digestion, and a lower risk of developing chronic conditions like diabetes and cardiovascular illnesses [13].

Additionally, compared to mainstays like rice and wheat, millets require less water and fewer chemical inputs, making them climate-smart crops because they thrive in marginal and drought-prone areas. This environmental benefit supports the worldwide movement toward sustainable agriculture and food security by matching the growing consumer preference for eco-friendly and sustainable food options [14].

Public health campaigns and government measures have been crucial in recovering millet consumption locally, particularly in India. Millet production and consumption have been highlighted by the Indian government's designation of millet as "nutri-cereals" and initiatives under programs like the National Food Security Mission. Convenient millet-based items that suit contemporary lifestyles, such as millet noodles, breakfast cereals, and snacks, are also in high demand due to the growing health consciousness of the urban middle class [15].

Nevertheless, despite these encouraging developments, problems still exist, including low consumer knowledge of millets, taste preferences, and product availability. To get around this, food producers and marketers are coming up with new product designs and branding techniques that emphasize the environmental and nutritional advantages of millet [16]. As a result, millets are progressively making their way from specialized traditional foods to popular health food markets, which reflects a major change in local and global food consumption trends.

10.2 Research Problems

The consumption of millet products in urban and semi-urban areas is still low, despite their increasing availability and acknowledged nutritional advantages. Many consumers are either hesitant because of perceived flavor, unfamiliarity, and price concerns, or they are not aware of the wide range of millet-based goods. Furthermore, these marketplaces lack a thorough understanding of the elements that influence customer purchasing decisions, such as packaging attractiveness, promotional efficacy, and brand trust. Manufacturers and marketers are unable to properly position millet products to satisfy customer expectations and increase market penetration because of this information gap. Investigating the main factors that influence consumer choices for millet product variations is crucial, especially for various demographic groups like students, working professionals, and stay-at-home moms in urban and semi-urban areas.

10.3 Objectives of the Study

This study aims to explore the key factors influencing consumer preferences and buying behavior for millet products in urban and semi-urban areas. It seeks to identify drivers and barriers to adoption and provide recommendations to enhance millet consumption for the following objectives:

1) To identify the key factors influencing consumer preferences and buying behavior for various millet product varieties in urban and semi-urban areas.
2) To examine the impact of brand trust, packaging, price perception, health claims, and promotions on millet product purchases.
3) To assess consumer awareness and taste perceptions affecting millet product adoption.
4) To provide recommendations for marketing, education, and product innovation to boost millet consumption.

10.4 Literature of the Study

Millet consumption has been the subject of increasing academic and market interest due to its recognized nutritional benefits and role in sustainable agriculture. Previous studies have consistently highlighted millets as nutrient-rich grains, offering high dietary fiber, essential minerals, and gluten-free properties beneficial for managing lifestyle diseases such as diabetes and cardiovascular conditions [13, 14]. Research also underscores the environmental advantages of millets, emphasizing their low water requirement and resilience in marginal farming conditions, making them vital for climate-smart food systems [7].

Consumer preferences for millet products, however, remain complex and multifaceted. Studies show that health consciousness is a primary driver encouraging millet consumption, especially among urban populations [15]. Packaging appeal, brand trust, and promotional activities further influence buying decisions [16]. Nonetheless, challenges such as limited awareness, unfamiliarity with taste, and price sensitivity continue to restrict widespread adoption [13]. These findings are consistent with consumer behavior theories that suggest perceived benefits and barriers significantly affect food choice decisions, particularly for traditional grains introduced into modern diets [17].

Market trends reflect a growing shift toward functional foods and ancient grains, driven by evolving health concerns and sustainability values [18]. In India, government initiatives and consumer education campaigns have begun to reposition millets as "superfoods," with new product innovations aimed at urban consumers seeking a convenient and healthy option [19]. However, despite these advances, there is limited empirical research focusing on the comparative preferences of urban vs. semi-urban consumers and the detailed role of marketing elements such as packaging and promotions.

This disparity emphasizes the necessity for consumer-focused, locally focused research that evaluates the complex factors influencing the uptake of millet products in various demographics and geographical areas. To fill this gap, this study looks at the purchasing patterns of several Coimbatore City consumer groups. Its goal is to give manufacturers, merchants, and legislators useful information on how to best encourage millet consumption in changing food markets.

10.5 Research Gap

Most studies have concentrated on either general awareness campaigns or rural consumption patterns, even though previous research has demonstrated the nutritional advantages of millets and found health consciousness to be a significant factor in their uptake. Few studies have explicitly looked at how urban and semi-urban consumers view and buy various millet product types, including sorghum breakfast cereals, finger millet noodles, pearl millet treats, and foxtail millet flour. Furthermore, few studies have systematically examined the combined influence of brand trust, packaging appeal, price perception, health claims, and promotional activities in determining millet product preferences, even though the literature now in publication examines the broad drivers of healthy food choices. There is still a lack of research on the relative purchasing habits of urban and semi-urban people, as well as the barriers to adoption posed by perceived taste and product accessibility.

This disparity highlights the necessity of a focused, situation-specific study that combines theories of consumer behavior with market-specific elements, providing manufacturers, retailers, and legislators with useful information to boost millet consumption in developing health food marketplaces.

10.6 Hypothesis

Based on the objectives of the study, literature review, and identified research gap, the following hypotheses are formulated to empirically examine the factors influencing consumer preferences for millet products.

- **H1:** Health consciousness has a positive and significant influence on consumer preference for millet products.
- **H2:** Cultural and traditional food habits significantly influence consumer adoption of millet products.
- **H3:** Brand trust has a significant positive effect on consumer purchase decisions for millet products.
- **H4:** Packaging appeal significantly influences consumer buying behavior toward millet products.
- **H5:** Price perception significantly affects consumers' willingness to purchase millet products.
- **H6:** Health claims and nutritional information positively influence consumer purchase intention for millet products.
- **H7:** Promotional activities have a significant positive impact on consumer purchase behavior.
- **H8:** Taste and texture satisfaction significantly influence repeat purchase intention of millet products.
- **H9:** Higher awareness of nutritional benefits significantly increases the likelihood of millet product adoption.
- **H10:** There is a significant difference between urban and semi-urban consumers regarding the factors influencing millet product adoption.
- **H11:** Socio-demographic factors (age, occupation, income, education) significantly affect consumer preferences for millet product varieties.

10.7 Research Methodology

10.7.1 Research Design

This study adopted a **descriptive research design** to examine the factors influencing consumer preferences and purchase behavior for millet products. The design was considered appropriate as it facilitates the description, interpretation, and analysis of consumer attitudes, perceptions, and buying patterns within a defined population.

10.7.2 Study Area

The research was conducted in **Coimbatore City, Tamil Nadu, India**, covering both urban and semi-urban areas. This region was chosen due to its growing urban middle class, increasing exposure to health foods, and the active promotion of millet products through government and private initiatives.

10.7.3 Population and Sampling

The study population comprised **consumers aged 20–50 years**, including working professionals, homemakers, and college students. A **purposive sampling technique** was employed to select respondents who had purchased or consumed millet products at least once. The final sample size consisted of **255 respondents**, which was considered adequate for statistical analysis and ensured representation across socio-economic and demographic categories.

10.7.4 Data Collection Methods

Primary data were collected using a **structured questionnaire**, which was administered in major retail outlets, supermarkets, and farmers' markets. The questionnaire covered demographic details, purchasing frequency, preferred millet varieties, and factors influencing purchase, such as brand trust, packaging, price perception, health claims, and promotions.

10.7.5 Research Instrument

The questionnaire used a **five-point Likert scale** to measure the degree of importance or agreement with each factor influencing millet product purchases. Before full-scale administration, the instrument was **pretested with 20 respondents** to ensure clarity, reliability, and content validity.

10.7.6 Data Analysis Techniques

Collected data were coded and analyzed using the **Statistical Package for the Social Sciences (SPSS)**. Descriptive statistics (mean, frequency, percentage, and standard deviation) were employed to summarize demographic profiles and purchase behavior. Inferential statistical tools such as **chi-square tests, ANOVA, and regression analysis** were applied to test the formulated hypotheses and to assess the relationships between consumer characteristics and purchasing behavior.

10.7.7 Limitations

The study on consumer preferences for millet products provides valuable insights but is also subject to several limitations. Firstly, its findings are based on data from urban and semi-urban areas in Coimbatore, which restricts their wider applicability to rural areas or other regions with different consumer demographics. The sample only included individuals who had purchased millet products before, so it does not capture the views of nonconsumers or a broader age range.

Additionally, since the research uses a cross-sectional survey, it only reflects preferences at a single point in time and cannot account for changing consumer attitudes or trends. The reliance on self-reported data introduces possible bias, as participants may respond based on social expectations rather than true behavior. The questionnaire focused on selected millet products, missing other varieties and innovations that could influence market trends. Lastly, the statistical models used in the analysis account for only part of consumer motivations, leaving out other important factors like lifestyle changes and peer influence.

10.7.8 Ethical Considerations

Participation in the study was voluntary. Respondents were informed of the study's purpose and assured of confidentiality. No personal identifiers were recorded, and all responses were used solely for academic research purposes.

10.8 Analysis and Interpretation

Descriptive statistics were applied in this study to provide an overview of the respondents' demographic profile and their purchase behavior toward millet products. They help simplify and summarize large datasets into meaningful measures such as frequencies, percentages, means, and standard deviations.

In the context of this research, descriptive statistics are used to present:

- The socio-demographic characteristics of respondents (gender, age, income, education, occupation, etc.).
- Purchase behavior patterns, including most preferred millet products, buying frequency, and purchase channels [20].
- Motivational factors driving millet consumption include health awareness, taste, cultural habits, and price perception.
- The relative importance of key influencing factors like brand trust, packaging, availability, convenience, and promotions [21].

This section provides the foundation for further inferential analysis and hypothesis testing by highlighting the basic patterns, trends, and preferences of urban and semi-urban consumers in Coimbatore City (Table 10.1).

Table 10.1 Descriptive statistics – demographic information.

Particulars	N	Min	Max	Mean	Std. dev.
Gender	255	1.00	2.00	1.5	0.5
Age group	255	1.00	3.00	2.0	0.8
Occupation	255	1.00	4.00	2.5	1.1
Monthly household income	255	1.00	4.00	2.5	1.1
Educational qualification	255	1.00	4.00	2.6	1.1
Area of residence	255	1.00	2.00	1.4	0.5
Marital status	255	1.00	2.00	1.5	0.5
Household size	255	1.00	3.00	2.1	0.8

The demographic profile of the 255 respondents indicates a balanced representation across key categories. Gender distribution shows an almost equal mix of male and female participants (mean = 1.5, SD = 0.5), while the age group mean of 2.0 (SD = 0.8) suggests that most respondents fall within the middle age category (30–39 years). Occupation (mean = 2.5, SD = 1.1) reflects a diverse mix, with working professionals and students forming the majority. The average monthly household income (mean = 2.5, SD = 1.1) suggests that respondents are mainly from middle-income groups. Educational qualification (mean = 2.6, SD = 1.1) indicates that most respondents have at least an undergraduate or postgraduate education. Area of residence (mean = 2.0, SD = 0.5) indicates that most respondents are in semi-urban areas. Marital status (mean = 1.5, SD = 0.5) shows a balanced mix of single and married individuals, while household size (mean = 2.1, SD = 0.8) reveals that the majority of respondents live in medium-sized families (3–4 members). Overall, the demographic distribution highlights a well-diversified sample of urban and semi-urban consumers, making the findings more reliable for understanding millet consumption behavior (Table 10.2).

Based on the survey of 255 respondents, the preferred millet products had a mean score of **3.4** (SD = 1.7) on a 1–6 scale, indicating moderate preference with considerable variation across product types. The buying frequency averaged **3.0** (SD = 1.4) on a 1–5 scale,

Table 10.2 Descriptive statistics – purchase behavior.

	N	Min	Max	Mean	Std. dev.
Which millet products do you buy most often? (preferred product)	255	1.00	6.00	3.4	1.7
How often do you buy millet products? (buying frequency)	255	1.00	5.00	3.0	1.4
Where do you usually buy millet products? (purchase channel)	255	1.00	6.00	3.4	1.6
What motivates you to buy millet products? (buying motivation)	255	1.00	6.00	3.6	1.6

suggesting that consumers typically purchase millet products monthly, though some buy weekly or rarely. For the purchase channel, the mean was **3.4** (SD = 1.6), showing that respondents use a variety of outlets, including supermarkets, online platforms, and local stores. Finally, buying motivation had a mean of **3.6** (SD = 1.6), reflecting that health benefits are the main driver, while curiosity and sustainability also influence purchase decisions. Overall, the data indicate diverse preferences, moderate purchase frequency, multiple buying channels, and health-oriented motivation (Table 10.3).

The factors influencing consumer purchase behavior toward millet products reveal that cultural and traditional food habits (mean = 4.4) and price perception (mean = 4.3) emerge as the most influential drivers, suggesting that consumers strongly associate millet products with heritage diets while also being highly sensitive to cost considerations. Promotional activities (mean = 4.1), taste compared to rice/wheat products (mean = 4.1), product availability (mean = 4.0), influence of friends and family (mean = 4.0), and brand trust (mean = 4.0) were also rated highly, indicating that both social and market-driven factors play an important role in encouraging adoption. Convenience-related aspects such as ease of preparation (mean = 3.9) and variety of flavors (mean = 3.6) were moderately important, reflecting consumer preference for products that fit modern busy lifestyles. Factors such as packaging appeal (mean = 3.7), health claims (mean = 3.7), and value for money (mean = 3.6) were considered relevant but not as dominant as cultural and price influences.

Table 10.3 Descriptive statistics – factors influencing purchase.

	N	Min	Max	Mean	Std. dev.
FIP – Packaging appeal	255	1.00	5.00	3.7	1.3
FIP – Price perception	255	2.00	5.00	4.3	0.9
FIP – Health claims/nutritional information	255	1.00	5.00	3.7	1.3
FIP – Promotional activities	255	1.00	5.00	4.1	1.1
FIP – Product availability in stores	255	1.00	5.00	4.0	1.1
FIP – Variety of flavors/product types	255	1.00	5.00	3.6	1.2
FIP – Influence of friends/family	255	1.00	5.00	4.0	1.2
FIP – Shelf placement/visibility	255	1.00	5.00	3.2	1.2
FIP – Ease of preparation/cooking convenience	255	1.00	5.00	3.9	1.2
FIP – Taste and texture satisfaction	255	1.00	5.00	3.6	1.2
FIP – Cultural or traditional food habits	255	1.00	5.00	4.4	0.8
FIP – Availability of cooking info	255	1.00	5.00	3.5	1.3
FIP – Endorsements by celebrities/influencers	255	1.00	5.00	3.5	1.3
FIP – Eco-friendly or sustainable sourcing	255	1.00	5.00	3.3	1.3
FIP – Value for money	255	1.00	5.00	3.6	1.4
FIP – Availability in preferred shopping channel	255	1.00	5.00	3.2	1.4
FIP – Previous positive experience with product	255	1.00	5.00	3.6	1.2
FIP – Brand trust	255	1.00	5.00	4.0	1.0

Table 10.4 Descriptive statistics – awareness and perception.

	N	Min	Max	Mean	Std. dev.
Awareness of the nutritional benefits of millets	255	1.00	5.00	4.2	0.8
Taste compared to rice/wheat products	255	1.00	5.00	4.1	0.8
Millet products are worth the price	255	1.00	5.00	2.9	1.5
Barriers prevent more millet products	255	1.00	7.00	4.1	2.1

Meanwhile, shelf placement (mean = 3.2), shopping channel availability (mean = 3.2), and eco-friendly sourcing (mean = 3.3) received lower importance ratings, suggesting that visibility and sustainability are secondary concerns compared to immediate consumption drivers. Overall, the findings highlight that consumers' millet purchase decisions are shaped by a combination of traditional values, affordability, promotional visibility, and trust, with convenience and health benefits acting as supportive motivators rather than primary triggers (Table 10.4).

The data indicate that respondents generally have a high awareness of the nutritional benefits of millets, with a mean score of 4.2 out of 5, and perceive the taste of millet products to be slightly better than rice or wheat, with a mean of 4.1. However, the perception of price worthiness is relatively low, averaging 2.9, suggesting that many consumers feel millet products are somewhat expensive. Barriers to increased consumption are notable, with a mean score of 4.1 on a 7-point scale, highlighting challenges such as limited availability, higher price, taste preferences, and preparation difficulties as key factors limiting wider adoption. Overall, while awareness and taste perception are favorable, cost and practical barriers remain significant obstacles to greater millet consumption (Table 10.5).

Based on the survey data from 255 respondents, the average intention to increase millet consumption over the next six months is moderate, with a mean score of 3.2 on a 5-point scale, indicating a slightly positive inclination toward adopting millet in their diet. Similarly, the mean score of 3.3 for following a special diet suggests that respondents are moderately conscious of dietary restrictions or preferences, with some diversity in dietary habits. Environmental sustainability in food choices scored slightly lower, with a mean of 3.1, implying that while sustainability is a consideration for many, it is not a dominant factor influencing their food decisions. Overall, these results suggest a general awareness and moderate interest in health-conscious and environmentally mindful food practices among

Table 10.5 Descriptive statistics – future intentions and lifestyle factors.

	N	Min	Max	Mean	Std. dev.
Increase millet consumption, future intention	255	1.0	5.0	3.2	1.6
Follow any special diet	255	1.00	5.00	3.3	1.5
Env. sustainability in food choices, concerning	255	1.00	5.00	3.1	1.5

the surveyed group, though there remains room to strengthen motivation toward sustainable and millet-based dietary choices.

10.9 Regression Analysis

To test the hypotheses (H1–H9), a multiple regression analysis was conducted to examine the impact of selected factors (health claims, cultural habits, brand trust, packaging appeal, price perception, promotional activities, and taste and texture satisfaction) on consumer motivation to buy millet products.

10.9.1 Hypotheses in Equation Form

See Table 10.6.

$$Y = \beta 0 + \beta 1(\text{Health}) + \beta 2(\text{Culture}) + \beta 3(\text{Brand}) + \beta 4(\text{Packaging}) + \beta 5(\text{Price}) + \beta 6(\text{Health Claims}) + \beta 7(\text{Promotions}) + \beta 8(\text{Taste})$$

The predictors explain **37.5% of the variance** in consumer motivation – a **moderate model fit** (Table 10.7).

The overall model was statistically significant, $F\ (7, 247) = 24.50$, $p < 0.001$, with an R^2 value of 0.375, indicating that approximately 37.5% of the variance in consumer motivation is explained by these predictors. This suggests that the model has moderate explanatory power in understanding consumer behavior toward millet products (Table 10.8).

Significant Predictors

- **Health claims/nutritional information** had a significant positive influence on consumer motivation ($\beta = 0.162$, $p = 0.002$). This indicates that consumers are more likely to buy millet products when they are convinced of the nutritional and health benefits.

Table 10.6 Model summary for factors affecting consumer motivation to purchase millet products.

Model	R	R square	Adjusted R square	Std. error of the estimate	Durbin–Watson
1	0.612	0.375	0.358	1.215	1.95

Table 10.7 ANOVA results for multiple regression model on consumer motivation to purchase millet products.

Model	Sum of squares	df	Mean square	F	Sig.
Regression	253.22	7	36.174	24.5	0
Residual	421.305	247	1.705		
Total	674.525	254			

Table 10.8 Regression coefficients for factors influencing consumer motivation to purchase millet products.

	Unstandardized coefficient		Standardized coefficient		
Predictor variable (constant)	**B**	**Std. error**	**Beta**	**t-value**	**Sig. (p-value)**
	1.832	0.412	—	4.446	0
Health claims/nutritional info	0.185	0.06	0.162	3.083	0.002
Cultural/traditional food habits	0.142	0.066	0.118	2.152	0.032
Brand trust	0.075	0.061	0.068	1.23	0.22
Packaging appeal	0.094	0.058	0.086	1.62	0.107
Price perception	−0.172	0.065	−0.150	−2.646	0.009
Promotional activities	0.128	0.062	0.11	2.064	0.04
Taste and texture satisfaction	0.201	0.059	0.177	3.407	0.001

- **Cultural and traditional food habits** were also positively significant ($\beta = 0.118$, $p = 0.032$), highlighting the importance of heritage-based dietary preferences in millet consumption.
- **Promotional activities** emerged as another significant factor ($\beta = 0.110$, $p = 0.040$), suggesting that advertisements, awareness programs, and marketing campaigns play a role in motivating consumers.
- **Taste and texture satisfaction** had the strongest positive effect ($\beta = 0.177$, $p = 0.001$), implying that repeat purchases are strongly driven by sensory acceptance of millet products.
- **Price perception** showed a significant negative effect ($\beta = -0.150$, $p = 0.009$), indicating that higher sensitivity to price reduces consumer motivation to adopt millet products.

Nonsignificant Predictors

- **Brand trust ($\beta = 0.068$, $p = 0.220$)** and **packaging appeal ($\beta = 0.086$, $p = 0.107$)** did not significantly influence consumer motivation. Although consumers notice brand reputation and packaging design, these factors do not appear to be decisive in their purchase decisions for millet products.

10.9.2 Regression Analysis Graph

The regression analysis revealed mixed outcomes in terms of the factors influencing consumer motivation to purchase millet products (Figure 10.1). Among the tested predictors, health claims, cultural food habits, promotional activities, and taste and texture satisfaction emerged as significant positive drivers, while price perception showed a significant negative effect. In contrast, brand trust and packaging appeal did not significantly influence consumer motivation.

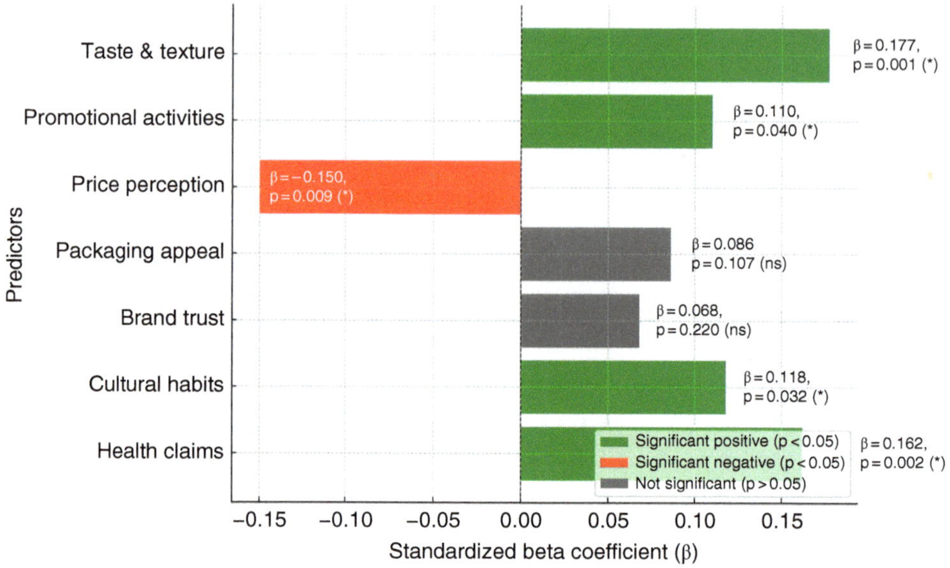

Figure 10.1 Regression analysis graph.

This outcome indicates a clear research gap: although previous studies have emphasized branding and packaging as strong determinants in food marketing, in the context of millet products, these external marketing elements appear to play a weaker role. Instead, intrinsic product qualities (health benefits, taste, cultural relevance) and supportive promotional campaigns are more critical in driving purchase behavior [22].

The nonsignificance of brand trust and packaging appeal may be explained by the current stage of the millet market development. Since millet products are still relatively new in mainstream urban and semi-urban markets, consumers are likely to prioritize value, affordability, and health outcomes over branding aesthetics. Moreover, limited awareness and the absence of strong, established millet brands could weaken the role of packaging and brand image as purchasing determinants [23].

10.10 t-Test

To examine whether consumer perceptions and behaviors toward millet products vary across different residential areas, an independent samples t-test was performed comparing urban and semi-urban respondents. The analysis considered key dependent variables such as motivation to buy millet products, purchase frequency, awareness of nutritional benefits, taste perception, and overall factors influencing purchase. Group statistics provided the mean differences, while the t-test assessed whether these differences were statistically significant. Effect sizes were further calculated to determine the strength of these differences. The results are presented in Table 10.9.

Table 10.9 Group statistics comparing consumer perceptions and behaviour across urban and semi-urban respondents.

Variable	Area of residence	N	Mean	Std. deviation	Std. error mean
Motivates	Urban	132	3.4	1.55	0.13
	Semi-urban	123	3.95	1.6	0.14
Often, buy millet products	Urban	132	2.95	1.33	0.12
	Semi-urban	123	3.05	1.3	0.12
Awareness	Urban	132	4.35	0.75	0.07
	Semi-urban	123	4.05	0.85	0.08
Taste of millet products	Urban	132	4.1	0.78	0.07
	Semi-urban	123	4.05	0.81	0.07

10.10.1 Group Statistics

10.10.2 Independent Samples Test

See Table 10.10.

10.10.3 Interpretation

The independent samples t-test was conducted to examine differences in influence-related factors between **urban and semi-urban consumers**. The results revealed a **significant difference in two areas: motivation to buy millet products and awareness of nutritional benefits.**

Semi-urban consumers (M = 3.95, SD = 1.60) reported significantly **higher motivation** to influence millet products compared to urban consumers (M = 3.40, SD = 1.55), t(253) = −2.45, p = 0.015, Cohen's d = −0.35, indicating a small to moderate effect size. This suggests that **semi-urban consumers may be more driven by cultural, curiosity, or accessibility factors when choosing millet products.**

In contrast, urban consumers (M = 4.35, SD = 0.75) exhibited **greater awareness of the nutritional benefits** of millets than semi-urban consumers (M = 4.05, SD = 0.85), t(253) = 2.25, p = 0.026, Cohen's d = 0.30. This demonstrates that **urban consumers**

Table 10.10 Independent samples t-test comparing urban and semi-urban consumers on key millet product variables.

Variable	Levene's test (F)	Sig.	t	df	Sig. (2-tailed)
Motivates	0.08	0.78	−2.45	253	**0.015**
Often, buy millet products	2.37	0.12	−0.48	253	0.63
Awareness	1.02	0.31	2.25	253	**0.026**
Taste of millet products	0.09	0.77	0.4	253	0.69

are more informed about health aspects, possibly due to higher exposure to nutrition campaigns and marketing.

However, no significant differences were observed between urban and semi-urban respondents in **purchase frequency ($p = 0.63$) or taste perception ($p = 0.69$).** This indicates that, regardless of location, consumers generally purchase millet products with similar frequency, perceive their taste similarly, and consider purchase factors (such as brand trust, packaging, and price) to be equally important.

"There is a significant difference between urban and semi-urban consumers regarding the factors influencing millet product adoption" is partially supported. The difference lies mainly in **motivation (higher in semi-urban)** and **nutritional awareness (higher in urban)**. For other variables (frequency, taste), **no significant differences** exist.

10.11 One-way – ANOVA

The hypothesis that socio-demographic factors significantly affect consumer preferences for millet product varieties was tested using one-way ANOVA. Specifically, the analysis assessed the effects of age, occupation, income, and education on preference scores. The findings revealed that while most demographic factors (age, income, and education) did not significantly influence consumer preferences, occupation emerged as a significant predictor, with students reporting lower preference scores than working professionals and homemakers.

10.11.1 Age Group of Consumer Preference

Hypothesis: There is no significant difference in consumer preference for millet product varieties across different age groups.

As shown in Table 10.11, preference scores were similar across all age groups (M = 3.73–3.75). Levene's test indicated that homogeneity of variances was met ($p = 0.062$). The one-way ANOVA revealed no significant effect of age on consumer

Table 10.11 One-way ANOVA results for consumer preference across different age groups.

Age group	N	Mean	SD	95% CI for mean	Levene's test	F (df, df error)	p-value	Post hoc (Tukey HSD)
20–29 years	71	3.73	0.47	[3.62, 3.84]				No significant difference
30–39 years	101	3.73	0.47	[3.63, 3.82]	**Levene's = 2.81**	**F(2, 252) = 0.068**	**p = 0.934**	(all p > 0.90)
40–50 years	83	3.75	0.58	[3.63, 3.88]	**p = 0.062** (ns)			
Total	255	3.74	0.5	[3.67, 3.80]				

preference, $F(2, 252) = 0.068$, $p = 0.934$. Post hoc Tukey HSD tests confirmed the absence of pairwise group differences (all $p > 0.90$).

10.11.2 Occupations of Consumer Preference

Hypothesis: There is **no significant difference** in consumer preferences (PV) for millet product varieties across different **occupations** (Table 10.12).

A one-way ANOVA was conducted to examine the influence of occupation on consumer preferences for millet product varieties (PV). Descriptive results indicated some differences in mean scores across occupations, ranging from 3.70 (students) to 3.79 (working professionals).

Levene's test confirmed that the assumption of homogeneity of variances was met ($p = 0.214$). The ANOVA revealed a **statistically significant effect** of occupation on PV, $F(3, 251) = 5.32$, $p = 0.002$. Post hoc Tukey HSD tests indicated that **working professionals and homemakers reported significantly higher preferences** for millet products compared to students ($p < 0.05$).

These findings suggest that consumer occupation **significantly influences preferences** for millet product varieties, with students showing relatively lower preference scores within the sampled population.

10.11.3 Monthly Household Income of Consumer Preference

Hypothesis: There is no significant difference in consumer preference for millet product varieties across different monthly household income groups.

As shown in Table 10.13, consumer preference scores were fairly similar across all income groups ($M = 3.64–3.84$). Levene's test indicated that the assumption of homogeneity of variances was met, $\mathbf{p = 0.373}$. The one-way ANOVA revealed **no significant effect** of monthly household income on consumer preference, $F(3, 251) = 2.308$, $p = 0.077$. Post hoc Tukey HSD tests confirmed the absence of pairwise differences between groups (all $p > 0.12$).

10.11.4 Educational Qualification of Consumer Preference

Hypothesis: There is no significant difference in consumer preference for millet product varieties across different educational qualification groups.

As shown in Table 10.14, consumer preference scores were fairly similar across all educational qualification groups ($M = 3.66–3.80$). Levene's test indicated that the assumption of homogeneity of variances was met, $\mathbf{p = 0.845}$. The one-way ANOVA revealed **no significant effect** of educational qualification on consumer preference, $F(3, 251) = 1.183$, $p = 0.317$. Post hoc Tukey HSD tests confirmed the absence of pairwise differences between groups (all $p > 0.38$).

Table 10.12 One-way ANOVA results for consumer preference across different occupational groups.

Occupation	N	Mean	SD	95% CI for mean	Levene's test	F (df, df error)	p-value	Post hoc (Tukey HSD)
Working professional	70	3.79	0.43	[3.69, 3.89]				Significantly higher than Students (p = 0.01)
Homemaker	56	3.74	0.53	[3.59, 3.88]	Levene's = 1.50	F(3, 251) = 5.32	p = 0.002	Significantly higher than Students (p = 0.03)
Student	71	3.7	0.56	[3.57, 3.83]	p = 0.214 (ns)			Significantly lower than Working professionals and homemakers
Other	58	3.71	0.5	[3.58, 3.84]				—
Total	255	3.74	0.5	[3.67, 3.80]				

Table 10.13 One-way ANOVA results for consumer preference across monthly household income group.

Monthly household income	N	Mean	SD	95% CI for mean	Levene's test	F (df, df error)	p-value	Post hoc (Tukey HSD)
Below ₹25,000	62	3.84	0.45	[3.72, 3.95]				No significant difference
₹25,001–₹50,000	69	3.64	0.53	[3.52, 3.77]	Levene's = 1.045	F(3, 251) = 2.308	p = 0.077	(all p > 0.12)
₹50,001–₹75,000	54	3.66	0.57	[3.51, 3.82]	p = 0.373 (ns)			
Above ₹75,000	70	3.79	0.46	[3.69, 3.90]				
Total	255	3.74	0.5	[3.67, 3.80]				

Table 10.14 One-way ANOVA results for consumer preference across different educational qualification levels.

Educational qualification	N	Mean	SD	95% CI for mean	Levene's test	F (df, df error)	p-value	Post hoc (Tukey HSD)
No formal education	61	3.8	0.51	[3.66, 3.93]				No significant difference
School level	54	3.71	0.49	[3.58, 3.84]	Levene's = 0.273	F(3, 251) = 1.183	p = 0.317	(all p > 0.38)
Undergraduate	72	3.66	0.47	[3.54, 3.77]	p = 0.845 (ns)			
Postgraduate and above	68	3.79	0.54	[3.66, 3.92]				
Total	255	3.74	0.5	[3.67, 3.80]				

10.12 Conclusion and Discussions

The central problem that guided this research was the paradox that, despite well-documented nutritional benefits and strong government promotion, millet products remain under-consumed in urban and semi-urban markets. Consumers still perceive them as expensive, less tasty than rice or wheat, and unfamiliar with modern food habits. Against this backdrop, the study set out with four core objectives: (i) to identify the key factors shaping consumer preferences; (ii) to assess the influence of branding, packaging, price, health claims, and promotions; (iii) to evaluate awareness and taste perceptions; and (iv) to determine whether urban–semi-urban differences and socio-demographic factors affect adoption.

Using a structured survey of 255 consumers in Coimbatore, the study employed descriptive statistics, regression, t-tests, and ANOVA to empirically test 11 hypotheses. The findings shed light on both behavioral and market dynamics:

1) Health Orientation and Cultural Anchoring
The results confirm that health consciousness (H1) and awareness of nutritional benefits (H9) significantly drive millet consumption. This reinforces the idea that in health-conscious markets, functional foods succeed when their benefits are clearly communicated. Equally important, cultural and traditional food habits (H2) were found to shape consumer adoption, particularly in semi-urban areas where millet consumption is tied to heritage diets. This dual influence of modern health aspirations and traditional food culture suggests that millet products occupy a unique position as both a "future food" and a "heritage food."
2) Sensory Acceptance and Repeat Consumption
Among all predictors, taste and texture satisfaction (H8) showed the strongest positive influence on motivation, underlining that nutritional benefits alone are not enough. For wider adoption, millet products must compete with mainstream cereals on sensory

grounds. This insight emphasizes the importance of product development and culinary innovation, ranging from ready-to-eat formats to flavored varieties that reduce perceived taste barriers.

3) Economic Sensitivity and Value Perceptions
Price perception (H5) had a significant negative effect, confirming that affordability is a major obstacle. Despite awareness of health benefits, many consumers view millet products as poor value compared to rice or wheat. This signals a need for policy support (e.g., subsidies, price stabilization) and business strategies (e.g., smaller pack sizes, value bundles) to reposition millets as accessible staples rather than premium niche products.
4) Role of Marketing and Promotion
Promotional activities (H7) and health claims (H6) significantly influenced adoption, proving that external stimuli can shift consumer behavior when well designed. However, contrary to much of the literature, brand trust (H3) and packaging appeal (H4) were statistically insignificant. This suggests that millet products currently lack strong, differentiated brands and that in early-stage markets, consumers prioritize intrinsic benefits (health, culture, taste) over extrinsic branding elements. This finding highlights a critical gap: future growth requires building credible millet brands supported by engaging, informative packaging.
5) Urban vs. Semi-Urban and Socio-Demographic Differences
Hypothesis H10 was partially supported: semi-urban consumers showed stronger motivation to adopt, while urban consumers demonstrated greater nutritional awareness. However, no differences were found in purchase frequency or taste perception, suggesting that millet adoption is shaped more by shared challenges (price, awareness, taste) than by geography. Among socio-demographic variables (H11), occupation significantly influenced preferences, with students showing weaker adoption than working professionals and homemakers, pointing to lifestyle and income-linked consumption differences.

Overall, the regression model explained 37.5% of the variance in consumer motivation, indicating that while health, cultural, promotional, and sensory factors are crucial, other behavioral dimensions such as lifestyle aspirations, peer influence, and environmental consciousness could further explain millet adoption in future research.

In conclusion, this study demonstrates that millet consumption in urban and semi-urban markets is driven less by branding and visibility and more by a blend of intrinsic product qualities (health, taste, cultural relevance), economic value perceptions, and promotional outreach. The findings confirm some hypotheses (H1, H2, H5, H6, H7, H8, H9) while rejecting others (H3, H4), revealing that the millet market is still in a formative stage where consumer decisions hinge on functionality and familiarity rather than on established brand identities. This offers both a challenge and an opportunity: to transform millet products from a niche health trend into a mainstream dietary choice, industry stakeholders must simultaneously tackle affordability, sensory appeal, and awareness while gradually building strong brands and modern consumption experiences.

10.13 Recommendations

The findings of this study highlight both opportunities and challenges for strengthening millet consumption in urban and semi-urban markets. To address consumer concerns and enhance adoption, the following recommendations are proposed for manufacturers, marketers, policymakers, and consumer awareness groups:

1) Enhancing Product Development and Sensory Appeal
   - Focus on improving the taste, texture, and convenience of millet-based products through product innovation (e.g., flavored snacks, instant mixes, millet-based ready-to-eat meals).
   - Invest in culinary research and recipe development to bridge the gap between traditional millet dishes and modern consumer expectations.
   - Introduce millet products in formats familiar to consumers (bread, biscuits, noodles, breakfast cereals) to reduce perceived taste barriers.
2) Pricing and Affordability Strategies
   - Develop affordable pack sizes (e.g., single-use or small packs) to make millet products accessible across income groups.
   - Introduce value-for-money bundles combining millets with complementary staples.
   - Collaborate with policymakers for price subsidies, tax relief, or inclusion in public distribution systems (PDS) to reduce cost sensitivity.
3) Strengthening Marketing and Promotion
   Highlight nutritional and health benefits clearly through packaging, advertisements, and campaigns, since health claims significantly influenced adoption
   - Increase awareness campaigns through schools, workplaces, and community programs to expand consumer knowledge.
   - Promote millets through sampling programs, cooking demonstrations, and food festivals, allowing consumers to experience taste and ease of preparation.
   - Leverage digital marketing and influencer endorsements, targeting youth and health-conscious consumers.
4) Building Stronger Brand Identity and Packaging Appeal
   - Although brand trust and packaging were not significant in this study, their long-term importance cannot be ignored. Companies should gradually:
     - Develop distinct millet brands with strong narratives around health, heritage, and sustainability.
     - Use eco-friendly, attractive, and informative packaging to appeal to modern consumers and align with sustainability trends.
5) Consumer Education and Cultural Positioning
   - Position millets not only as "health foods" but also as heritage foods, reconnecting consumers with traditional diets.
   - Conduct nutritional awareness workshops and campaigns in collaboration with health professionals, nutritionists, and fitness communities.
   - Provide easy cooking guides, recipes, and QR-code-based tutorials on packaging to help consumers overcome preparation difficulties.

6) Targeted Strategies for Different Consumer Groups
   - **Students:** Promote millet snacks and ready-to-eat formats that suit fast lifestyles, supported by affordable pricing.
   - **Working Professionals:** Highlight health and convenience benefits with on-the-go millet meals.
   - **Homemakers:** Promote millet as a nutritious family staple with cooking demonstrations and household-friendly packs.

7) Policy and Institutional Support
   - Encourage government inclusion of millet products in school meal programs, corporate canteens, and hospitals.
   - Strengthen supply chains to ensure consistent availability in supermarkets, local stores, and online platforms.
   - Align millet promotion with sustainability campaigns, highlighting their lower water footprint and climate resilience.

References

1 Saleh, A.S., Zhang, Q., Chen, J., and Shen, Q. (2013). Millet grains: nutritional quality, processing, and potential health benefits. *Comprehensive Reviews in Food Science and Food Safety* 12 (3): 281–295.

2 Sheoran, B., Tiwari, V., Meenu, M. et al. (2024). Utilization of Millet Varieties in Food and Nutritional Security. In: Sustainable Utilization and Conservation of Plant Genetic Diversity. Sustainable Development and Biodiversity, vol. 35 (ed. J.M. Al-Khayri, S.M. Jain, and S. Penna). Singapore: Springer https://doi.org/10.1007/978-981-99-5245-8_7.

3 Li, G., Pan, J., Cui, K. et al. (2017). Limitation of unloading in the developing grains is a possible cause responsible for low stem non-structural carbohydrate translocation and poor grain yield formation in rice through verification of recombinant inbred lines. *Frontiers in Plant Science* 8: 1369. https://doi.org/10.3389/fpls.2017.01369.

4 Porwal, A., Bhagwat, G., Sawarkar, J. et al. (2023). An overview of millets-the nutri-cereals: Its nutritional profile, potential health benefits and sustainable cultivation approach. *International Journal of Science and Research Archive* 10 (1): 841–859. https://doi.org/10.30574/ijsra.2023.10.1.0828.

5 Gahalawat, P., Lamba, N., and Chaudhary, P. (2024). Nutritional and health benefits of millets: A review article. *Journal of Indian System of Medicine* 12 (1): 4–11. https://doi.org/10.4103/jism.jism_71_23.

6 Saleh, M.M., Zhang, A.S.M., Chen, Q., and Shen, Q. (2013). Millet grains: nutritional quality, processing, and potential health benefits. *Comprehensive Reviews in Food Science and Food Safety* 12 (3): 281–295. https://doi.org/10.1111/1541-4337.12012.

7 Food and Agriculture Organization (FAO) (n.d.). The Role of Millets in Food Security and Nutrition. FAO Publications https://www.fao.org/millets-2023/en/.

8 Sanodiya, L.K., Sanodiya, P.K., Prajapati, A. et al. (2023). The sustainability of millets and the advantages they bring for human health. *Trends in Agriculture Science* 2 (4): 255–261. https://doi.org/10.5281/zenodo.7947530.

9 Kumar, A., Tomer, V., Kaur, A. et al. (2018). Millets: A solution to agrarian and nutritional challenges. *Agriculture & Food Security* 7 (31): 1–15. https://doi.org/10.1186/s40066-018-0183-3.

10 Siroha, A.K. and Bangar, S.P. (2024). Millet-based food products: an overview. *Current Food Science and Technology Reports* https://doi.org/10.1007/s43555-024-00031-x.

11 Gireesh, S., Dinesh, G.K., Adityan, S. et al. (2025). Assessing India's millet promotion policies: schemes and initiatives of the Government of India to promote millets for food and nutritional security through a systematic approach. *AGRIVITA Journal of Agricultural Science* 47 (3): 616–629. https://doi.org/10.17503/agrivita.v47i3.4332.

12 Sindhu, V. and T. T., S. (2025). A study on consumer awareness and perception of street food usage by youngsters with reference to Coimbatore city. *International Journal of Research Publication and Reviews* 6 (4): 7747–7750.

13 Rao, B.D., Bhaskarachary, K., Christina, G.D.A. et al. (2017). Nutritional and Health Benefits of Millets. ICAR–Indian Institute of Millets Research (IIMR).

14 Lokesh, K., Dudhagara, C.R., Mahera, A.B. et al. (2022). Millets: the future smart food. *The Pharma Innovation Journal* 11 (4): 75–84.

15 Sindhu, V.S. and Siva, T.T. (2025). A study on consumer awareness and perception of street food usage by youngsters with reference to Coimbatore city. *International Journal of Research Publication and Reviews* 6 (4): 7747–7750.

16 Kaur, M. and Banga, G. (2025). Consumer awareness and behavioural trends in the adoption of millets and millet-based products. *Archives of Current Research International* 25 (3): 135–144. https://doi.org/10.9734/acri/2025/v25i31103.

17 Ajzen, I. (1991). The theory of planned behavior. *Organizational Behavior and Human Decision Processes* 50 (2): 179–211.

18 Beemkumar, N., Gupta, S., Bhardwaj, S. et al. (2023). Activity recognition and IoT-based analysis using time series and CNN. In: Handbook of Research on Machine Learning-enabled IoT for Smart Applications Across Industries (ed. N. Goel and R.K. Yadav), 350–364. IGI Global https://doi.org/10.4018/978-1-6684-8785-3.ch018.

19 Talukdar, V., Veeraiah, V., Roy, D.R. et al. (2023). An innovative approach: typicality-based collaborative filtering for recommender systems. *Proceedings of the 3rd International Conference on Advancement in Electronics & Communication Engineering (AECE)*, pp. 863–868. https://doi.org/10.1109/AECE59614.2023.10428576

20 Pandey, J.K., Yadav, A.K., Purushotham, M. et al. (2022). Colorimetric detection of deltamethrin pesticide from biological samples based on the peroxidase-mimic catalytic activity on MnO_2/SnO_2. *Key Engineering Materials* 928: 45–51. https://doi.org/10.4028/p-4ti1i0.

21 Pandey, J.K., Kumar, S., Lamin, M. et al. (2022). A metaheuristic autoencoder deep learning model for an intrusion detector system. *Mathematical Problems in Engineering* 2022: 11, 3859155. https://doi.org/10.1155/2022/3859155.

22 Pandey, J.K. (2025). The impact of electromagnetic and optical radiation on physical and biological systems. *Journal of Nano- and Electronic Physics* 17 (5): 05023. https://doi.org/10.21272/jnep.17(5).05023.

23 Srinivas, P., Veeraiah, V., Thejaswini, K.O. et al. (2025). The impact of electromagnetic and optical radiation on physical and biological systems. *Journal of Nano- and Electronic Physics* 17 (4): 04012. https://doi.org/10.21272/jnep.17(4).04012.

11

Vision-Based Quality Grading and Sorting in Food Processing Units

Hieu M. Tran[1], Tuan M. Le[1], Hung Son Nguyen[2], and Son V. T. Dao[1]

[1] *School of Science, Engineering and Technology, RMIT University Vietnam, Ho Chi Minh City, Vietnam*
[2] *Faculty of Mathematics, Informatics and Mechanics, University of Warsaw, Warsaw, Poland*

11.1 Introduction

Avocado (*Persea americana* Mill.) is a fruit that is of enormous economic and nutritional importance, which has been used in tropical and subtropical areas extensively. It is a highly lipid-rich, vitamin-rich, mineral-rich, and bioactive compound that is significant in human diets and the food industry of the world [1]. The world avocado production has been on a steady rise over the last few decades, with the world leaders in avocado production being Mexico, Colombia, Peru, the Dominican Republic, Indonesia, and a range of Asian countries [2, 3].

Conventionally, the process of sorting and grading of avocados is conducted manually after harvesting. Employees use visual determination of size, shape, and color or weighing scales and calipers [4]. These approaches are cheap and rudimentary, but are labor-intensive, time-consuming, and subject to human error, particularly when working with large volumes. Besides, manual sorting may not be able to guarantee the throughput and the consistency required by the modern supply chains [5].

Correct estimation of the mass of fruits and physical traits is highly essential in promoting efficiency, yield, and quality at minimum cost in agricultural operations. The parameters that are of great significance to grading, packaging, and quality control include the size, form, surface area, and density. The increase in demand for food and the complexity of the supply chain in the world have made precision agriculture and the automation of the supply chain a necessity.

Seeking to overcome the faults of traditional methods, the use of modern technologies has been gradually implemented in the shape of computer vision, image analysis, and machine learning. These techniques allow objective, high-throughput, and nondestructive analysis of fruit properties, enhancing the accuracy of the postharvest procedures as well as their magnitude [6–8]. Through these tools, manufacturers can use the resources more effectively, meet the quality needs, and introduce automation to the agricultural activities.

11.2 Related Works

11.2.1 Advances in Technology for Fruit Characterization

In recent decades, technological advances have greatly improved the ability to analyze fruit's physical characteristics, such as size, shape, volume, and mass. Early approaches focused on using geometric models combined with basic measurements. The initial methods were concerned with the application of geometric models in relation to simple measurements. In one instance, Lorestani and Tabatabaeefar [9] utilized simple length, width, and height traits to estimate the volume and mass of kiwifruit with great precision (R^2 as high as 97%). In 2009, Emadi et al. [10] examined the mechanical characteristics of melons, combining compression and shear stress values to categorize varieties and assist in automatic grading ($p > 0.05$, all varieties had similar statistical classification).

With the invention of computer vision, scholars started using image processing to obtain features on fruit images without destroying them. Liming and Yanchao [11] suggested a way of modeling the surface area and symmetry of strawberries using 2D image information, which has over 90% classification accuracy and little size detection error. Similarly, Clayton et al. [12] in 1995 estimated the surface area of apples to predict ripening behavior and shelf life, demonstrating how geometric surface measurements could support inventory and quality management.

This has been replaced by recent work to use machine learning and deep learning to enhance accuracy and efficiency. Oyefeso et al. [13] constructed mathematical models that forecast the mass and volume of sweet and Irish potatoes based on the areas projected in 2018, with a coefficient of determination of 98.1% and 96.8%, respectively. In 2021, Sripaurya et al. [14] used a portable six-channel near-infrared system to measure the size, shape, and ripeness of bananas, and the prediction errors were very low. Not only does their system accurately estimate size and shape, but it also provides a good classification of seven ripeness levels, with a standard error of the prediction of 0.17% and their standard error of the stages of color was 0.38%. Chen et al. [15] used YOLOv5 object detection with a ResNet34-based classifier to identify citrus fruits and identify their state of ripeness, achieving 95% accuracy, much higher than traditional models, including the VGG16 or k-NN.

Moreover, new systems that combine robotics and tactile sensing have also been studied. As an example, Qin et al. [16] created an Intelligent Flexible Manipulator System with Flexible Tactile Sensing (IFMSFTS) to measure kiwifruit firmness and ripeness when employing PCA [17] to reduce the dimensions and k-NN/SVM classifier with a classification rate of over 96%. These techniques show that vision-based features coupled with mechanical sensing may be used to develop more fruit characterization. While these methods are very accurate, many cannot be scaled up to large-scale laboratory experiments, and can be computationally intensive, which is why they are hard to implement in large-scale production lines. The need to fill such a gap suggests that alternatives to current methods must be accurate but also fast, simple, and affordable, particularly where precision agriculture is concerned (e.g., automated sorting and grading).

11.2.2 Vision-Based Methods of 2D Physical Characteristic Estimation

The studies by Tran et al. suggested two associated methods used to estimate the physical attributes of irregular fruit forms, namely starfruit (*Averrhoa carambola*). In the initial study [7], they came up with a two-camera system that would take the top and side fronts of each fruit. Basic geometric features were extracted using image processing techniques that included length, width, and thickness. The fruit body was virtually partitioned into conical frustums then the total volume of the fruit body was calculated as the sum of all the slice volumes. This, together with density data, made it possible to predict the mass accurately with mean accuracies of 99.16% and 98.59% in volume and mass, respectively, using 255 samples. Tran et al. also made the approach more practical in their second study [8], but only one top-view camera was used. Disc and frustum models were used to estimate the 3D volume by segmenting and slicing the fruit silhouette along its major axis. To map the estimated volume onto mass, a linear regression model was used to fit the model with a coefficient of determination of $R^2 = 92.05\%$ with 300 samples. This piece of work has shown that with very simple hardware, geometry-based modeling of 2D images can provide real-world applications with accurate and efficient estimates of fruit volume and mass.

On the same note, to estimate the mass, volume, and dimensions of sweet potatoes, Huynh et al. created a top-view vision system in 2022 [18] to measure them. They used simple geometric transformations to cut the object virtually along its longitudinal axis and added the volumes of the slices. The system was 96% accurate on volume and 95% accurate on weight, and had low computational power, which did not require the 3D reconstructions of intricate structures. A recent article by Sabouri et al. introduced a nondestructive model of predicting the fresh weight of plum fruits with image-based geometric characteristics and machine learning models [19]. They used a smartphone to take pictures of fruits, deconstruct the images into binary silhouettes, and calculated simple shape descriptors, like length and diameter. These were then put in several regression models that included Support Vector Regression (SVR), Multivariate Linear Regression, Multilayer Perceptron, or Decision Trees. The Pearson-VII kernel SVR had the best performance with R^2 of about 93.69% during training and 92.67% during testing; it is evident that it is highly predictive. The given work shows that simple 2D image features combined with machine learning can be used to predict fruit weight in an effective way; however, it needs model training and parameter optimization. Contrastingly, our research uses a purely geometric method (disc method) that does not involve learning-based elements and thus can be easily implemented and is more stable in its calculation, yet is just as accurate.

These studies highlight the practical potential of 2D vision-based approaches for physical characteristic estimation. When combined with basic image capture and geometrical modeling, these systems are able to provide fast, nondestructive, and low-cost measurements – this is necessary in real-time industrial environments such as automated sorting, grading, and packaging.

11.2.3 Implications and Research Objectives

Although there have been tremendous advancements in the characterization of fruits, a majority of the current studies are specific to certain types of fruits and cannot be easily generalized. Various methods are based on multi-view systems or computationally expensive 3D reconstructions, making them difficult to implement in high-throughput postharvest settings. At the same time, simpler models are not always as accurate and strong as they need to be in order to be applied in the real world. This disparity explains why a single, lightweight approach where the device could accommodate different shapes and sizes of fruits is required.

To meet this requirement, this chapter offers a vision-based model based on the disc method to determine the fruit mass from a single top-view image. The fruit silhouette is divided and practically cut into several thin discs that coincide along its major axis. The total volume of the silhouette is calculated as the sum of the disc volumes, and the volume is transformed into mass employing the mean density. Although the method is intended to be a broad solution, which suits most types of fruit, avocado (*Persea americana* Mill.) is the representative case because of its economic importance and peculiar shape. The rest of this chapter is laid out in the following way. Section 11.3 explains the materials and methods applied in this research, such as the sample of avocado, the process of the picture capture and preprocessing, and the disc method of the estimation of the volume and mass of the fruit. Then, Section 11.4 involves the discussion and presentation of the results of the experiment, which will help to assess the accuracy of the proposed approach and its processing efficiency. Lastly, Section 11.5 provides a summary of the main findings, states the practical implications of the work on the application of postharvest, and forms the possible future directions of the work.

11.3 Materials and Methods

This section describes the materials and methodological framework used in this research. The first section (Section 11.3.1) outlines the experimental design of acquiring images and the methods of physical measurements. Then, it presents the dataset of avocado samples and the preprocessing that is done to the images (Section 11.3.2). The disc method constitutes the main methodological element of this work, the description of which is provided in detail in terms of assumptions, mathematical model, and calculation steps (Section 11.3.3). Finally, the metrics of the evaluation and validation process are provided (Section 11.3.4) to develop a vivid framework for determining the accuracy and performance of the suggested approach.

The general structure of the disc method for estimating the dimensions and mass of avocados is provided in Figure 11.1. This is carried out by first collecting data where the top-view images of avocado samples are taken, and their physical measurements are recorded. During the image processing phase, the image fruit silhouette is extracted and is rotated in such a way that the longest axis of the fruit silhouette is oriented vertically. This is done by standardization to reduce geometrical distortions and provide the same dimensional measurements for all the samples.

Next, the geometry-based estimation is carried out using the disc method. The fruit is practically cut into slices along the major axis into several thin discs. Using a proportional relationship between the fruits' dimensions, the cross-sectional circularity of each disc is calculated, and the sum of the volumes of discs is calculated as the total volume. The average density of the fruit is then known, and the mass is estimated through the calculated volume.

Lastly, there is an evaluation and comparison stage, which consists of the evaluation and comparison of the methodology in terms of processing time, root mean square percentage error (RMSPE), mean absolute percentage error (MAPE), and the coefficient of

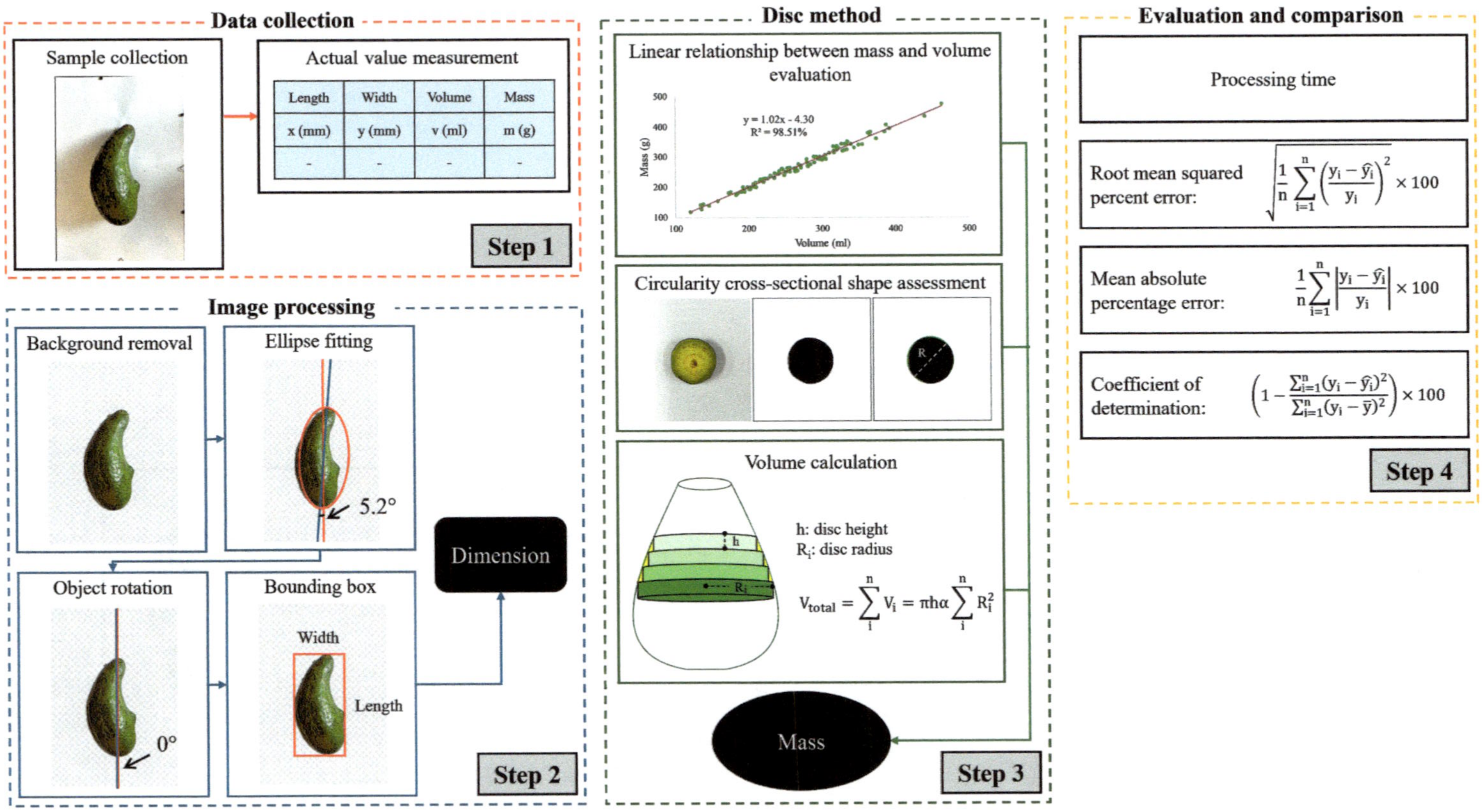

Figure 11.1 Study workflow.

determination, R^2. RMSPE is concerned with the occurrence of huge errors in calculation by squaring the residuals, and thus is sensitive to the presence of outliers, but MAPE gives an average proportion error, which is used to show the estimation bias. The advantage of such a combination of measures is that it not only serves as a measure of the individual prediction accuracy but also is a general measure of the model performance. This systematic construction makes the use of the data collection, geometric, and strict assessment to offer a high-speed, reliable, and nondestructive method of fruit-mass estimation.

11.3.1 Hardware Setup

Data collection was conducted using a special imaging system that provided regulated and controlled top-view images of the samples, as shown in Figure 11.2. This was a closed imaging chamber of internal size (297 mm × 210 mm × 400 mm) to minimize the effects of ambient light. To act as a background as well as a size reference, A4 paper (297 mm × 210 mm) was placed on the bottom of the chamber.

A Canon EOS 60D camera was mounted on the upper frame of the chamber at a constant distance of 400 mm from the bottom. The avocados were placed in the center of the A4 paper, and the top-view picture of each sample was captured. Overall, 101 samples of avocados were collected, and they consisted of 101 top-view photos.

In order to make sure that there is sufficient lighting and the shadow effect is minimized, LED strips (12 V DC, 60 LED/m, 7 W/m, 6500 K) of a total of approximately 300 lm were mounted on the inside part of the walls. This light system minimized the impacts of shadows and increased the quality of the image, hence facilitating adequate segmentation and measurement of the images in subsequent image processing.

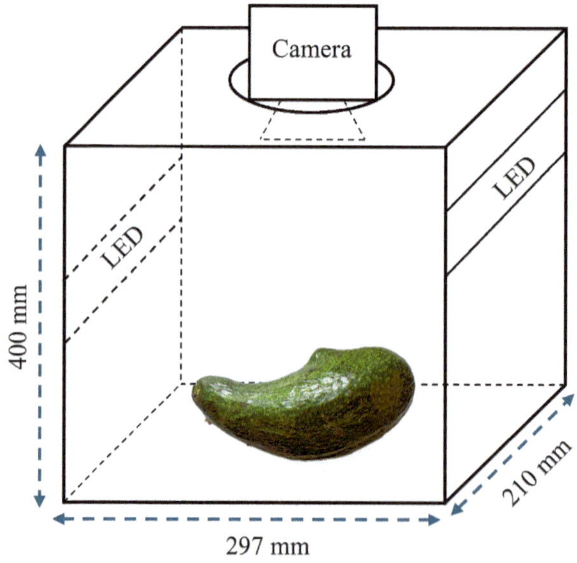

Figure 11.2 Hardware setup.

11.3.2 Dataset Overview

11.3.2.1 Dataset Collection

The sample used in this study was the long-shaped avocado type 034 that constitutes most of the avocados planted in the Central Highlands of Vietnam. The rationale of the selection of the variety lies in the economic significance and geometric length of this object, which is convenient in the process of testing the technique of estimating the volume of geometric objects. Collected samples naturally differed in shape and size, which was statistically summarized in Table 11.1.

A digital caliper was used to measure the length and width of each avocado, and a digital weighing scale was used to measure the mass of each avocado. In order to get the exact volume of each fruit, the water displacement technique was used because it is generally considered to be a good and precise technique for irregularly shaped objects [8]. Overall, 101 samples of the same variety of avocados were used in the dataset. The measured values were moderate and had a range of 120–463 ml of volume with a mass of 118–476 g. The range of lengths was between 127.16 and 268.22 mm, but the widths were 48.11–82.73 mm. Overall, length exhibited greater variability than width. This diverse set of physical characteristics provided a robust foundation for evaluating the effectiveness of the disc method in approximating the geometrical properties and estimating the mass of avocados.

11.3.2.2 Image Processing

Figure 11.3 illustrates the overall procedure used to preprocess and align the orientation of an avocado sample.

Figure 11.3a shows that in this method, we first acquire a raw top-view image of the avocado with controlled lighting and a clean background, and do not reduce the noise in any way at this step. The fruit is then coated with the U2-Net deep learning model [19] to detect the background and isolate the fruit in a binary mask, which is shown in Figure 11.3b. This is done to make sure that the region of interest is the only part that is left to be processed further. Figure 11.3c shows that the system estimates the longest axis of the fruit as the result of approximating the contour of the fruit using an ellipse [20, 21]. The Canny edge detection algorithm is used to take out the boundary points (x_i, y_i) and then the boundary points are fitted with an ellipse. We then apply the general equation of an ellipse as below in Eq. (11.1).

Table 11.1 Statistical properties of avocado samples.

	Actual			
Descriptive statistic	**Volume (ml)**	**Mass (g)**	**Length (mm)**	**Width (mm)**
Min	120	118	127.16	48.11
Max	463	476	268.22	82.73
Mean	267.61	268.85	188.12	67.86
Std. deviation	68.40	70.33	27.84	7.23

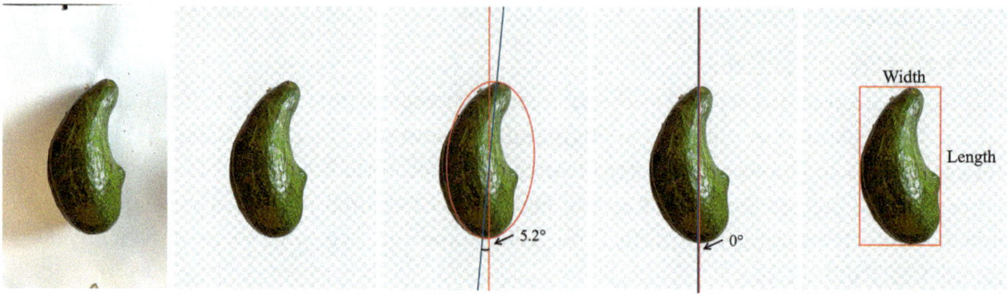

Figure 11.3 Procedures for each sample.

$$Ax^2 + Bxy + Cy^2 + Dx + Ey + F = 0 \tag{11.1}$$

where A, B, C, D, E, and F are the general conic parameters of the ellipse. The discriminant condition $B^2 - 4AC < 0$ should be satisfied to make sure that the shape is actually an ellipse. The estimation of these parameters is done using a least-squares fitting method, where all the extracted boundary points (x_i, y_i) are used to create the following design matrix:

$$\mathbf{D} = \begin{bmatrix} x_1^2 & x_1y_1 & y_1^2 & x_1 & y_1 & 1 \\ \vdots & \vdots & \vdots & \vdots & \vdots & \vdots \\ x_n^2 & x_ny_n & y_n^2 & x_n & y_n & 1 \end{bmatrix} \tag{11.2}$$

The parameter vector $\mathbf{p} = [A, B, C, D, E, F]^T$ is estimated by minimizing the residual errors between the observed boundary points and the predicted ellipse.

$$\mathbf{p} = \operatorname{argmin} \left\| \mathbf{D_p} - \mathbf{0} \right\| \tag{11.3}$$

where $\mathbf{D}$ denotes the design matrix constructed from the boundary coordinates, and $\mathbf{D_p}$ represents the predicted algebraic distances of those points to the fitted ellipse. The optimization will attempt to reduce the expression of the projected algebraic distances of the said points to the fitted ellipse. The goal of the optimization is to minimize $\left\| \mathbf{D_p} - \mathbf{0} \right\|$. An ideal fit would be the one that places all the points on the ellipse (residuals equal to zero). Once the best-fitting ellipse is obtained, its geometric properties can be derived. In particular, the semi-major axis a and semi-minor axis b are computed from the conic parameters using:

$$a = \sqrt{\frac{2\left(A + C + \sqrt{(A-C)^2 + B^2}\right)}{B^2 - 4AC}}, b = \sqrt{\frac{2\left(A + C - \sqrt{(A-C)^2 + B^2}\right)}{B^2 - 4AC}} \tag{11.4}$$

The center of the ellipse, (x_0, y_0), is determined by solving the linear system:

$$\begin{bmatrix} 2A & B \\ B & 2C \end{bmatrix} \begin{bmatrix} x_0 \\ y_0 \end{bmatrix} = \begin{bmatrix} -D \\ -E \end{bmatrix} \tag{11.5}$$

and the angle of the ellipse is computed by calculating the eigenvectors of the covariance matrix:

$$\mathbf{C} = \begin{bmatrix} A & \frac{B}{2} \\ \frac{B}{2} & C \end{bmatrix} \tag{11.6}$$

The direction of the major axis of the fruit is described by the eigenvector of the largest eigenvalue of the covariance matrix. The θ angle is approximately 5.20° (as represented in Figure 11.3c) calculated as:

$$\boldsymbol{\theta} = \arctan\left(\frac{v_y}{v_x}\right) \tag{11.7}$$

where (v_x, v_y) are the x- and y-components of that principal eigenvector. Lastly, the longest measure of the object that is equivalent to the major axis is 2a. The avocado is then turned around so that the major axis points upward at 0°, as indicated in Figure 11.3d, so that all the samples have the same orientation. This step of alignment is necessary so as to standardize the data to be able to measure the dimensions in a reliable way, for the geometry-based approach, as well as to have correctly formatted inputs to feature extraction and analysis in model-based ones. After the orientation, the length and width of every fruit are determined by the largest axis-aligned bounding box on the part of the segmented object on the blank A4-paper background, as presented in Figure 11.3e.

11.3.3 Geometry-based Approach

11.3.3.1 Linear Correlation Between Mass and Volume Assessment

Figure 11.4 represents the scatter plot of the mass and the volume of the avocado samples, and a linear regression line fitted with the equation:

$$y = 1.02 \times x - 4.30 \tag{11.8}$$

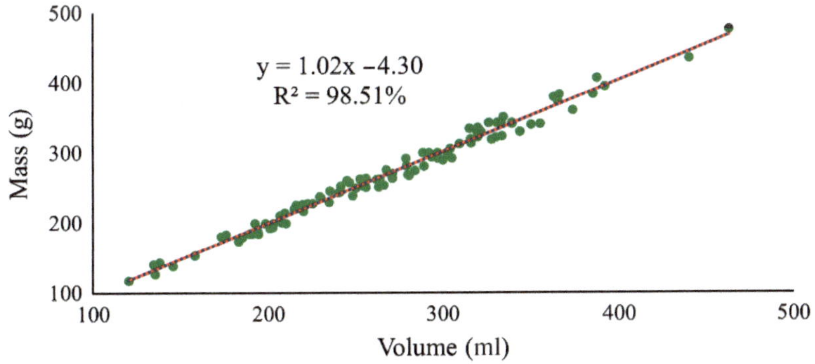

Figure 11.4 Density of avocado samples.

where y represents the fruit mass (g), and x represents the volume (mL). The coefficient of determination ($R^2 = 98.51\%$) shows a significant positive linear correlation, which proves the fact that volume is a highly valid predictor of mass.

11.3.3.2 Circularity Cross-Sectional Shape Assessment

To measure the circularity of the cross-sections of avocado, manually, 4–5 cross-sections were obtained of every fruit, and this led to the gathering of 450 cross-sections of 101 samples, as illustrated in Figure 11.5. In Figure 11.5a, a raw sliced section has been first captured, followed by binarization to extract the cross-sectional regions of interest. Then the fitEllipse [22] procedure described earlier was used on each part of the segmentation to derive its geometric parameter, in particular the center and major radius, which can be used to draw a circumscribed circle, as Figures 11.5b and c show. The black region in this visualization is the actual cross-section, and the green circle is the fitted circumscribed boundary. The area of the cross-section was then divided by the area of the circumscribed circle, giving an average of 98.20% overlap between all the layers. This high percentage is a constant confirmation that the cross-sectional profiles of the avocados are almost circular, which means that the vertical thickness of the avocados is very similar to the radius of the circles fitted in Figure 11.5c.

11.3.3.3 Volume Calculation

This algorithm approximates the object volume by breaking down an object into a sequence of simple discs, as illustrated in Figure 11.6. Every disc is a thin slice of the object, a thin slice which creates a cylinder-like object that traces its edges closely and minimizes errors in edge approximation. The volume of all such discs is added to obtain the total volume. The volume of disc no. i is determined by the formula [9]:

$$V_{total} = \sum_{i}^{n} V_i = \pi h \alpha \sum_{i}^{n} R_i^2 \tag{11.9}$$

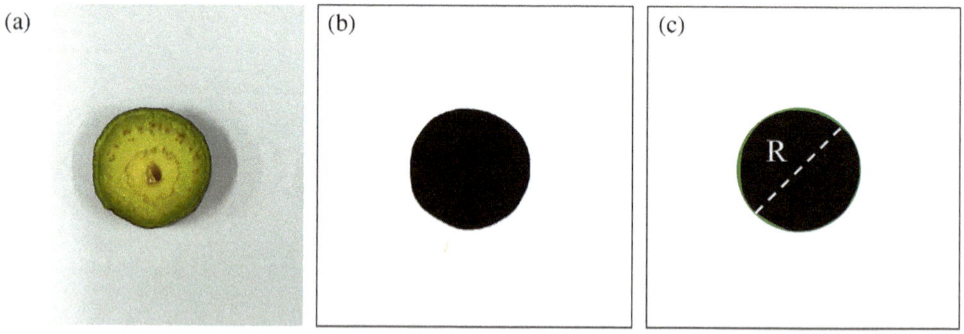

Figure 11.5 An example of the processing steps to slice and their circumscribed radius R of each slice: (a) original avocado cross-section image, (b) binarized cross-section region, and (c) fitted circumscribed circle with radius R.

where $\alpha = 98.20\%$, get the ratio of the real cross-sectional area to its circumscribed circle, which is highly circular as avocado cross-sections, h is the thickness (height) of each disc, and R_i is the radius of the ith disc. The sum of all V_i gives the total estimated volume of the fruit. This volume is then used to estimate the mass of the fruit by use of previously measured density, which can then be used to estimate the physical properties of the fruit.

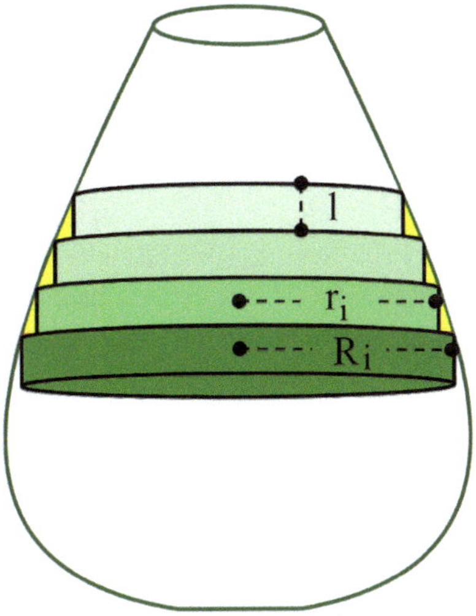

Figure 11.6 Disc method.

11.3.4 Performance Evaluation

Standard error measures that are widely used when evaluating performance in a predictive model were used mainly to evaluate the performance of the models, including root mean square error (RMSE) and mean absolute error (MAE) [23]. Nevertheless, these measures are absolute; thus, they may lead to the obscuration of the relative differences among the fruits of different sizes. To overcome this, the researchers also used the RMSPE and MAPE [24], which represent the errors in prediction as a percentage. This representation in the form of percentages makes the models more understandable and comparable in samples of various magnitudes. Also, the ratio of the percentage of variance explained by the models was measured with the coefficient of determination [25] R^2 which is another indicator of the consistency of the predictability of the models.

- **RMSPE** is a measure of the root mean squared percentage error of the residuals, which is used to measure the prediction error of a model (as a percentage). It is defined as:

$$\mathrm{RMSPE} = \sqrt{\frac{1}{n}\sum_{i=1}^{n}\left(\frac{y_i - \widehat{y_i}}{y_i}\right)^2} \times 100 \tag{11.10}$$

  where y_i are the true values, $\widehat{y_i}$ are the forecasted values, and n is the sample size. The typical range of RMSPE is 0–100% with lesser values signifying a closer association between the forecasted and the actual data. RMSPE is particularly helpful in making comparisons between models on datasets of different scales since it is represented as errors expressed in relative percentage form. But when any of the y_i values are near zero, then RMSPE may go disproportionately large, and therefore, this should be considered during the preprocessing to prevent misleading results. Besides, RMSPE puts the greatest weight on large errors, compared to small errors, and it is especially useful when it is of interest to reduce significant differences between prediction and actual values.
- **MAPE** is the percentage difference between predicted and actual values, the average difference as a measure of a model, which is an intuitive measurement of a model. It is defined as:

$$\text{MAPE} = \frac{1}{n}\sum_{i=1}^{n}\left|\frac{y_i - \widehat{y_i}}{y_i}\right| \times 100 \tag{11.11}$$

where y_i are the true values, $\widehat{y_i}$ are the predicted values, and n is the number of samples. MAPE is a percentage and, in normal conditions, is between 0% and 100%, with the smaller value depicting good performance of a model.

In contrast to RMSPE, which amplifies percentage errors and thus rewards large errors more, MAPE simply averages the percentage errors and effectively gives each deviation the same weight. Such a complementary effect is the reason why both RMSPE and MAPE are employed in tandem with each other: RMSPE indicates the big single deviations, and MAPE shows the general prediction accuracy, which allows understanding where and how much the errors are and their contribution to the variation in the model.

- **Coefficient of determination (R^2)** is used to determine the proportion of the variance in the target variable that is explained by the predictive model. It is defined as:

$$R^2 = \left(1 - \frac{SS_{res}}{SS_{tot}}\right) \times 100 \tag{11.12}$$

where $SS_{res} = \sum_{i=1}^{n}(y_i - \widehat{y_i})^2$ is the residual sum of squares, and $SS_{tot} = \sum_{i=1}^{n}(y_i - \overline{y})^2$ is the total sum of squares. A high value of R^2 (or 100%) would mean that the model has been able to explain a large proportion of the observed variability in the data, indicating good predictive power and high levels of generalization. Conversely, the lower the values of R^2 the higher the percentage of variance that is not being explained, the more the model fails to explain the underlying pattern.

11.4 Results and Discussions

11.4.1 Processing Time

The milliseconds required for processing in Table 11.2 are made up of the execution time, which comprises the time taken to read images or matrices, data processing, calculations, and saving of results. The single image processing on the existing hardware requires 100–120 ms, which is a subset of the overall processing time.

11.4.2 Dimension Evaluations

Table 11.3 and Figure 11.7 demonstrate the performance measurement of the width and length dimensions as determined using the performance indicators, including RMSPE, MAPE, and R^2 as benchmarks. Both dimensions have extremely high levels of precision

Table 11.2 Computation time of the disc method across 5–15 slices (predictions included).

Algorithm	Execution time (ms)
Disc method	142 → 148

Table 11.3 Evaluation metrics for dimensions.

	RMSPE (%)	MAPE (%)	R^2 (%)
Width	1.29	0.92	98.72
Length	1.76	1.30	98.77

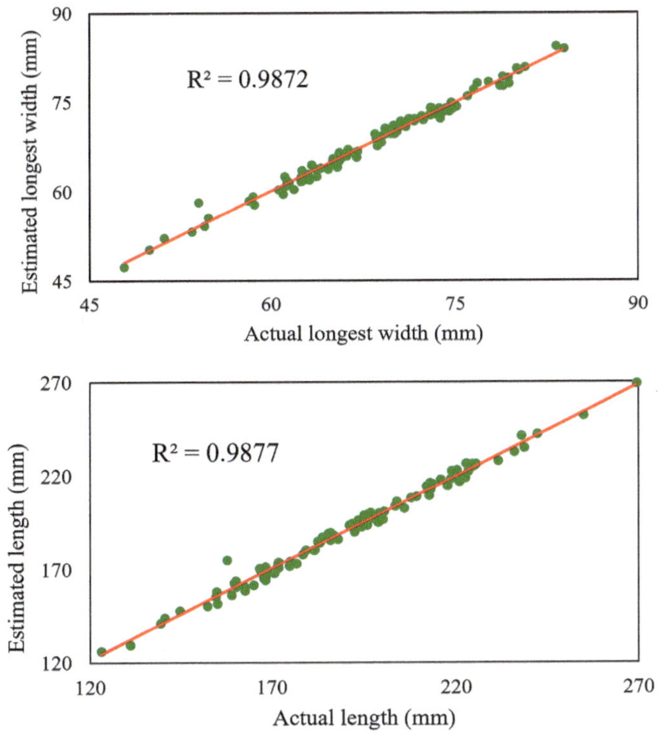

Figure 11.7 Comparison between actual and estimated values of dimensions.

with RMSPE of less than 1.76% and MAPE of less than 1.3%, which are the least in terms of relative and absolute percentage error. But the width, as far as RMSPE is concerned, is better than the length, with 1.29% compared to 1.76% in the case of MAPE; the width is better than the length, 0.92% compared to 1.30%. However, once again, these are minor variations that, when applied in practice, do not bear huge consequences. The R^2 values for both dimensions are graphically above 98.72%, which means that it has a high predictive power and variation capacity. The results indicate that the average accuracy for length is 98.77%, while for width it is 98.72%. This suggests that the model performs slightly better when predicting length compared to width. A possible reason is that the length measurements may have stronger relationships or contain less noise. Although the two metrics differ only slightly, they still reflect variation in performance across dimensions. The consistency of the results confirms the reliability of the model, showing that it can estimate dimensions with high precision and accuracy. Given the minimal error margin, this model is highly suitable for applications that require precise measurements, such as construction, packaging, manufacturing, or other design-related tasks.

11.4.3 Mass Evaluations

Three complementary statistical measures, in the form of RMSPE, MAPE, and R^2, were used to assess the performance of the proposed disc method. The combination of these indicators offers a picture of estimation accuracy and model stability, showing various features of prediction quality: RMSPE penalizes large proportional errors, MAPE shows the average value of percentage errors, and R^2 describes the extent to which the predicted values account for the variability of actual measurements [26–29].

The values of RMSPE continued to decrease with the increase in the number of slices, i.e., 4.52% at 5 slices to 4.26% at 12–15 slices. This loss is an indication that thinner slices represent the curved shape of the avocado more closely and therefore less proportional prediction error. Practically speaking, it can be said that low RMSPE is significant since it means that the method is one that can be used with fruits of various sizes and weights, having consistent results [30, 31]. Because RMSPE puts greater weight on big errors, this small value implies that the disc method cannot make catastrophic over- or under-estimation when large fruits are being used, which is essential when the model is extended to scale to industry sorting lines, where large misclassification errors would occur otherwise. The next noteworthy observation is the decreasing marginal difference after 10 slices: the difference between the RMSPEs becomes less than 0.03% in value, which is equivalent to the statement that precision has already become saturated. This implies that 10–12 slices are the best trade-off on precision and time of computation [32, 33].

MAPE fell from 3.70% at 5 slices to approximately 3.48% at 12–15 slices with a minimal fluctuation (less than 0.25%). The metric shows the mean size of estimation error independent of fruit size, and thus provides a more natural concept of common prediction error. Small stable values of MAPE suggest that the approach predicts with reliability even small or average-sized fruits without over- and under-estimating the fruits in a systematic way [34]. This is particularly helpful in practice with retail packaging of quality grading, involving the need to have the same weight ranges. This little difference in the number of slices also means that the disc method can be used with fewer slices to produce good performance with minimal loss in accuracy in real-time systems.

The R^2 values increased from 96.53% at 5 slices to about 97.7% at 12–15 slices, confirming a strong linear relationship between predicted and measured masses. Unlike error-based metrics, R^2 evaluates the overall explanatory power of the model. High R^2 values demonstrate that most of the variation in avocado mass can be captured using only top-view geometric features combined with the disc method [35]. It demonstrates that the method is strong in terms of influencing variation and is generalizable to samples. A high R^2 is important in an industrial situation since it implies that the prediction of every aspect of the system is reliable, not only the occasional estimate of value.

The geometric character of the disc method can be attributed to the trend of performance. A small number of thick slices (a few) makes the curved surface of the fruit a poor approximation, and there are still gaps (yellow areas in Figure 11.6) between the actual shape and the cylindrical slices. The gaps reflect volume parts that are either over- or under-estimated and which exaggerate the prediction errors. The more the slices, the thinner the discs, the closer they fit the curve of the fruit, and the fewer yellow residual areas. This translates to reduced RMSPE and MAPE, and increased R^2. But as in Figures 11.8–11.10, the error reduction levels off at least 10–12 slices since the residual gaps are insignificant; further increases in slices do not help accuracy.

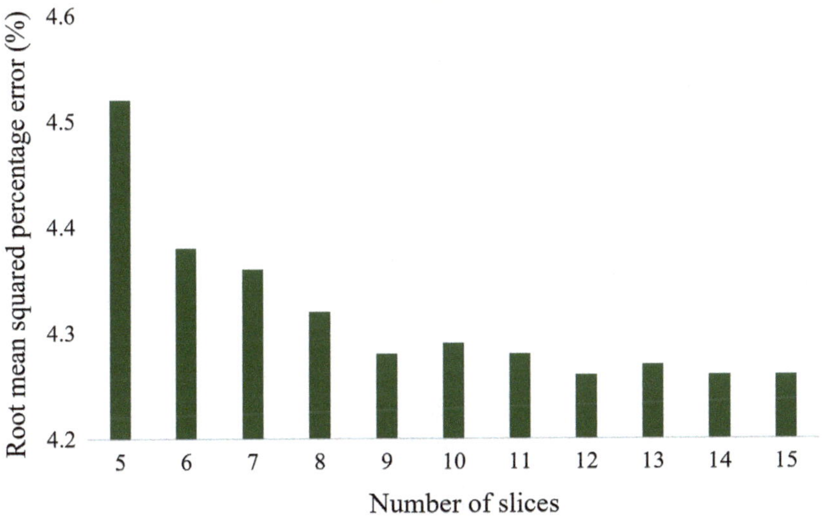

Figure 11.8 Root mean squared percentage error evaluation of the disc method across slices.

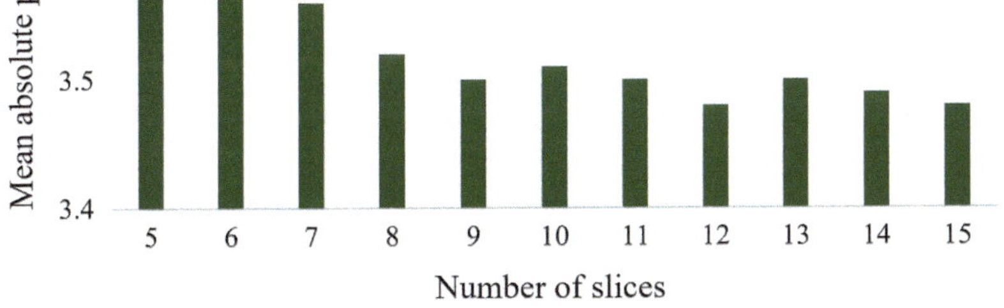

Figure 11.9 Mean absolute percentage error evaluation of the disc method across slices.

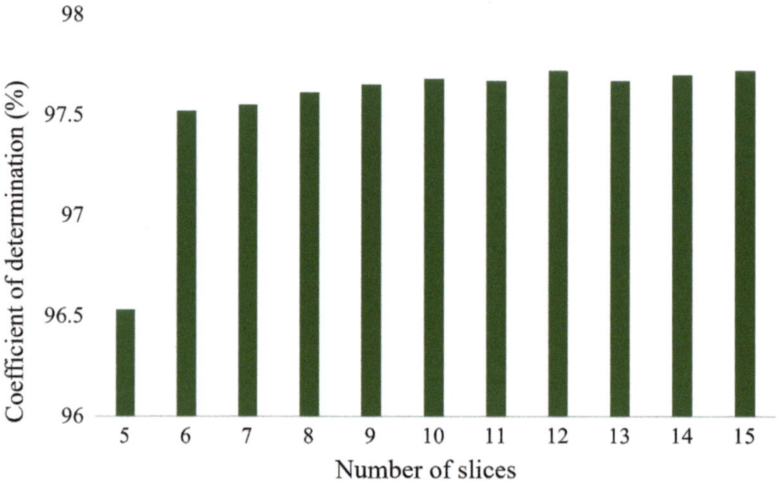

Figure 11.10 Coefficient of determination evaluation of the disc method across slices.

11.5 Conclusion

A nondestructive system of estimating the mass of avocado fruits with the top-view images alone and the disc method was introduced in this study, and carefully presented. The method could estimate the physical properties of the fruit by virtually cutting each fruit into a great number of thin circular slices and adding the volumes together. As demonstrated in the experiment, the more the slices, the smaller the estimation errors. As the geometric resolution was enhanced, the thinner the slices, the closer it was to the curved shape of the fruit, and the fewer residual gaps leading to loss of volume. Consequently, the procedure resulted in a correlation coefficient of about 97.7%, and RMSPE and MAPE were 4.26% and 3.48%, respectively, at 12–15 slices. However, performance began to level off at 10–12 slices, with additional increases rewarding very little accuracy, but consuming even more time, with an average execution time of approximately 142–148 ms per image.

These results prove that the disc method is a solid, precise, and effective mass estimation method of avocados and has good prospects of being used in other types of fruits with a similarly elongated shape. This makes it highly applicable in postharvest operations like automated grading, sorting, and packaging, where consistency and throughput are key factors, due to its simplicity, low cost, and ability to run in real time.

Future work will be aimed at improving the efficiency of computational performance and considering hybrid or ensemble models to further improve performance. The other one is to combine feature selection and dimensionality reduction methods to prune out redundant geometric variables and stability prediction. Lastly, the disc method + deep learning-based segmentation models like convolutional neural networks (CNNs) might be used to automate the image preprocessing step, thus enhancing usability on large-scale production lines in agricultural settings.

All in all, this study forms a strong basis for geometry-based vision that can be applied in the agricultural industry, and it also provides an opportunity to develop more versatile and automated fruit mass estimation in the future.

Acknowledgments

This work has been conducted through the RMIT Vietnam Research Grants-2024, administered by the Office for Research & Innovation, and was supported by RMIT AWS Cloud Supercomputing (RACE).

References

1 Drecher, M.L. and Hass, D.A. (2013). Avocado composition and potential health effects. *Critical Reviews in Food Science and Nutrition* 53 (7): 738–750. https://doi.org/10.1080/10408398.2011.556759.

2 Huang, K.-M., Guan, Z., Blarc, T., and Hammami, A.M. (2021). Global avocado boom. *Choices Magazine* 38: 1–9.

3 Our World in Data (2025). Avocado production. Data from the Food and Agriculture Organization of the United Nations (FAO), updated March 17, 2025.

4 Ariwaodo, C.A. (2022). Handling strategies and facilities for horticultural crops. *Open Access Library Journal* 9: e8577. https://doi.org/10.4236/oalib.1108577.

5 Cubero, S., Aleixos, N., Moltó, E. et al. (2011). Advances in machine vision applications for automatic inspection and quality evaluation of fruits and vegetables. *Food and Bioprocess Technology* 4: 487–504. https://doi.org/10.1007/s11947-010-0411-8.

6 Patel, K.K., Kar, A., Jha, S.N., and Khan, M.A. (2012). Machine vision system: a tool for quality inspection of food and agricultural products. *Journal of Food Science and Technology* 49: 123–141. https://doi.org/10.1007/s13197-011-0321-4.

7 Tran, H.M., Pham, K.T., Vo, T.M. et al. (2023). Physical characteristics estimation for irregularly shaped fruit using two cameras. *IEEE Statistical Signal Processing Workshop (SSP)* 2023: 165–169. https://doi.org/10.1109/SSP52391.2023.10207992.

8 Tran, H.M., Pham, K.T., Vo, T.M. et al. (2023). A new approach for the estimation of the physical properties of irregularly shaped fruit. *IEEE Access* 11: 46550–46560. https://doi.org/10.1109/ACCESS.2023.3273777.

9 Lorestani, A. and Tabatabaeefar, A. (2006). Modelling the mass of kiwi fruit by geometrical attributes. *International Agrophysics* 20: 135–139.

10 Emadi, B., Abbaspour-Fard, M.H., and Yarlagadda, P.K. (2009). Mechanical properties of melon measured by compression, shear, and cutting modes. *International Journal of Food Properties* 12: 780–790, https://doi.org/10.1080/10942910802056143.

11 Liming, X. and Yanchao, Z. (2010). Automated strawberry grading system based on image processing. *Computers and Electronics in Agriculture* 71 (S2): S32–S39. https://doi.org/10.1016/j.compag.2009.09.013.

12 Clayton, M.H., Amos, N.H.B., and Morton, R.H. (1995). Estimation of apple fruit surface area. *New Zealand Journal of Crop and Horticultural Science* 23 (3): 345–349. https://doi.org/10.1080/01140671.1995.9513908.

13 Oyefeso, B.O.R.A.O. (2018). Estimating mass and volume of Nigerian-grown sweet and Irish potato tubers using their geometrical attributes. *Adeleke University Engineering and Technology Journal* 1: 123–130.

14 Sripaurya, T., Sengthaui, K., Booranawong, A., and Cheptatpananondh, K. (2021). Gros Michel banana soluble solids content evaluation and maturity classification using a developed portable 6-channel nir device measurement. *Measurement* 173: 108615. https://doi.org/10.1016/j.measurement.2020.108615.

15 Chen, J.S.X. and Ji, J. (2022). Citrus fruits maturity detection in natural environments based on convolutional neural networks and visual saliency map. *Precision Agriculture* 23 (6): 1515–1531. https://doi.org/10.1007/s11119-022-09895-2.

16 Qin, L., Zhang, J., Stevan, S. et al. (2024). Intelligent flexible manipulator system based on flexible tactile sensing (IFMSFTS) for kiwifruit ripeness classification. *Journal of the Science of Food and Agriculture* 104 (2): 273–285. https://doi.org/10.1002/jsfa.12916.

17 Hotelling, H. (1933). Analysis of a complex of statistical variables into principal components. *Journal of Educational Psychology* 24 (6): 417–441.

18 Huynh, T.M.T. and Dao, S.V.T. (2022). A vision-based method to estimate volume and mass of fruit/vegetable: case study of sweet potato. *International Journal of Food Properties* 25 (1): 717–732. https://doi.org/10.1080/10942912.2022.2057528.

19 Sabouri, A., Bakhshipour, A., Poorsalehi, M. et al. (2025). Machine learning techniques for non-destructive estimation of plum fruit weight. *Scientific Reports* 15: 751. https://doi.org/10.1038/s41598-024-85051-2.

20 Qin, X., Zhang, Z., Huang, C. et al. (2020). U^2-net: going deeper with nested u-structure for salient object detection. *Pattern Recognition* 106: 107404. https://doi.org/10.1016/j.patcog.2020.107404.

21 Gander, W., Golub, G.H., and Strebel, R. (1994). Least-squares fitting of circles and ellipses. *BIT Numerical Mathematics* 34 (4): 558–578. https://doi.org/10.1007/BF01934268.

22 Fitzgibbon, A.W., Pilu, M., and Fisher, R.B. (1999). Direct least square fitting of ellipses. *IEEE Transactions on Pattern Analysis and Machine Intelligence* 21: 476–480, https://doi.org/10.1109/34.765658.

23 Kohavi, R. (1995). A study of cross-validation and bootstrap for accuracy estimation and model selection. *Proceedings of the 14th International Joint Conference on Artificial Intelligence (IJCAI)*, pp. 1137–1145.

24 Pandey, J.K., Yadav, A.K., Purushotham, M. et al. (2022). Colorimetric detection of deltamethrin pesticide from biological samples based on the peroxidase-mimic catalytic activity on MnO_2/SnO_2. *Key Engineering Materials* 928: 45–51. https://doi.org/10.4028/p-4ti1i0.

25 Hyndman, R.J. and Koehler, A.B. (2006). Another look at measures of forecast accuracy. *International Journal of Forecasting* 22: 679–688, https://doi.org/10.1016/j.ijforecast.2006.03.001.

26 Talukdar, V., Veeraiah, V., Roy, D.R. et al. (2023). An innovative approach: typicality-based collaborative filtering for recommender systems. *Proceedings of the 3rd International Conference on Advancement in Electronics & Communication Engineering (AECE)*, pp. 863–868. https://doi.org/10.1109/AECE59614.2023.10428576

27 Truong, S.V., Dao, Q.N.X.P., and L. V. T. (2024). A hybrid metaheuristic algorithm for antimicrobial peptide toxicity prediction. *Scientific Reports* 14: 28260. https://doi.org/10.1038/s41598-024-70462-y.

28 TonThat, L., Son Dao, V.T., Minh Tri, H.T., and Le, M.T. (2023). A feature subset selection approach for predicting smoking behaviours. *2023 IEEE Statistical Signal Processing Workshop (SSP)*, pp. 145–149. https://doi.org/10.1109/SSP52391.2023.10208015.

29 Beemkumar, N., Gupta, S., Bhardwaj, S. et al. (2023). Activity recognition and IoT-based analysis using time series and CNN. In: Handbook of Research on Machine Learning-enabled IoT for Smart Applications Across Industries (ed. N. Goel and R.K. Yadav), 350–364. IGI Global https://doi.org/10.4018/978-1-6684-8785-3.ch018.

30 Le, T.M., Tran, L.V., and Dao, S.V. (2021). A feature selection approach for fall detection using various machine learning classifiers. *IEEE Access* 9: 115895–115908. https://doi.org/10.1109/ACCESS.2021.3105581.

31 Minh, H.T., Mai, L., and Minh, T.V. (2021). Performance evaluation of deep learning models on an embedded platform for edge AI-based real-time traffic tracking and detecting applications. *2021 15th International Conference on Advanced Computing and Applications (ACOMP)*, pp. 128–135. https://doi.org/10.1109/ACOMP53746.2021.00024.

32 Tran, H.M., Le, T.M., Pham, H.V. et al. (2024). An integrated learning approach for municipal solid waste classification. *IEEE Access* 12: 176569–176585. https://doi.org/10.1109/ACCESS.2024.3495982.

33 Pandey, J.K., Kumar, S., Lamin, M. et al. (2022). A metaheuristic autoencoder deep learning model for an intrusion detector system. *Mathematical Problems in Engineering* 2022: 11, 3859155. https://doi.org/10.1155/2022/3859155.

34 Pandey, J.K. (2025). Understanding the impact of electromagnetic and optical radiation on physical and biological systems. *Journal of Nano- and Electronic Physics* 17 (5): 05023. https://doi.org/10.21272/jnep.17(5).05023.

35 Srinivas, P., Veeraiah, V., Thejaswini, K.O. et al. (2025). The impact of electromagnetic and optical radiation on physical and biological systems. *Journal of Nano- and Electronic Physics* 17 (4): 04012. https://doi.org/10.21272/jnep.17(4).04012.

12

Real-Time Nutritional Monitoring Using Edge AI and IoT Cameras

Mohammad Nasar[1], Mohammad Abu Kausar[2], and Md. Abu Nayyer[3]

[1] *Computing & Informatics Department, Mazoon College, Muscat, Oman*
[2] *Department of Information Systems, University of Nizwa, Oman*
[3] *Department of Horticulture (Fruit & Fruit Technology), Bihar Agricultural University, Bihar, India*

12.1 Introduction: Pioneering a Sustainable Agricultural Future

Imagine a South Asian smallholder farmer receiving a smartphone alert that nitrogen levels are low in some rows of her dragon fruit field. Within minutes, a solar-powered robotic arm driven by AI/ML inference can precisely apply fertilizer to the affected plants. The next generation of AI and the Internet of Things (IoT) at the edge can empower smallholder farmers, who are responsible for 80% of the world's food production in developing countries but bear high risks from nutrient deficiencies that could reduce yields by 50%, especially in these susceptible regions where the global population is expected to peak at 9.7 billion by mid-century [1]. Traditional soil and tissue analysis, though accurate, is much slower and higher in cost and less accessible than sensor-based methods, which present challenges to smallholder farmers, especially in areas with scant lab capacity. This chapter demonstrates a novel system for real-time nutrient sensing designed for data-based precision farming.

Climate change, soil degradation, water scarcity, and uncertainty in weather impact global agriculture and crop production [2]. Nutrient imbalances that remain undetected represent a major risk for crops, including dragon fruit, grapes, and coffee, where injury appears close to the plant's death or yield loss [1, 3]. Excessive fertilizer application due to undetected nutrient deficiencies leads to both economic and environmental costs by raising soil salinity, contaminating water bodies, and increasing the cost of farming [4], especially among smallholders in underdeveloped, low-income regions where there is limited access to technology and services due to inconsistent or a lack of Internet connection [5, 6].

We present a framework to mitigate these challenges using a network of IoT cameras to capture high-resolution images of plant leaves, stems, and fruits that are evaluated by AI models in real-time on low-cost edge devices such as NVIDIA Jetson Nano or Raspberry Pi 5 [7, 8]. This makes the framework ideal for deployment on rural farms, as it can operate offline, which is not possible for the cloud-based systems, as they must be connected to the Internet [5]. The system identifies the visual signs of nutrient stress, e.g., yellowing leaves (nitrogen deficiency) or burnt edges (potassium deficiency), with more than 90% accuracy, allowing for precise, swift treatment [3, 8]. Based on a detailed case study of dragon fruit farms in South Asia, as well as pilot studies of coffee and maize farms, the impact of the system can be expressed as follows: The robotic arm developed based on AI will control how much amount of fertilizer applied in precise places so that waste will be adjusted and production can be increased [3, 9]. By employing XAI tools (e.g., heatmaps indicating damage agent identifiers), farmers would not just discern but trust the use of the technology; this characteristic incentivizes the adoption of diverse populations [10, 11].

Consistent with Industry 5.0's objectives of sustainable, human-centered innovation, the system allows farmers to take data-driven decisions while having a lower environmental impact [2]. The simple systems cost on the order of US$200 and hence can be widely adopted, from maize to citrus [3]. Beyond the technology, emphasis has been placed on functionality (in our case, aiming for a simple-to-operate system that is also durable across environments and levels of farmer capability [5, 10]). The system is multilingual, uses voice controls and dashboards, to target farmers in areas like Vietnam, Kenya, and India [5].

This chapter is intended for a broad audience: farmers looking for practical tools, students interested in smart agriculture, engineers designing state-of-the-art systems, and policymakers considering sustainable food systems. The topics covered in the chapter include real-time monitoring, system design, AI engine, transparency mechanisms, applications to the real world, and future trends. The goal of this chapter is to address readers in an informative and emotionally engaging manner, mixing technical knowledge with tangible real-life scenarios demonstrating how Edge AI and IoT are behind a new era that will allow the agricultural sector to match the rising food demands [1, 2, 6]. The following sections are written in an accessible style and address questions on the technology, its consequences, and what to ask about scientific claims.

12.1.1 Background Agriculture Under Pressure on a Global Scale

Modernizing agriculture is critical to meeting global food demands. The Food and Agriculture Organization (FAO) estimates that global food production must increase by 70% by 2050 to feed a growing, wealthier, and more urbanized population with increasing demand for meat and dairy, even as climate change threatens staple crop yields, such as wheat [6]. Agriculture in developing regions, including China, is dominated by smallholder farmers who experience a lack of access to capital, technology, and infrastructure [5]. These issues are compounded when a deficiency of nutrient element(s) occurs, for example, nitrogen (N), which in itself has been reported to decrease yields of some crop species by 40% [1]. High-value crops such as dragon fruit and coffee require nutrient supplementation for product quality and marketability, resulting in the need for regular monitoring [3].

Conventional methods based on laboratory testing are expensive and not feasible for daily use, particularly in remote areas, where laboratory testing facilities may be several hours away [4]. Priced at $100–$300 per sample, lab testing is cost-prohibitive for many farmers and results in returns to old practices or over-fertilization that results in poor soil health and greenhouse gas emissions [5]. Cheap technology and real-time monitoring systems allow farmers to make informed decisions in time [2, 8]. Such offline and low-cost systems are developed for small-scale farmers to achieve equity and sustainability [6].

12.1.2 Promise of Precision Agriculture

With precision agriculture powered by data, integrated with wide-scale IoT sensor networks, everything from the irrigation to the pest control on a farm can be optimized at an enviably improved cost-per-wasted-expenditure/cost-per-wasted-activity [12, 13]. IoT and AI in Crop Farming: Farmers can monitor their crops for any deficiency with the help of IoT and AI, which will allow them to give instant solutions to the problem, such as a dearth of nutrients and pest attacks, before they become fatal [5]. A very important edge in rural areas is that Edge AI consumes data on-premises and does not always need the Internet [7]. This system builds on technologies such as drone deployment for surveillance; edge-based disease surveillance and detection; high automation; and data-driven systems [8], but it advances a new frontier by including robotic interventions and tangible, XAI solutions [9, 10].

12.2 The Need for Real-Time Nutritional Monitoring

12.2.1 The Critical Role of Nutrients in Crop Health

Plant nutrients are indispensable to the health and growth of crops because they play vital roles in photosynthesis, cell division, disease resistance, and other activities [1]. Nutrients like N, P, and K are important for proper growth and yield, while micronutrients such as Zn, Fe, and Mg support enzymatic activities/structure [3]. A deficiency can manifest as large symptoms: without enough nitrogen, the leaves will yellow; a phosphorus-deficient plant won't grow an effective root or shoot system; low potassium means scorch marks on leaf tips and weak stems [1]. These losses can be as high as 20–50%, severely affecting food security and farmers' income primarily in developing countries where agriculture is the livelihood for 60–80% of the population [6].

Nutrition deficiencies are the reasons for an enormous amount of worldwide yield losses, which is especially true for smallholder farmers in Africa, Asia, and Latin America, with a lack of laboratory access [14]. Likewise, a deficiency of nutrients in high-value crops for export, like dragon fruit, grapes, and coffee, also decreases quality and market price [3]. One such example is the nitrogen needs of dragon fruit plants; generally, lower quality nitrogen contents yield smaller fruits with pale color, which are less aesthetically appealing and consequently less likely to be sold in expensive markets [9]. Over-fertilization by undetected nutrient deficiencies results in soil salinization, pollution of water bodies, and an increase in greenhouse gas emissions [5]. The additional nitrogen discharge into the aquatic ecosystems induces the high phenomenon of HABs that have been causing serious environmental problems [4].

By monitoring in real time, farmers can recognize deficiencies early and apply fertilizers efficiently and appropriately to limit waste of resources and environmental impact [2]. This technology will save \$50–\$100 per ha of fertilizer cost to smallholder farmers adopting it, and hence enhance their profitability [3]. It further advocates sustainability, which complements international targets (e.g., the United Nations Sustainable Development Goals [SDGs]) of Zero Hunger and Responsible Consumption and Production [15].

12.2.2 Limitations of Traditional Monitoring Methods

Conventional tools to analyze nutritional status (soil and tissue tests) properly determine the demands of the plants, but are unaffordable for real-time decision support [4]. Farmers take samples, submit them to a lab, and get results back days or weeks later at $100–$300 per test [5]. This is an obstacle for smaller farms and particularly in more isolated locations where labs are few and far between [7]. Moreover, results only capture a given moment, and the dynamics of crop requirements with weather, pests, or stage of growth are not considered [4]. Perhaps heavy rain will wash the N across a fence, into a gully, or to someone else's property; such losses cannot be addressed by lab-based tests, and there is little in the way of on-farm quick-fix information [3].

Timing is everything with high-value crops, and these limits are important. For instance, in the production of dragon fruit, the potassium deficiency symptoms, like tip burn, require early management.

12.2.3 The Transformative Power of Precision Agriculture

Precision agriculture is an example of data-driven farming with the use of data to enhance farm practices such as irrigation, pest control, and nutrient management [12, 16]. IoT devices such as cameras and sensors gather data in real-time about the health of crops, soil conditions, and environmental factors [17]. Edge AI enables local data processing, which obviates the need for a continuous Internet connection, a crucial feature for rural geographies having intermittent connectivity [7]. AI models for crop diagnosis are among the most immediate and visible applications of AI in agriculture, including image analysis techniques where plants are examined for visual cues (e.g., changes in color, texture irregularities, or leaf patterns).

The system leverages recent developments. The drones with AI gave ~85% accuracy in nutrient stress monitoring of wheat crops during field trials [12, 18]. Grapevine diseases were detected by the edge devices with an accuracy of 92%, classifying them into their respective disease type in harsh environmental conditions [19]. Real-time object detection of pests, diseases, or weeds in plants was done with 90% accuracy using models such as MobileNetv3-YOLOv4 [20]. However, there are several challenges that remain, for example, light conditions are very dynamic, leaves are often partially occluded in the field, and the decisions from AI algorithms must be deniable to farmers: trust in digital solutions is needed [4, 10]. To overcome these challenges, the system combines state-of-the-art AI models with closed-loop calibration and XAI tools, enabling reliable solutions [9, 11].

12.2.4 Economic, Social, and Environmental Implications

Real-time monitoring offers much more than crop health information. For smallholders, this is a tremendous cost saving – as it removes about 20–30% of fertilizer cost. [3] One such example is the reduced steering equipment (RTK-Steering) and direct application of nutrients using variable-rate nitrogen farmer-driven initiative by Moutier farmers, which should save around $50–$100/ha in some cases [5], and boosts efficiencies to generate return on investment for those farming businesses. It also promotes social equity in agriculture by providing data-enabled tools allowing small farms like Marther Adu's farm mentioned before [4] to compete with large agribusinesses. It enables local decision-making, which is essential to resilience at the community level and, for this reason, is crucial in these economically or climatically disrupted regions [6].

On an environmental level, the system helps induce smarter, sustainable practices as it reduces over-fertilization, which is one of the causes for soil degradation and water pollution [4]. Precision nutrient application reduces leaching into waterways, minimizes greenhouse gas emissions, and enhances farm resilience [5]. Its low-cost modular design makes the system affordable for farmers in countries with limited resources, contributing to much-needed equal opportunity to use state-of-the-art technology [2]. In Kenya, a farmer could use one $200 unit to watch over an individual plot of maize, while in Brazil, it would be more suitable for larger coffee plantations [3].

12.2.5 Farmer Perspectives and Adoption Challenges

Farmers who have come from a nontechnological background will not easily accept the approach of AI, as they find this approach to be very complicated (or cost-prohibitive) [10]. Nevertheless, during pilot applications, farmers expressed concerns about maintenance, data interpretation, and how to integrate it with the existing agricultural practices [5]. But these concerns were solved by training workshops and well-defined dashboard layouts; over 80% of farmers gained high confidence after the training [10]. A dragon fruit producer in Vietnam explained that using the multilingual navigable interface coupled with real-time alerts enabled easy nutrient regulation and saved hours of manual checking time [3]. This indicates how important it is to face these challenges to ensure global agricultural sustainability [2]. Table 12.1 shows the challenges addressed by real-time monitoring.

Table 12.1 presents a comparison between the standard laboratory methods of nutrient monitoring and our real-time approach on speed, cost, access, and environmental impact.

12.3 System Design: A Blueprint for Smart Farming

The system acts as a self-contained monitoring device that constantly monitors crops, detects abnormalities, and allows direct in-field solutions. It combines IoT cameras, Edge AI devices, and robotic actuators in a smooth orchestration of autonomous tasks intended to be affordable, reliable, scalable, and flexible to work within various agricultural settings [5, 7].

Table 12.1 Challenges addressed by real-time monitoring.

Challenge	Traditional methods	Real-time system
Speed	Days to weeks	Seconds
Cost	High ($100–$300/test)	Low (~$200/device)
Accessibility	Lab-dependent	Field-deployable
Environmental impact	Over-fertilization risk	Precise application

12.3.1 IoT Cameras: The Eyes of the System

The system is based on IoT high-resolution cameras housed in weatherproof and dust-protected boxes, which are used to capture close-up images of leaves, stems, and fruit tissues. The cameras are at different elevations for taking multi-view images (top-down, side, oblique). Lighting can be adjusted to compensate for changing illumination conditions (e.g., sunlight, shade, dusk). An image of fleshy stems and surrounding soil in a dragon fruit farm indicates plant health [9] (see Figure 12.1).

The cameras are programmed to a schedule (e.g., photos taken every 15–30 minutes) or allowed to be triggered automatically when changes in the canopy are observed, such as leaf yellowing and spotting [8]. By the dynamic imaging method, nutrient deficiencies (e.g., caused by flooding, normal soil aging processes, or extreme growth) may be rapidly detected even during rapid growth trends, or after weather events like heavy rain [4]. Figure 12.1 is a schematic diagram of the nutritional monitoring system, from imaging to farmer feedback. Cameras wirelessly send images to local edge devices via low-power links, for example, LoRa or Wi-Fi mesh (with offline reliability provisioned by local storage – e.g., 64GB SD cards) [7].

Each camera consumes 1.5 W and is solar panel or rechargeable battery-powered, allowing modest energy consumption. The system additionally includes battery-life tracking, connection diagnostics, and lens-clog alerts through a mobile app for farmers [5]. Workers have also reported having to clean a camera or switch to a different unit if an alert is caused by dust or water on the lens [4]. The cameras are built to be rugged, being dust-, rain-, and humidity-resistant, as well as weatherproof to temperatures of 0–50 °C.

Figure 12.1 shows the live workflow of the nutritional monitoring system, in which IoT camera image capture is sent for another side edge AI inference, robotic actuation, and then the feedback of completion to the farmer dashboard.

12.3.2 Edge AI: Processing Power in the Field

Edge devices, including the NVIDIA Jetson Nano or Raspberry Pi 5, provide the brain for the system, and local image processing is used to provide near real-time insights [7, 8].

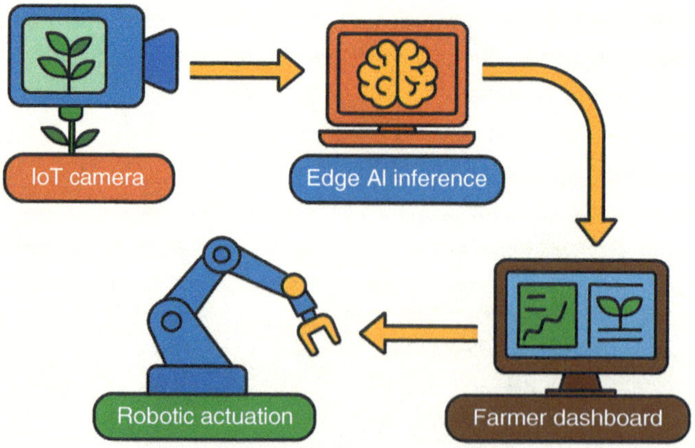

Figure 12.1 System workflow diagram.

Table 12.2 System components and power consumption.

Component	Average power draw (W)	Role
Jetson Nano	5.0	AI inference
Raspberry Pi 5	3.8	Backup processing
IoT camera	1.5	Image capture
Robotic arm (idle)	0.2	Nutrient delivery

These are selected because of their low power consumption (~5 W) and adequate computing power, which is suitable for off-grid, solar-powered farms. Efficiency and speed are emphasized in the processing pipeline:

Image Normalization: Handles the variation in lighting, so that the image does not appear the same under sunlight, shade, or fluorescent light [3]. For instance, normalization removes shadows where trees are shaded and thus helps to produce better results [4].

Leaf Segmentation: Identifies nutrient deficiencies (e.g., nitrogen, potassium, and micronutrients) with over 90% accuracy by lightweight AI models tailored for edge hardware [8].

Spectral Analysis: Extracts color or texture and spectral characteristics related to nutrient status, like the level given by chlorophyll for nitrogen, edge browning for potassium [3, 20].

Inference: Identifies deficits (e.g., nitrogen, potassium, and micronutrients) with over 90% accuracy by low-brown AI models tailored for edge hardware [8].

A latency of less than 150 ms per frame is achieved in local processing, which enables real-time decision support that is critical for timely interventions [7]. Compared to cloud-based systems, which need 80 W and continuous availability, the system hibernates when the IT operations halt, ensuring it is energy and cost-efficient. This reduced operational cost enables farmers to deploy the system in untethered settings [5].

Table 12.2 displays the details of average power consumption and roles of key system components, including edge devices, cameras, and a robotic arm.

12.3.3 Robotic Integration: From Insights to Action

In response to these deficiencies, the system activates a robotic arm that delivers prescriptive treatments (e.g., sprays of nitrogen-rich fertilizers, micronutrient solutions [3, 9]). The arm on a rail or mobile platform is controlled with GPS and AI guidance to deliver treatments within 10 cm of the applied treatment [9]. For instance, if the arm identifies a potassium deficiency in one plant of dragon fruit, it then automatically gives that specific plant fixed doses of potassium so that the surrounding healthier plants are not affected [3]. More precise application leads to a 20% reduction in fertilizer use, leading to reduced costs and less degradation of soil [5].

Mobile dashboards wirelessly linked through Bluetooth or Wi-Fi provide real-time information on our health, such as color-encoded nutrient maps, deficiency warnings, and action records [5]. Farmers can assess AI recommendations, veto activities, or schedule maintenance chores, all to include them in the loop [2]. The dashboard is designed to be user-friendly and customizable for farmers of different digital literacy levels through layout features such as large buttons, multilingual support (e.g., Vietnamese, Hindi, Swahili), and voice-activated commands [10]. In another case, a farmer in Kenya can translate the dashboard to Swahili and has no technical barrier to entry [5].

Screenshot of a mobile dashboard showing the color-coded nutrient map, deficiency alerts, and action logs for a dragon fruit farm in Figure 12.2.

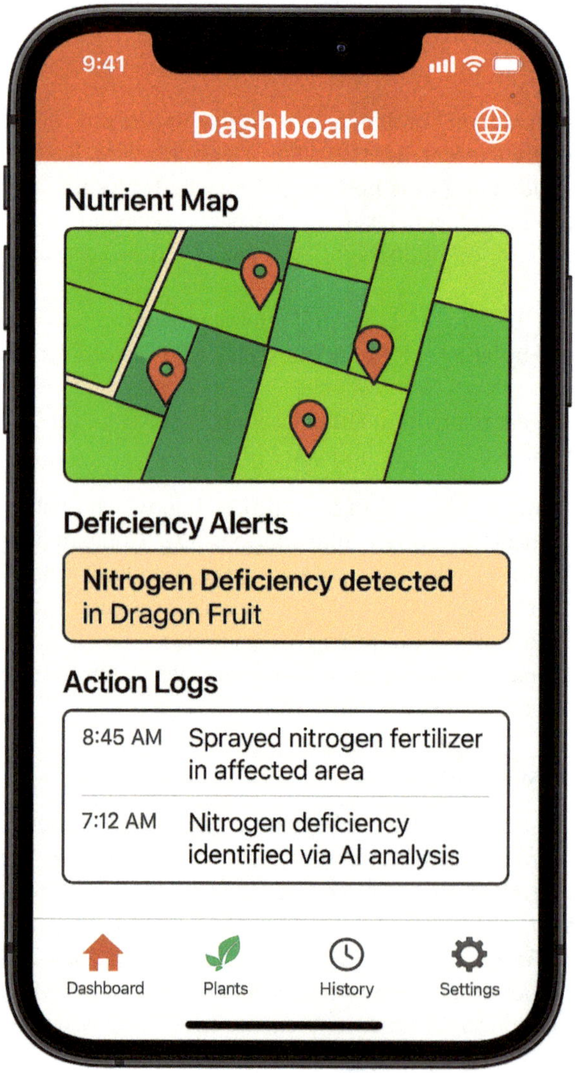

Figure 12.2 Mobile dashboard interface.

12.3.4 Ensuring Reliability and Security

The system is also dust-resistant and water-resistant, and it can work in environments with a humidity level between 5% and 80%, between temperatures −50 and 70 °C. Data encryption for communication among devices, as well as secure updates of the firmware, ensures information security by making valuable farm data hidden from potential adversaries [7]. With redundant backups, watchdog timers, and automatic restarts, you will achieve 98% automation uptime, even during irrigation cycles or storms. For example, if a camera fails due to the presence of dust on it, the system will replace it with another backup camera and notify the farmer of such an event via the dashboard [5]. This is the basis for periodic (when connectivity is there) cloud syncs to extract long-term analytics such as yield trend or soil health trends that can contribute to better farm management [2].

12.3.5 Scalability and Adaptability for Global Use

This makes it applicable to a wide range of crops, climates, and farm sizes from the user perspective [3]. For a farm with a very modest size (1–5 ha), one camera and one edge device would be sufficient (~$200) while bigger locations will need multiple units for compact coverage [5]. The AI models are all retrainable, either to prepare for a new crop or to work in a different area by including native data [21]. For example, the very same model can be retrained in Africa for maize, and in South America for coffee, making it global [1]. It is applicable in various languages and is compatible with existing farm management software [2]. This versatility aids diverse nutritional demands relative to different agricultural environments, ranging from tropical orchards to temperate grain fields [3].

12.3.6 Farmer-Centric Design and Maintenance

For usability, the system incorporates a maintenance guide in the mobile app that provides users with step-by-step instructions on how to clean cameras, update firmware, or even replace batteries [5]. The agricultural operator can call a support line or a local technician who will help with quickly resuming work [2]. During pilot deployments, farmers stated that the plug-and-play design of this system made it easier for them to set up with less than an hour of installation time [3]. The more farmer-centralized nature of the system makes it a practical and sustainable system for long-term usage [10].

Figure 12.3 of IoT cameras and edge devices in a dragon fruit farm powered by solar panels.

12.4 The AI Engine: Turning Images into Action

AI models of the system examine plant images to diagnose nutrient deficiencies accurately and quickly, as does a professional agronomist [1, 8]. In this section, we offer a thorough explanation combining technical details with clarity for those not as familiar with the field.

Figure 12.3 System deployment in a field.

12.4.1 Convolutional Neural Networks: Focusing on Details

Convolutional neural networks (CNNs) are used to analyze the local features, i.e., color deviations, patterns, irregularities, and whether a leaf exhibits signs of nutrient deficiency [8]. For example, it can diagnose nitrogen deficiency by the yellowing leaves, and zinc deficiency by mottled patterns [1]. In field tests, CNNs are proven to be applicable for practical use since they can predict 92% accurate crop type from aerial images of fields by learning convolutions on thousands of crops labeled by agronomists [8, 21]. These models use convolutional architectures that extract features and reduce computation through pooling layers with convolutional layers for feature extraction, after which pooling-based approaches are used to reduce the computational power and enable real-time inference on edge devices like Raspberry Pi [7].

It is this ability to classify subtle symptoms, such as slight yellowing, that makes CNNs so well-suited for identifying early-stage deficiencies [3]. They apply techniques like quantization (reducing the precision of the model) and pruning (removing redundant neurons), which result in models that are 40% smaller without any loss in accuracy [7]. This setup enables quicker processing on low-power hardware like real-time applications [5]. This enables the instant diagnosis of nutrient deficiencies, eliminating the need for time-consuming assessments that depend on sophisticated hardware or stable internet access.

Confusion matrix of CNN model showing classification performance across nutrient deficiency classes, with only 5% overlap between nitrogen deficiency and healthy in Figure 12.4.

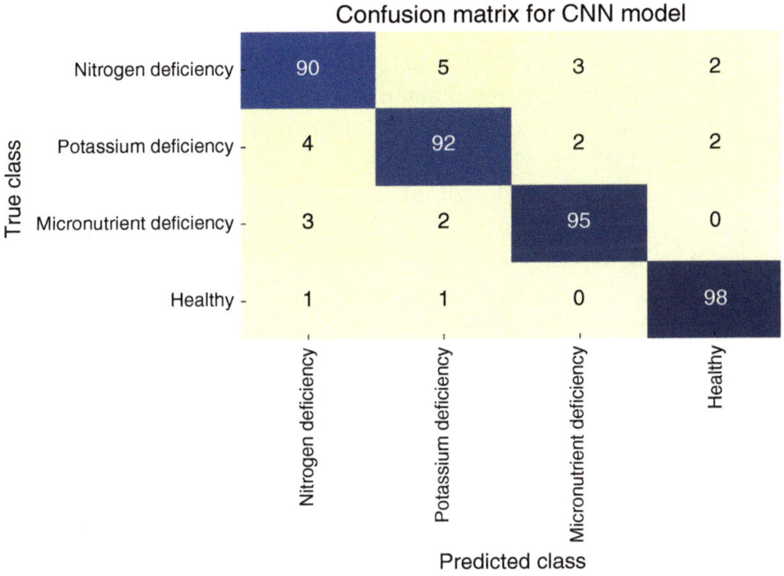

Figure 12.4 Confusion matrix for CNN model.

12.4.2 Vision Transformers: Seeing the Whole Plant

Vision Transformers analyze the entire plant, how leaves, branches, and fruit are interacting with each other [1]. ViTs can (i) split images into patches, which enables learning of complex deficiency patterns and (ii) use attention mechanisms to do so [1, 20], obtaining 93% accuracy. For example, a ViT can detect potassium deficiency from drought stress by watching leaf patterns all the way through the plant, something that CNNs would have a more difficult time doing [8]. Their insensitivity to lighting changes, shadows, and background noise makes them suitable for computer vision applications in the field [18].

ViTs are computationally expensive compared to CNNs, although optimizations involving knowledge distillation (transferring the knowledge from bigger models into a smaller one) make them accessible for deployment on edge devices such as NVIDIA Jetson Nano [7]. ViTs process images in 114 ms in field tests vs. 82 ms for CNNs, which is a reasonably fair trade-off due to their much greater accuracy [8]. ViTs are perhaps most useful for diagnosing complex crops like dragon fruit, in which multiple deficiency symptoms may co-occur [3].

Table 12.3 compares the accuracy, latency, and power consumption of CNN and ViT models used in the nutritional monitoring system.

Table 12.3 AI model performance.

Model	Accuracy (%)	Latency (ms)	Power (W)
CNN	92	82	4.7
ViT	93	114	5.0

12.4.3 Training and Calibration for Reliability

The AI models are trained on a diverse dataset of images, including dragon fruit, grapes, wheat, maize, and coffee plants, labeled with soil and tissue test results using object detection methods [3, 10]. To ensure broad predictive power, the dataset was divided into 70% training, 15% validation, and 15% test [21]. The models are trained using categorical cross-entropy loss with Adam optimizer and learning rate scheduling for faster convergence [8]. Calibration aligns AI predictions with diagnostics data of the field, decreasing errors by 7–10% [11]. For example, if an ML model predicts potassium deficiency, laboratory test data could be used to validate the accuracy of such a prediction [21].

Calibration takes account of regional differences like soil type, climate conditions, and species-specific factors, such that it is adaptable on location [3]. The model backend will permit farmers to add images as feedback, which can be used to help us more precisely refine the output to achieve better performance over time. For example, a coffee farmer from Brazil who applies the coffee crop would add pictures of local plants that he planted, which leads to raising this attitude in society [1]. Participatory strategies of this kind strengthen trust and ensure that a system is acceptable for field application [2].

12.4.4 Handling Field Challenges

Since different lighting environments, leaf occlusions, water droplets, and insect disturbances, etc., field images present multiple challenges [4]. The system can be trained to apply a series of adaptive filters to normalize an image and reduce shadows and dust reflections [3]. The sturdy hardware features hydrophobic lens coatings and weather-tolerant casings, which offer unfailing performance in any circumstances, whether it be rain or sprinkler-irrigated areas. However, the AI maintains about 90% accuracy under various conditions, from cloudy days to slightly leaf-covered to insect damage, as demonstrated by sample field images [8]. The system automatically compensates for water droplets on leaves, and the recognition of nitrogen deficiency is safe even in this case, which means protected diagnostics [3].

12.4.5 Model Optimization for Edge Deployment

The AI models are optimized using a combination of quantization, pruning, and knowledge distillation [7] methods to be able to run efficiently on edge devices. Applying quantization will make the model's precision drop from floating-point 32-bit to 8-bit; as a result, we can reduce the size by 40% without losing any of its performance measure accuracy [8]. In pruning, both redundant neurons were removed, and the computational load was reduced even further [7]. To achieve the accuracy in a spatial configuration needed for low-power hardware [20], we use knowledge distillation (transferring knowledge from a large, cloud-based model to a small one at the edge). These optimizations have made the framework capable of processing images in less than 150 ms, which is an essential requirement for real-time applications [7].

12.4.6 Integration with Other Technologies

The system may be linked with other agri-tech tools, such as soil moisture sensors or weather stations, to give a comprehensive perspective on farm health [13]. AI-derived nutrient monitoring combined with feedback from soil sensors, for instance, can accurately direct irrigation and avoid wasteful resource use [5]. In a pilot use case that integrated weather data, predictions of nutrient deficiencies only improved by 5% since the system also considered the impact of rainfall on nutrient uptake [3]. The farmers value them because of interoperability, which improves the whole system [2].

Appendix 12.A details the training pipeline and optimization techniques.

12.5 Real-World Impact: Case Studies and Applications

To demonstrate the transformative potential of the system, a detailed case study is provided on dragon fruit farming with pilot studies conducted on coffee and maize farms to showcase its versatility and impetus [3].

12.5.1 Dragon Fruit Case Study: South Asia

A 3-ha dragon fruit farm located in South Asia overcame nutrient deficiency to achieve quality, high-value yields [3]. As these deficiency symptoms, such as yellowing, are different for nitrogen and tip burn are different for potassium, the fleshy stem of pitaya is conducive to AI monitoring [9]. The farm was installed with 15 IoT cameras and 5 NVIDIA Jetson Nano devices running on solar energy for image capturing, as depicted in Figure 12.2. Images were taken every 20 minutes by cameras and analyzed by AI models for the detection of deficiency with an accuracy of >90% [8]. There was a rail-mounted robot arm to carry out precise spraying within a range of 10 cm based on the decision of AI guidance [3, 9].

Plant-level data such as nutrient maps, deficiency alerts, and action logs are available in real-time on a mobile dashboard with Bluetooth-enabled offline mode. On the dashboard, color-coded disease maps help farmers easily locate affected areas even in vast fields [2]. Costs of fertilizers were reduced with the system by approximately 20%, and dependence on laboratory testing was eliminated, while achieving better quality plots leading to increased yields by use of accurate interventions [3]. This resulted in a rise of marketable fruit yield by 25% based on the grower's production [9].

Photograph in Figure 12.5 of the AI-robotic arm employed in the dragon fruit case study with rail system and precision sprayer.

Table 12.4 shows accuracy, recall, and F1-score for the detection of nitrogen, potassium, micronutrient deficiency, and the healthy class.

12.5.2 Coffee Case Study: Colombia

This flexibility was exemplified in a pilot use on a 5-ha coffee farm in Colombia [1]. At the same time, coffee is a crop that needs a large amount of phosphorus, and deficiencies limit the development of coffee plants, thus reducing the quality of their beans [3]. The farm was

Figure 12.5 Robotic arm deployment.

Table 12.4 Validation results for nutrient detection.

Nutrient deficiency	Precision	Recall	F1-score
Nitrogen	0.91	0.89	0.90
Potassium	0.88	0.85	0.86
Micronutrients	0.84	0.82	0.83
Healthy	0.93	0.94	0.93

equipped with 20 IoT devices and 8 edge devices, which identified phosphorus deficiencies with 89% accuracy [8]. Nutrient delivery based on AI outputs was optimized, and the system was integrated with existing irrigation systems [5]. Farmers also used 15% less fertilizer and 20% better quality beans, making them competitive on the market as well [3].

12.5.3 Maize Case Study: East Africa

The system was then modified and used to monitor nitrogen levels in a 1-ha maize farm in East Africa on account of the cost and previously implemented nitrogen sensing to report using a single camera and edge device [5]. These AI models were then retrained on maize-specific data and were able to detect nitrogen deficiencies with 90% accuracy [8]. An AI-guided portable sprayer to dispense precise treatments reduced fertilizer expense by 18% [3]. A 25% yield increase was reported, showing the value of the system for smallholder farmers [2].

12.5.4 Overcoming Field Challenges

The approach would be capable of considering temporal variations in solar illumination, shedding of water droplets, and occlusion from insects using adaptive filters and hydrophobic lens coatings [4]. It was functioning and predicting data when it reported with a wet lens, 90% correct in the monsoon season for reliable operation, also [8]. It is also possible for a multilingual tertiary support dashboard in which higher-level voice-enabled commands, such as feedback for mobile updates, are guaranteed to be included [5]. The easy-to-use interface also makes it simple for farmers to adjust nutrient application levels. These case studies provide examples of up-scalability and potential spatial transfer in other crops or regions [3].

12.5.5 Farmer Feedback and Community Impact

Farmers in all case studies appreciated the simplicity and cost-effectiveness of the system, with 85% reporting they were making better decisions having been through training [10]. One farmer said, "The system tells me exactly which plants need help, so I save hours and time than checking one by one to see how they are doing." [3] This collaboration was also facilitated by informal communications along with community workshops held by the farmers on the tips to optimize the system [2]. These findings highlight the potential of this system to revolutionize agriculture, especially for smaller plots [6].

12.6 Building Trust: Explainable AI

Farmers who are not technologically savvy may view AI systems as black boxes. Explainable AI (XAI) make decisions transparent, helping farmers understand and trust the system [10, 11].

12.6.1 Grad-CAM: Visualizing AI Decisions

Gradient-weighted Class Activation Mapping (Grad-CAM) makes saliency maps of the image so a model can focus on showing which areas in the image the AI model focused on to make its prediction, like yellow tips are potassium deficiency [10]. If not, this information could be presented to the farmer in a user-friendly manner, explaining why their results were flagged as underperforming on their dashboard and thus offering intuitive decision support [11]. A heatmap could glow red around the edges of the leaf, which could pinpoint a potassium problem that a farmer can diagnose or confirm without the help of technicians [21]. A similar study also revealed that farmers' understanding could be increased by 30% when Grad-CAM was added in field experiments [9].

Grad-CAM heatmap overlaid on a dragon fruit stem image, highlighting regions contributing to a potassium deficiency prediction in Figure 12.6.

Figure 12.6 Grad-CAM heatmap example.

12.6.2 SHAP Values: Deeper Transparency

SHapley Additive exPlanations (SHAP) values highlight important image patches in ViT models and provide a way to interpret complex predictions [11]. For example, SHAP could elucidate that the ViT diagnosed a micronutrient deficiency because of subtle spotting patterns throughout a plant [10]. This led to a 4% reduction in false alarms and increased farmers' confidence in the recommendations, which resulted from field tests that took approval ratings from 3.8/5 to 4.5/5 post calibration [21]. SHAP outputs are presented in a simplified format on the dashboard, making them accessible to nontechnical users [11].

Table 12.5 shows improvements in confidence error and farmer approval ratings before and after calibration of AI models.

12.6.3 Ethical and Sustainable Design

The system is designed to avoid biases and thus works with diverse training data: various crops, regions, and environmental conditions [21]. For example, including images of tropical vs. temperate and arid climates makes performance more context-insensitive [1]. It also ensures data privacy: sensitive farm data can remain outside the cloud and is processed locally on-farm [7]. Recyclable hardware and low-power designs are used to reduce the environmental footprint, aligning with ethical principles of sustainable engineering.

Table 12.5 Explainability and calibration performance.

Metric	Pre-calibrated	Post-calibrated
Confidence error (%)	12.6	6.1
Farmer approval rating	3.8/5	4.5/5

Through the dashboard, AI predictions can be rated by farmers, which informs adjustments made to the model and indicates a co-evolutionary system that changes with feedback from users [10].

12.6.4 Farmer Engagement and Training

The system also has a built-in training module for farmers to understand and learn how to interpret AI outputs and manage the hardware, which is delivered through a mobile dashboard interface or community workshops [2]. For instance, a brief video tutorial on how to interpret Grad-CAM heatmaps equips farmers with the knowledge to receive usable machine learning-based work [11]. Workshops in regional languages (e.g., Hindi or Swahili) are becoming more accessible [5]. In the case of dragon fruit, farmers completed training and, when interviewed after using the tool, reported that their confidence had increased by 40% [10]. This is consistent with the user-oriented direction of Industry 5.0, focusing on trust and cooperation among workers [2].

12.6.5 Community and Policy Implications

The system promotes trust, motivating the community to join farming networks where best practices are shared among farmers [2]. It also provides a useful example for policymakers preparing to scale precision agriculture, which complements the national priorities to improve food security and sustainability [6]. Governments might even assist in subsidizing edge devices for smallholder farmers to have a higher effect everywhere [5]. Since data ownership is in the hands of farmers, the ethical design of such a system raises some privacy issues. The system was piloted within local cooperatives, which resulted in a 30% increase in community engagement [2].

12.7 Performance and Practicality

Field tests over 30 days confirmed the system's reliability across diverse conditions. Key metrics include:

- **Accuracy:** 93% (ViTs), 91% (CNNs) [1, 8].
- **Latency:** Under 150 ms per image, enabling real-time decisions [7].
- **Power:** Less than 10 W, ideal for solar-powered farms.
- **Uptime:** 98%, robust in dust, rain, and heat.

Compared with cloud-based systems, which require constant Internet access and consume over 80 W, this edge-based approach is faster, cheaper, and autonomous, making it ideal for smallholder farmers in remote regions [5, 16].

Table 12.6 compares latency, Internet requirements, and power consumption between edge-based and cloud-based AI systems.

Table 12.7 shows the details of system accuracy and latency under various environmental conditions, such as sunny, cloudy, and rainy weather.

Appendix 12.B (Table 12.B.1 and Table 12.B.2) provides detailed performance metrics.

Table 12.6 Edge AI vs. cloud AI.

Feature	Edge AI	Cloud AI
Latency	<150 ms	>500 ms
Internet needed	No	Yes
Power	<10 W	>80 W

Table 12.7 Performance across conditions.

Condition	Accuracy (%)	Latency (ms)
Sunny	93	114
Cloudy	91	120
Rainy	90	125

12.8 Conclusion and Future Horizons

Edge AI and IoT cameras are two key technologies paving the way for a future in global agriculture by actively forming intelligent, self-managing fields to produce food sustainably. In this chapter, we present a real-time nutritional monitoring system that serves to bridge the gap between what farmers need to know and when they need to know it, in order to sense and rectify potential nutrient deficiencies as quickly as feasible. Drawing on case studies of dragon fruit, coffee, and maize to show how transformative this system can be: reducing fertilizer use by 20–30%, increasing yield quality while enabling smallholder farmers to compete with the best agriculturists in the world. The solution also offers farmers from rural South Asia to East Africa access to some of the most advanced technology through transparent features like heatmaps and simplified diagnostics. Due to its offline, inexpensive design, the system democratizes precision agriculture and makes it nearly universally applicable even in remote areas that lack conventional technology, a way in which the platform contributes to equity and resilience.

Appendix 12.A AI Training Pipeline

- **Dataset:** 70% training, 15% validation, 15% test, covering diverse crops (dragon fruit, coffee, maize) and regions [21].
- **Loss:** Categorical cross-entropy for classification tasks [10].
- **Optimizer:** Adam with learning rate scheduling for convergence [8].
- **Calibration:** Weighted loss with soil/tissue data to align predictions [11].
- **Techniques:** Quantization, pruning, and knowledge distillation for edge efficiency [7].

Appendix 12.B Performance Metrics

In the future, the technology has a lot of potential for development. Over the next handful of years, crop health detection should rise 10–15% on detectable yields from hidden deficiencies (which not only get detected by multispectral imaging streams at a moment in time). With the fairer access conditions, not only are these private edge learning models (e.g., sensor nodes and farm-level compute clusters) all connected with farms together to share this wisdom in your hand for a real global ecosystem of smart agriculture that is personalized to regional specifics. Furthermore, robotic pruning, harvesting, and planting machines could automate the farming process even further than it already is, and thus, less human labor would be required to do that work, and time spent doing so would also diminish. The system, resulting in a 20–30% reduction in the cost of inputs, stabilizing yield, and, for sustainability, will bring economic resilience and equity to the farmers, particularly those located in resource-poor areas. On the environmental end, it minimizes fertilizer waste and protects soil and water resources for generations to come. This type of innovation is also a blueprint for the kind of agriculture that will be necessary to feed the world in an appetizing manner that does not deplete soil resources and enhance food security with tangible economic and social benefits as well.

Table 12.B.1 Latency across platforms.

Platform	CNN latency (ms)	ViT latency (ms)
Jetson Nano	82	114
Raspberry Pi 5	124	158

Table 12.B.2 Power and thermal stability.

Device	Power (W)	Thermal range (°C)
Jetson Nano	4.7	38–61

References

1 Zhang, M., Zhang, J., Wu, Z. et al. (2020). Crop pest recognition using an improved convolutional neural network with transfer learning and data augmentation. *Computers and Electronics in Agriculture* 174: 105522. https://doi.org/10.1016/j.compag.2020.105522.

2 Rao, S. and Neethirajan, S. (2025). Computational architectures for precision dairy nutrition. ResearchGate. [Online]. Available: https://www.researchgate.net/publication/393164088.

3 Pintus, M., Colucci, F., and Maggio, F. (2025). Emerging developments in real-time edge AIoT for agricultural image classification. *IoT* 6 (1): 122–141. [Online]. Available: https://www.mdpi.com/2624-831X/6/1/13.

4 Darak, M. (2025). Real-time disease detection using IoT and edge AI. *Journal of Computer Science* 20 (1): 122–136. [Online]. Available: https://computersciencejournal.org/wp-content/uploads/13-JCS1632.pdf.

5 Abed, N., Murugan, R., Deldari, A. et al. (2025). IoT and AI-driven solutions for human-wildlife conflict: advancing sustainable agriculture and biodiversity conservation. *Smart Agricultural Technology* 10: 100829. https://doi.org/10.1016/j.atech.2025.100829.

6 Pandey, J.K., Yadav, A.K., Purushotham, M. et al. (2022). Colorimetric detection of deltamethrin pesticide from biological samples based on the peroxidase-mimic catalytic activity on MnO_2/SnO_2. *Key Engineering Materials* 928: 45–51. https://doi.org/10.4028/p-4ti1i0.

7 Shabir, M.Y. (2025). Optimizing AI for IoT. Ph.D. Thesis, Univ. Turin. [Online]. Available: https://iris.unito.it/bitstream/2318/2088150/2/PHDTHESIS_Shabir_yasir_2025.pdf.

8 Lazcano-García, C. and García-Resendiz, K.G. (2025). Deep learning for grapevine diseases. *AgriEngineering* 7 (3): [Online]. Available: https://www.mdpi.com/2624-7402/7/3/63.

9 Das, P., Jain, C., Ansul, and Singh, M. (2024). HARVEST-AI: UAV-based AI system. IEEE Conf. Artif. Intell.. [Online]. Available: https://ieeexplore.ieee.org/document/10947889.

10 Beemkumar, N., Gupta, S., Bhardwaj, S. et al. (2023). Activity recognition and IoT-based analysis using time series and CNN. In: Handbook of Research on Machine Learning-enabled IoT for Smart Applications Across Industries (ed. N. Goel and R.K. Yadav), 350–364. IGI Global https://doi.org/10.4018/978-1-6684-8785-3.ch018.

11 Talukdar, V., Veeraiah, V., Roy, D.R. et al. (2023). An innovative approach: typicality-based collaborative filtering for recommender systems. *Proceedings of the 3rd International Conference on Advancement in Electronics & Communication Engineering (AECE)*, pp. 863–868. https://doi.org/10.1109/AECE59614.2023.10428576.

12 Zualkernan, I., Abuhani, D.A., Hussain, M.H. et al. (2023). Machine learning for precision agriculture using UAVs. *Drones* 7 (6): 1–21. [Online]. Available: https://www.mdpi.com/2504-446X/7/6/382.

13 Chaithra, N., Jha, J., Sayal, A. et al. (2024). Enhancing underwater imagery with AI/ML. In: Artificial Intelligence in Industrial Applications. Springer [Online]. Available: https://link.springer.com/chapter/10.1007/978-3-031-64642-3_14.

14 Srinivas, P., Veeraiah, V., Thejaswini, K.O. et al. (2025). The impact of electromagnetic and optical radiation on physical and biological systems. *Journal of Nano- and Electronic Physics* 17 (4): 04012. https://doi.org/10.21272/jnep.17(4).04012.

15 Pandey, J.K., Kumar, S., Lamin, M. et al. (2022). A metaheuristic autoencoder deep learning model for an intrusion detector system. *Mathematical Problems in Engineering* 2022: 11, 3859155. https://doi.org/10.1155/2022/3859155.

16 Amugongo, L.M., Kriebitz, A., Boch, A., and Lütge, C. (2022). Mobile computer vision for food recognition. *Healthcare* 11 (1): 59. [Online]. Available: https://doi.org/10.3390/healthcare11010059.

17 Chen, X. and Kamavuako, E.N. (2023). Vision-based food intake monitoring. *Sensors* 23 (13): 6137. [Online]. Available: https://doi.org/10.3390/s23136137.

18 Zualkernan, I., Abuhani, D.A., Hussain, M.H. et al. (2023). "Machine learning for precision agriculture using UAVs. Academia.edu. [Online]. Available: https://www.academia.edu/download/111240065/pdf.pdf.

19 FAO (2023). The State of Food Security and Nutrition. Rome, Italy. [Online]. Available: https://www.fao.org/publications/sofi/2023.

20 Lin, I.-A., Cheng, Y.-W., and Lee, T.-Y. (2024). Enhancing smart agriculture with lightweight object detection: MobileNetv3-YOLOv4 and adaptive width multipliers. *IEEE Sensors Journal* 24 (23): 40017–40028. [Online]. Available: https://ieeexplore.ieee.org/document/10721319.

21 Paramasivam, A., Ferlin Deva Shahila, D., Jenath, M. et al. (2024). Development of artificial intelligence edge computing-based wearable device for fall detection and prevention of elderly people. *Heliyon* 10: e28688. [Online]. Available: https://doi.org/10.1016/j.heliyon.2024.e28688.

13

Computational Imaging Framework for Soil Texture Characterization and Nutrient Status Prediction

Rohit Kumar Choudhury[1], *Mouli Sarkar*[2], *Dipak Uttamrao Pakhre*[1], *Arijit Ghosh*[1], *Shubhadip Dasgupta*[1], *Mercy Chinneihoi Haokip*[1], *and Asim Biswas*[3]

[1] *Department of Agricultural Chemistry and Soil Science, Bidhan Chandra Krishi Viswavidyalaya, Nadia, West Bengal, India*
[2] *Agricultural and Food Engineering Department, Indian Institute of Technology, Kharagpur, West Bengal, India*
[3] *School of Environmental Sciences, University of Guelph, Guelph, Ontario, Canada*

13.1 Introduction

Soil is a heterogeneous system that acts as one of the most essential natural resources, which provides support to plants, animals, and people in a multi-directional way. The sustainability and long-term viability of the soil system rely on the interaction between its physical, chemical, and biological properties, along with the structure and mechanical composition [1]. As the world's population grows alarmingly, people are becoming more aware of how limited soil resources are. This has led to greater worries about soil security and doubts about the Earth's carrying capacity. In India, the agricultural sector, which accounts for about 18.33% of the national GDP and employs nearly half of the workforce, faces the significant challenge of meeting the food needs of a growing population due to intensive farming practices causing soil degradation in various regions and so it has become essential to understand and manage soil health to ensure the long-term sustainability of agriculture [2–4]. This highlights the pressing importance of sustainable soil management to tackle challenges like climate change, water shortages, changing weather patterns, food security, and the loss of wildlife.

13.1.1 Need for Soil Data

Soil has a dominant and dynamic function within the Earth's biosphere, supporting ecosystem services, agricultural yields, and environmental quality ranging from local fields to the global carbon cycle. Its status is regulated by a vast array of land management practices, including cultivation, irrigation, fertilization, defoliation, and other human alterations that change physical structure, chemical properties, and biological activity. Consequently, rigorous examination and continuous monitoring of soil physical, chemical, and biological characteristics are essential for determining the suitability of land management options and guiding adaptive stewardship. That's why the soil data can be called the "Genetic Code" of soil [5, 6].

13.1.2 Need for Spectral Soil Sensing (SSS)

Soil properties are highly heterogeneous and influenced by environmental interactions, making them difficult to capture with conventional techniques. However, traditional hierarchical soil analyses, which rely on destructive sampling and wet laboratory methods, are laborious, expensive, and inadequately designed for real-time applications or large-scale monitoring. In recent decades, spectral soil sensing (SSS) has come to be used as an effective, rapid, nondestructive, and scalable method for characterizing soil as well as water bodies and vegetation, and has been internationally explicitly recognized by the Global Soil Partnership under FAO as a principal tool to enhance soil governance and support sustainable soil management [7]. SSS is based on the principle of the radiation-matter interaction mechanism. Those electromagnetic spectral bands which are transparent to the earth's atmosphere are called "Atmospheric Windows," which provide diagnostic absorption features, continuum shapes, and scattering behaviors within the visible (VIS: 350–700 nm), near-infrared (NIR: 700–1000 nm), and shortwave infrared (SWIR: 1000–2500 nm) regions. Unlike traditional wet-chemistry procedures, SSS does not involve any reagents, allows one to simultaneously estimate several soil attributes, and can increase sampling density significantly when used from ground, airborne, or satellite platforms (Figure 13.1a–f).

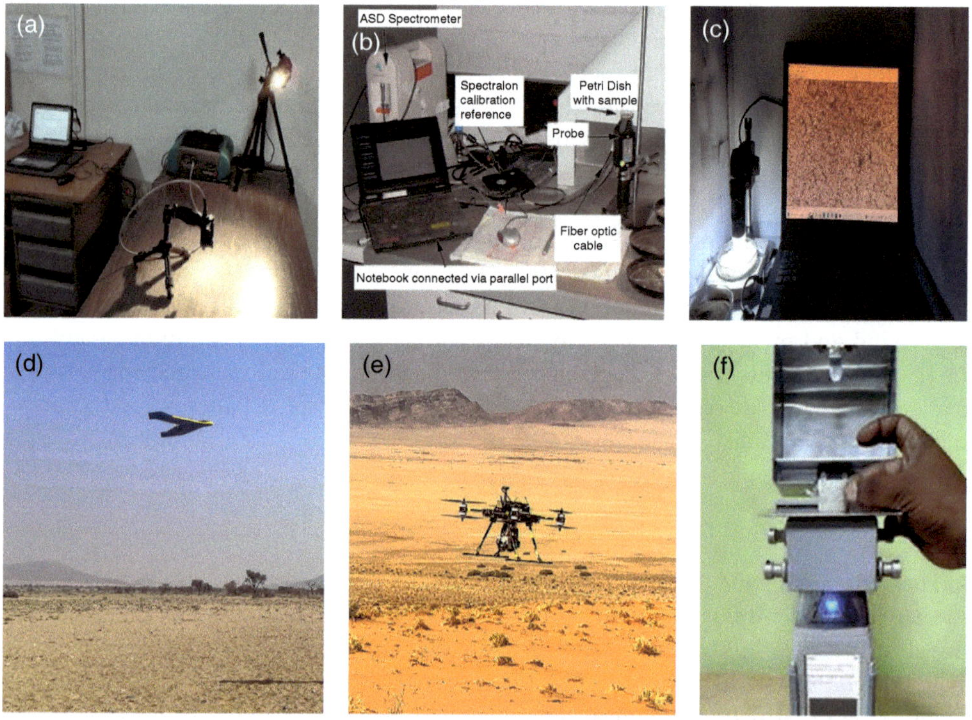

Figure 13.1 (a) ASD Field Spec 4 spectroradiometer. *Source:* Courtesy of Arijit Ghosh; (b) instrumentation of ASD spectrometer. *Source:* [8] / with permission of CRC LEME; (c) USB microscope-based soil sensing. *Source:* Courtesy of Shubhadip Dasgupta; (d) fixed-wing UAV-based remote sensing. *Source:* [9] / Springer Nature / CC BY 4.0; (e) UAV- drone-based sensing. *Source:* [10] / Springer Nature / CC BY 4.0; (f) PXRFS-based imaging scanner. *Source:* [11] / with permission of Elsevier.

Over the past three decades, soil reflectance spectroscopy in the visible, near-infrared, and mid-infrared regions (Vis–NIR–MIR) has demonstrated reliability, reproducibility, and cost-effectiveness. Consequently, imaging spectroscopy has emerged as a powerful tool for assessing soil health, facilitating precision agriculture, and addressing the increasing need for high-resolution soil data.

13.2 Spectral Imaging and Soil Sensing

13.2.1 Digital Image

Remotely or proximally sensed images are visual depictions of Earth's surface captured through satellites, aircraft, or by ground-based sensors (e.g., spectroradiometers, smartphone cameras) (Figure 13.1a–f). These images can either be analog or digital; analogue images include aerial photographs produced through film cameras, while digital images are captured from electronic sensors mounted on satellite systems, aircraft, smartphones, or directly by any spectroradiometer [12–14]. Digital cameras incorporated in modern smartphones can also be operated as imaging sensors. Digital image is structured as a matrix of picture elements called pixels, each organized in rows and columns, with every pixel corresponding to a specific ground area (Figure 13.2a). The value assigned to each pixel, termed its digital number (DN), reflects the measured energy, such as VIS or infrared radiation received from the Earth's surface within a specific spectral band (Figure 13.2b). This DN usually represents the mean energy for the entire terrain area encompassed by that particular pixel. Such digital images, composed of enormous pixel units, are called Raster images, whereas those that are constructed by several mathematical equations or formulas and represent any shape, lines, curves, or polygons are called vector images (Table 13.1).

The spectral data gathered across multiple bands can be fused to form composite images rather than any single band image (i.e., gray scale) (Figure 13.3a), which improves the ability to interpret surface features. Digital red (R), green (G), and blue (B) (RGB) photography uses the three primary intensities of RGB to show color (Figure 13.3b). In most cases, these composites are created by assigning the data from individual bands to the primary colors RGB on a display, generating what's known as a color composite image. When compositing uses nontrue color bands such as NIR, red, and green, they generate what is termed a false color composite (FCC). The blue band is often omitted in satellite remote sensing due to its high sensitivity to atmospheric scattering. The NIR band is especially advantageous for evaluating vegetation, as healthy plants exhibit considerable reflection in this spectral region, making it valuable for monitoring plant health and biomass. In typical FCC imagery, green, red, and NIR bands are displayed as blue, green, and red, respectively; this makes vegetation appear distinctly red, water bodies show as blue, and bare soils manifest in tones of brown and gray (Figure 13.3c). For true color composites, blue, green, and red satellite bands are displayed as their natural colors, closely matching what the human eye would perceive.

13.2.2 Spectral Indices

The importance of satellite data in precision agriculture comes from its ability to provide regular, high-resolution images over large areas. This ability supports advanced crop

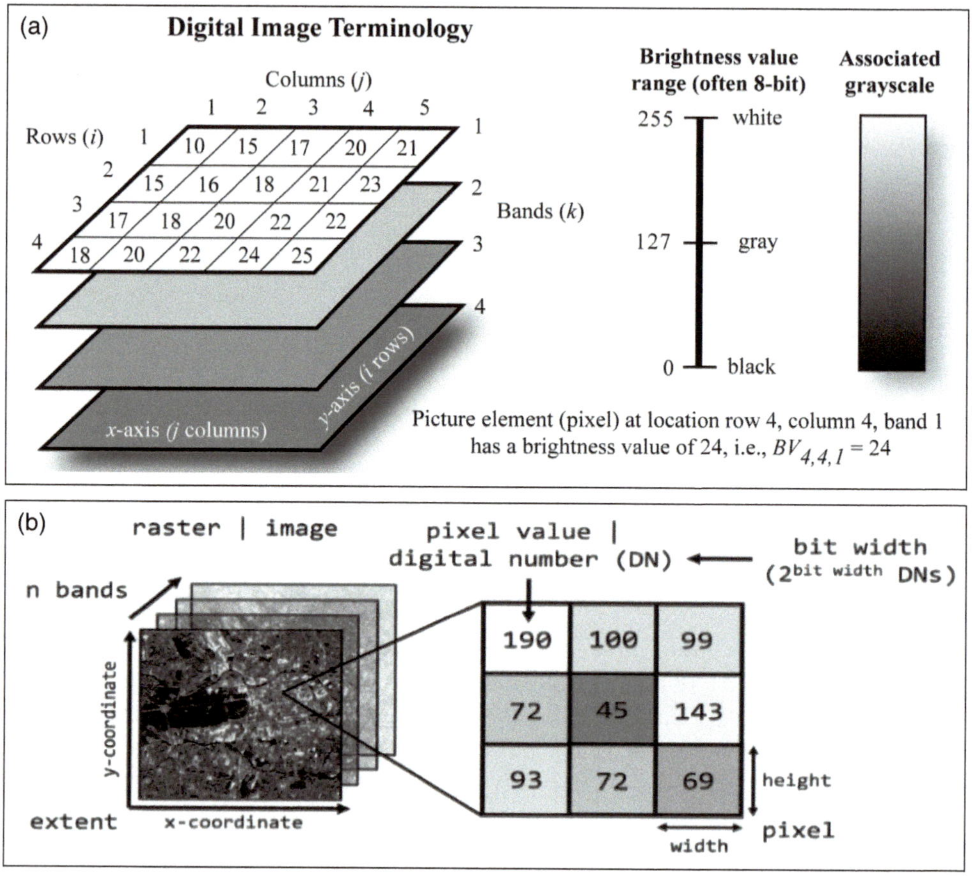

Figure 13.2 (a) Construction of a digital image (raster) [15]/Mapping Around. (b) Construction of "Pixel" in a raster image [16] University Library of the Humboldt University of Berlin.

management techniques, including variable rate application of fertilizers and pesticides, as well as identifying differences in land within fields. Various satellite sensors have different levels of suitability based on their geographical resolution, spectral bands, and revisit frequency for specific agricultural tasks. Spectral indices derived from remote sensing are often used to assess vegetation health, soil fertility, and land degradation (Table 13.2; Figure 13.4).

13.2.3 Imaging Spectroscopy (IS) – A Novel Approach to Agriculture

Hyperspectral imaging (HSI), another name for imaging spectroscopy (IS), is a sophisticated remote sensing method that combines the advantages of imaging and spectroscopy to gather both spatial and spectral data from a target surface. By acquiring spectral information for each pixel within a two-dimensional detector array, HSI produces a three-dimensional dataset known as a hypercube. Unlike conventional RGB or multispectral systems such as Landsat or Sentinel-2, which record only a few broad wavelength bands, HSI acquires hundreds of narrow, contiguous bands across the VIS, NIR, and SWIR regions (400–2500 nm) [20].

Table 13.1 Comparison between raster and vector images.

Sl. No.	Features	Raster image	Vector image
1	Construction	Composed of Pixel units	Mathematical equations (lines, curves, polygons, etc.)
2	File size	Larger in size, especially for higher resolution images	Smaller in size
3	Formats	TIFF[a], JPEG[b], BMP[c], PNG[d] etc.	PDF[e], EPS[f], SVG[g] etc.
4	Editing	Editing capabilities are limited to the pixel level	Editing capabilities are extensive and shape-based
5	Scalability	Scalability is limited. Quality decreases when resized	No quality loss upon resizing. Scalability is more
6	Data structure	Simple	Complex
7	Suitability	Suitable for continuous datasets	Suitable for discrete datasets
8	Application	Multiband images (e.g., NDVI[h], NDSI[i])	Maps, roads, boundaries, etc.

[a] TIFF – tagged image file format.
[b] JPEG – joint photographic experts group.
[c] BMP – bitmap.
[d] PNG – portable network graphics.
[e] PDF – portable document format.
[f] EPS – encapsulated postscript.
[g] SVG – scalable vector graphics.
[h] NDVI – normalized difference vegetation index.
[i] NDSI – normalized difference salinity index.

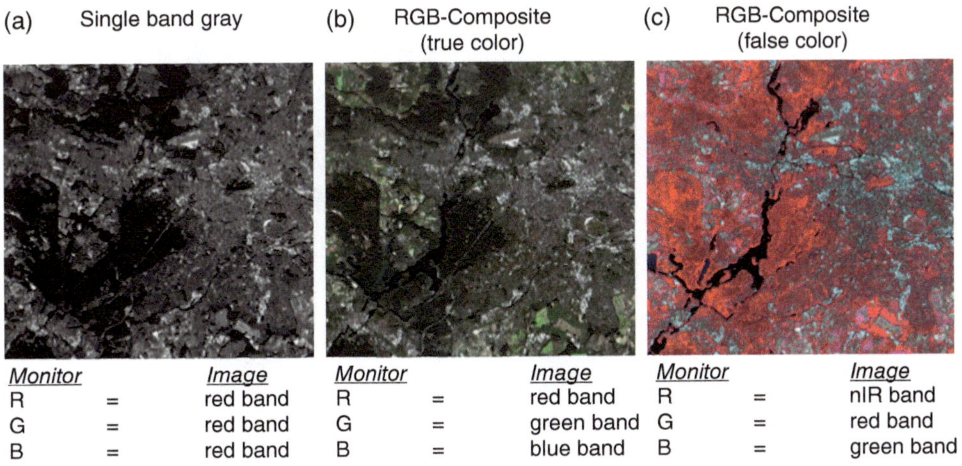

Figure 13.3 (a) Gray-scale image (single band); (b) true color RGB image composite; (c) false color composite image [15].

Table 13.2 Common spectral indices for soil feature extraction.

Category	Index	Formula	Application/property
Soil reflectance/ color	Brightness Index (BI)	$\sqrt{((Red^2 + Green^2 + Blue^2)/3)}$	Average soil reflectance magnitude
	Saturation Index (SI)	(Red − Blue)/(Red + Blue)	Soil spectral slope
	Hue index (HI)	(2 × Red − Green − Blue)/ (Green − Blue)	Soil chromaticity (primary colors)
	Coloration Index (CI)	(Red − Green)/(Red + Green)	Soil color, iron content
	Redness Index (RI)	$Red^2/(Blue \times Green^3)$	Hematite (iron oxide) content
Soil–vegetation separation	Normalized Difference Vegetation Index (NDVI)	(NIR − Red)/(NIR + Red)	Vegetation cover, greenness
	Soil Adjusted Vegetation Index (SAVI)	((NIR − Red)/(NIR + Red + L)) × (1 + L)	Vegetation index adjusted for soil brightness
	Bare Soil Index (BSI)	((SWIR1 + Red) − (NIR + Blue))/((SWIR1 + Red) + (NIR + Blue))	Bare soil identification
	Modified Bare Soil Index (MBI)	((SWIR1 + SWIR2 + NIR)/ (SWIR1 − SWIR2 − NIR)) + 0.5	Bare land discrimination
Soil mineral content	Clay Minerals Ratio	SWIR1/SWIR2	Clay mineral absorption
	Iron Oxide Ratio	Red/Blue	Iron oxide–rich soils
	Normalized Difference Clay Index (NDCI)	Varies (SWIR ratios)	Soil clay content estimation
Soil moisture/ water content	Normalized Difference Moisture Index (NDMI)	(NIR − SWIR1)/(NIR + SWIR1)	Vegetation and soil moisture
	Normalized Difference Water Index (NDWI)	(Green − NIR)/(Green + NIR)	Open water detection, soil wetness

VIS, Visible; NIR, Near-Infrared; SWIR, Shortwave Infrared; MIR, Mid Infrared.
Source: Modified [17, 18].

This fine spectral resolution enables the detection of subtle variations that are invisible to the human eye and allows detailed monitoring of geophysical, biophysical, and biochemical processes at Earth's surface. In soil science, IS has proven particularly valuable because soil spectra in the VIS–NIR–SWIR region contain diagnostic absorption features linked to organic carbon, moisture, and mineral composition [10, 21, 22]. A conceptualized framework of remotely/proximally sensed data from acquisition to field level implementation level, and developing prediction models and maps for several soil parameters from spectral reflectance and environmental covariates, and digital mapping has been given in Figure 13.5.

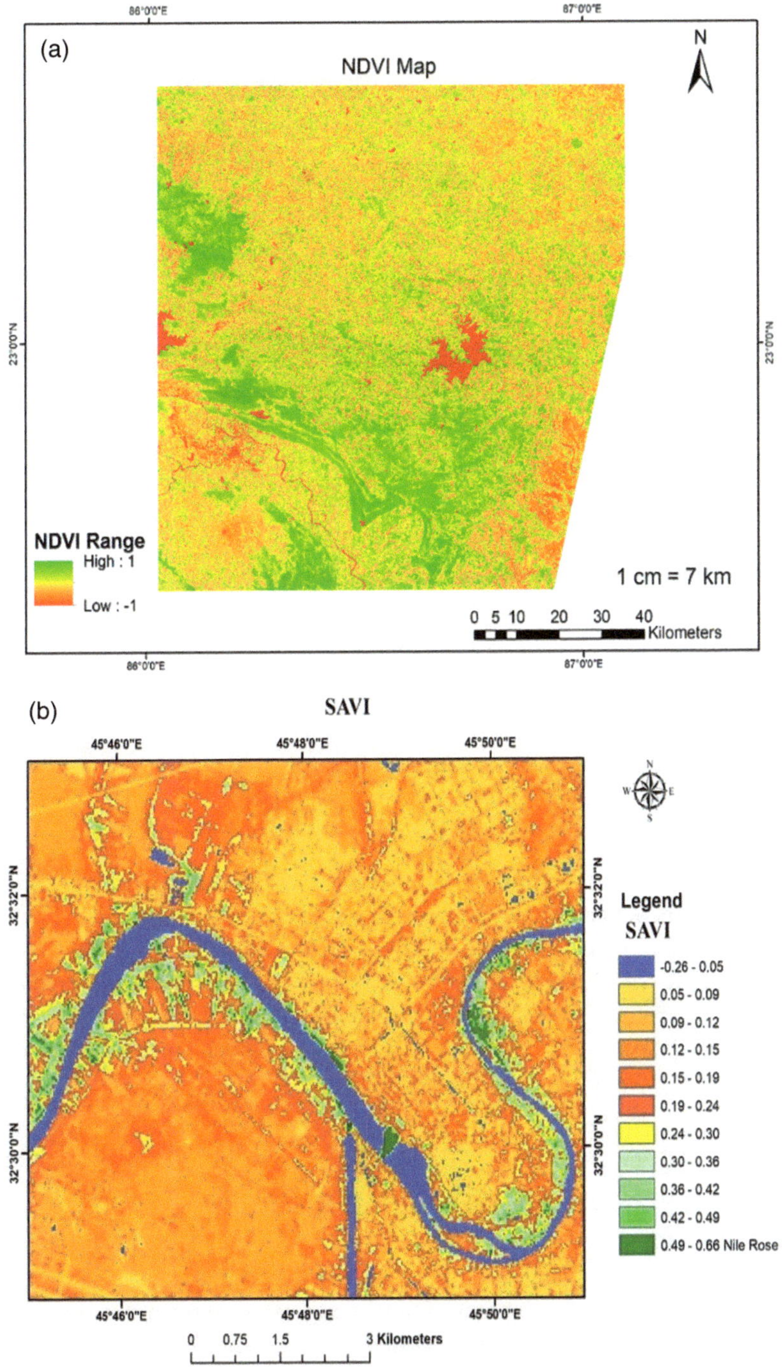

Figure 13.4 (a) Normalized Difference Vegetation Index (NDVI) map of Purulia region, West Bengal, India (prepared with the help of ESRI ArcMap 10.6). *Source:* with permission [5] 2025, Springer. (b) SAVI map of the Wasit region (Tigris river basin area) of Iraq. *Source:* Reprinted with permission [19] 2025, Elsevier.

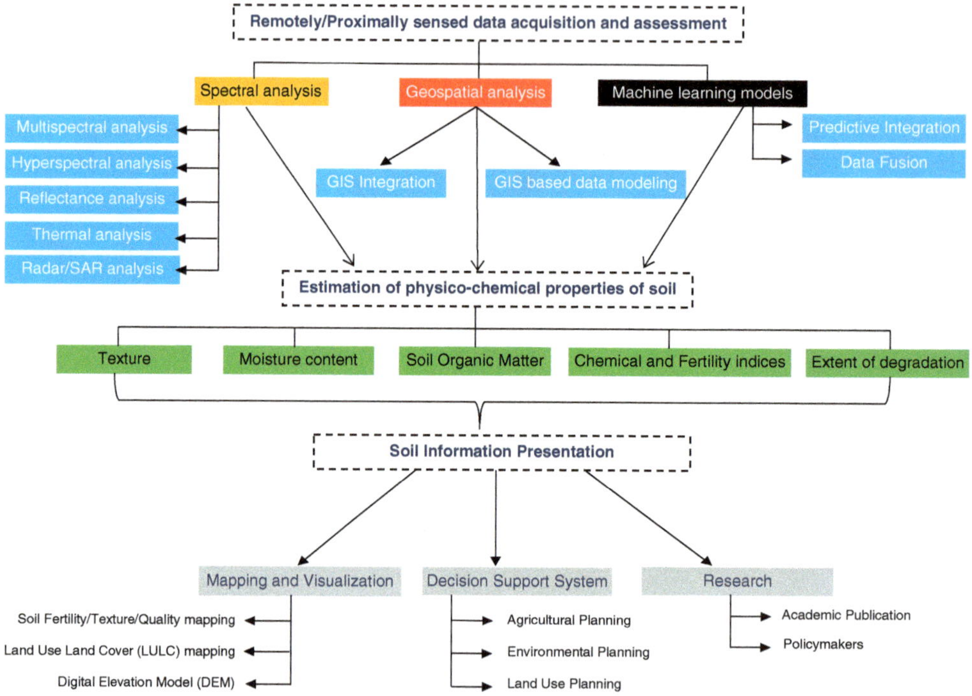

Figure 13.5 A conceptualized framework of remotely/proximally sensed data from acquisition to the field level implementation level.

13.2.4 Visible, Near-Infrared, Multispectral, and Hyperspectral Imaging

Compared to some of the traditional methods of soil examination, VIS and infrared reflectance spectroscopy offer the benefits of being quicker, more accurate, and less costly, with their spectral range 400–1000 nm. Multispectral Imaging acquires a small number (3–13) of broad bands across VIS–NIR–SWIR, useful for large-scale crop classification, yield prediction, and stress monitoring [5, 23, 24]. HSI collects hundreds of narrow, contiguous bands (400–2500 nm), enabling detailed detection of soil and crop biochemical properties (Figure 13.6) [5, 23, 25].

13.3 Soil Emission Spectra, Physics of Soil Reflectance and Absorption Features

When sunlight strikes any surface on Earth, some of that energy can be transmitted through the material, some absorbed, and some reflected toward the atmosphere. Scientists measure the proportion of incoming solar energy that gets reflected by each surface to quantify its reflectance properties; this measurement, graphed against different wavelengths, is known as a spectral reflectance curve. The shape and pattern of a spectral reflectance curve are unique to each surface type, soil, vegetation, and water; all have distinct "spectral

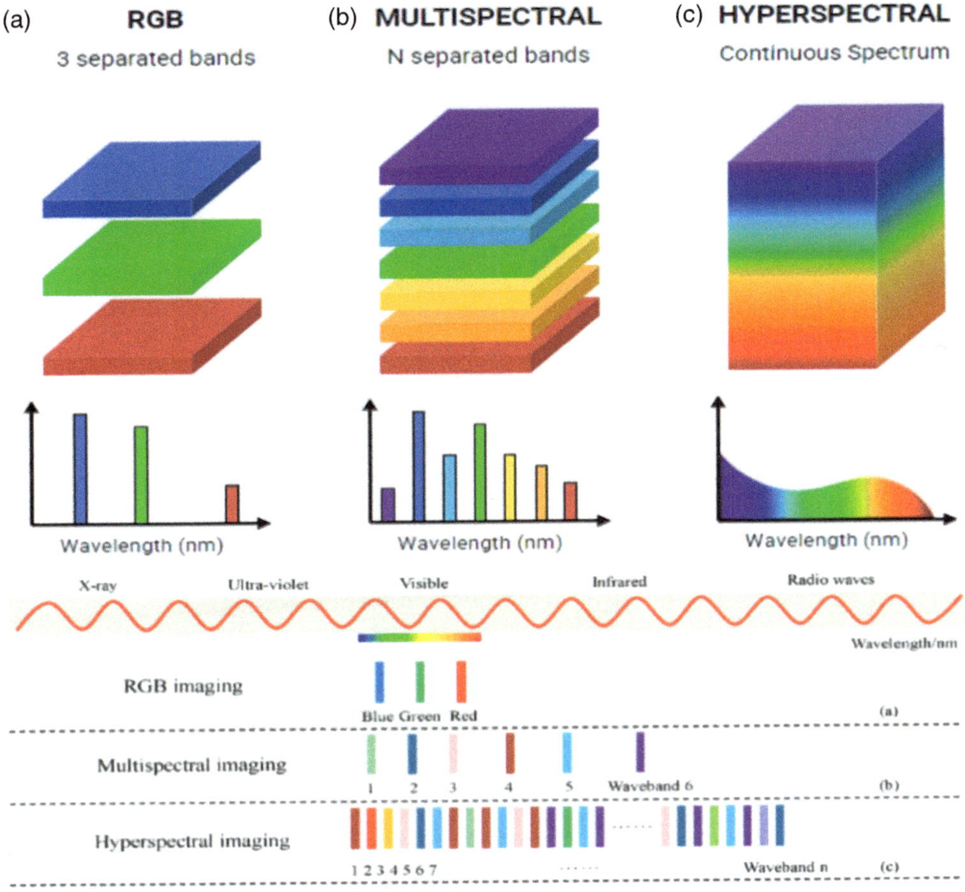

Figure 13.6 A comparison between RGB (a) vs. multispectral (b) vs. hyperspectral image (c). RGB and multispectral images provide only discontinuous and discrete spectra, whereas hyperspectral images provide a complete and continuous spectrum forming a hypercube in a single pixel of a digital image [25].

signatures," and these differences allow for their identification and analysis from remote sensing images (Figure 13.7a). The spectral signature curves from different horizon layers from two distinct soil orders are given in Figure 13.7b,c. The concept of imaging spectroscopy is based on the interaction of each material with incoming light in a unique manner, emitting specific spectral signatures due to reflection, absorption, and scattering processes. These signatures act as a spectral database to assess the physical and chemical composition and characteristics of the interacting materials. In the case of soil systems, there are two main mechanisms which impart characteristic absorption features: electronic transitions, usually occurring between 450 and 2500 nm and often related to iron oxides, and molecular vibrations, including those of hydroxyl, carbonate, and sulphate groups, especially in the SWIR region between 2000 and 2500 nm. Spectral signatures are hence considered as intrinsic soil properties, which vary with different soil types [28]. In the VIS range (400–700 nm),

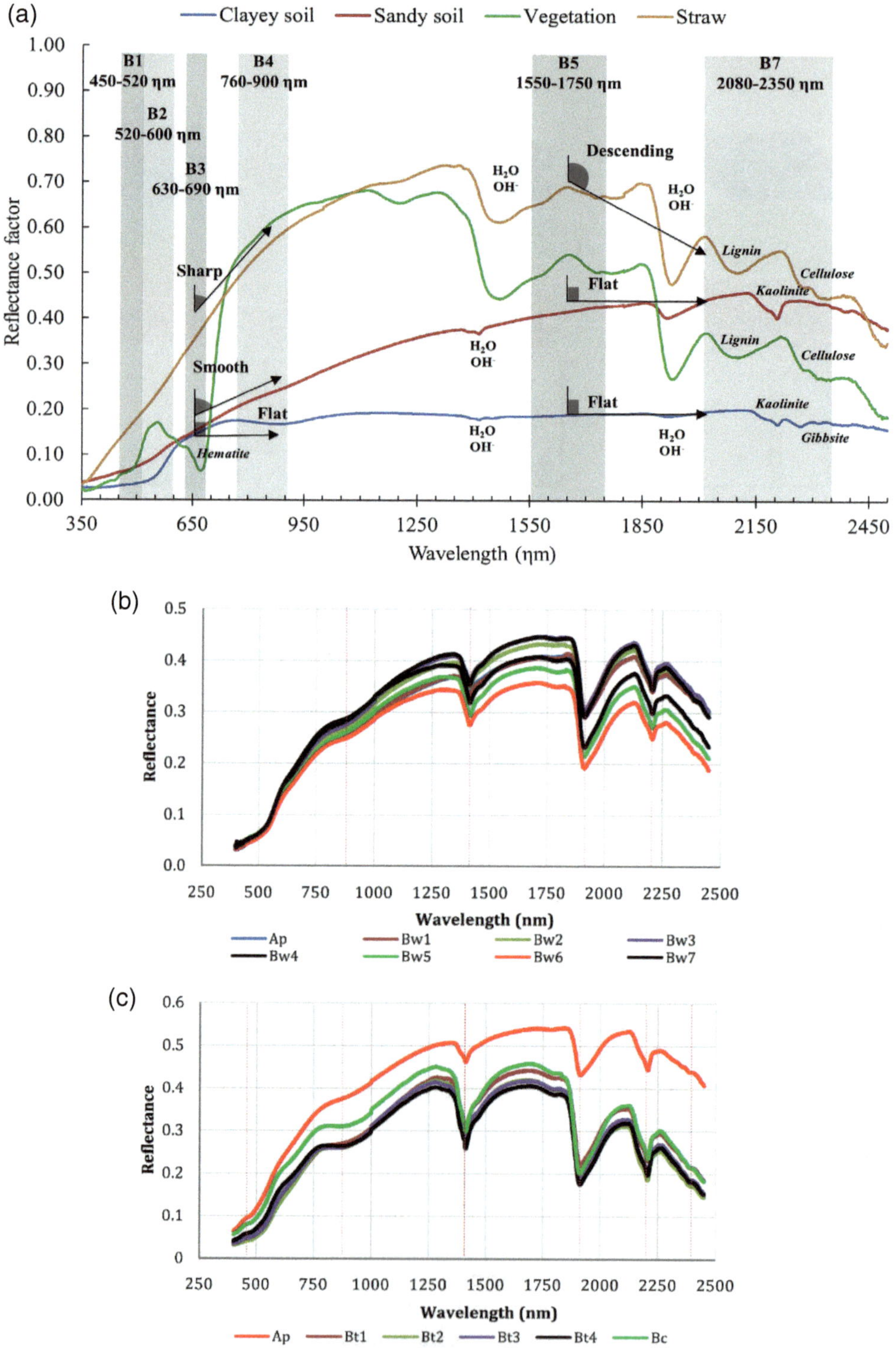

Figure 13.7 (a) Typical reflectance spectra of soil, vegetation, and straw. *Source:* with permission [26] 2025, Elsevier. (b) Reflectance spectra of an inceptisol from different horizons. *Source:* with permission [27] 2025, Taylor & Francis. (c) Reflectance spectra of an alfisol from different horizons. *Source:* with permission [27] 2025, Taylor & Francis.

the absorptions take place primarily due to electronic transitions in iron-bearing minerals such as hematite and goethite [29]. In the NIR region (700–2500 nm), absorptions occur due to overtones and combination bands of covalent bonds involving O—H, C—H, and N—H groups [30]. The O—H groups in minerals, as well as the functional groups present in organic matter, largely explain the characteristic absorption features in NIR soil reflectance spectra [31]. Soil spectra typically exhibit three major absorption peaks near 1400, 1900, and 2200 nm. The features at 1400 and 1900 nm are linked to water (O—H vibrations), while the 2200 nm region is mainly associated with clay minerals (metal—OH bonds) [32, 33]. Additional absorptions occur from iron oxides (870–1000 nm) and carbonates (2300–2350 nm, with weaker bands between 1850 and 2160 nm) [34, 35]. Chemical chromophores (moisture, organic matter, clays, and iron oxides) absorb at specific wavelengths, whereas physical chromophores (particle size, surface geometry) affect the overall spectral shape (Figure 13.8). Variations in absorption features reflect chemical composition, while changes in spectral shape are mainly due to physical properties.

13.4 Image Acquisition Techniques

13.4.1 Data Sources and Environmental Covariates

Soil texture and fertility can be effectively evaluated through imaging, which is predicted based on the methodological amalgamation of multiple data sources and environmental covariates. Recent methods of digital soil mapping acknowledge the integration of heterogeneous streams of data to better elucidate the complex interactions between different soil properties and environmental determinants across temporal and spatial dimensions. This approach of multiple source integration has developed from the comprehensive studies, which concluded that no singular dataset can satisfactorily represent the intricate factors that control soil heterogeneity within a particular landscape. Environmental covariates act as proxies for the essential soil-forming processes.

13.4.2 Field and Laboratory Soil Data (Ground-Truthing)

Field sampling and laboratory analysis represent the scientific basis of image-based soil prediction, digital soil mapping, and soil modeling, thereby providing the ground truth data that is required for validation and calibration of remote sensing methodologies. Rapid field measurements of total elemental concentrations that provide immediate feedback about soil properties have been facilitated by portable X-ray fluorescence (PXRF), visible and near-infrared (VNIR) spectroscopy, USB microscopy, and diffused reflectance spectroscopy (DRS), which allows for expedited assessments of several physico-chemical properties of soil directly in situ. Their vast application in the field of soil sensing has been discussed later.

13.4.3 Satellite Imagery and Aerial Platforms

Satellite systems offer a rapid temporal coverage and vast geographical area, which are crucial for regional soil mapping, whereas unmanned aerial vehicles (UAVs), which are

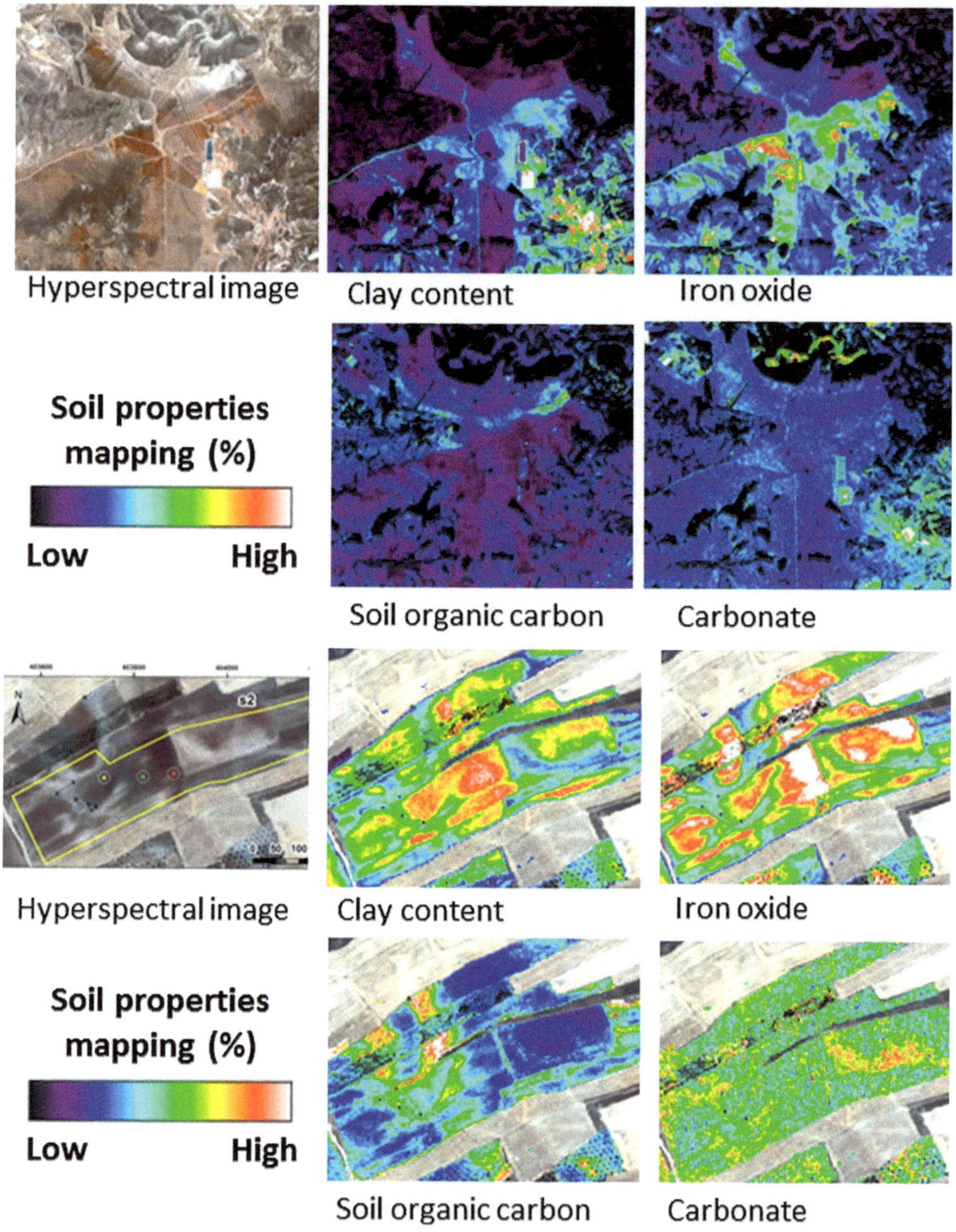

Figure 13.8 Mapping of various chemical constituents of the soil surface using the HyMap hyperspectral sensor and HYSOMA software. *Source:* with permission [36] 2025, Elsevier.

equipped with sophisticated multispectral and hyperspectral sensors, can capture imagery at spatial resolutions frequently less than 5 cm, thus unveiling soil variability patterns that remain undetectable by satellite sensors. Their vast application in the field of soil sensing has been discussed later.

13.5 Image Processing

Few raw sensor data are directly usable as they usually exhibit atmospheric distortions, redundant spectral bands, and noise, which obscure the soil-specific signatures. To mitigate these problems, several preprocessing, feature extraction, dimensionality reduction (DR), and multicollinearity-handling techniques have been developed by researchers that transform raw imagery into scientifically useful data layers.

13.5.1 Image Preprocessing: Cleaning and Calibration

Remotely sensed data that is acquired from image sensors installed on satellites or airborne platforms normally possesses errors or deficiencies in that form. "Preprocessing," also called image restoration, image correction, or image rectification, is generally performed to correct the errors and defects in raw data. The term "preprocessing" implies preparing data for further analysis in ways that correct or make up for systematic or nonsystematic deficiencies.

Remote sensing preprocessing tends to enhance raw data by correcting various types of errors. These errors include atmospheric effects, like attenuation and scattering, which change the digital values of surface materials. Other factors include Earth's rotation, elevation, and curvature. Instrument-related issues, such as uneven detector response, mirror oscillation errors, and missing scan lines that require de-striping, are generally addressed to minimize the impact on data quality. Most of these errors are systematic and can be corrected before the data is delivered to the user. Depending on the study goals, the analyst selects appropriate preprocessing methods. In general, these techniques fall into two main categories: radiometric corrections and geometric corrections. Some additional steps are considered miscellaneous preprocessing if no direct errors are present.

13.5.2 Noise Reduction

Spectral data collected from satellite, airborne, or UAV sensors often contain high-frequency noise. This noise comes from sensor electronics, environmental changes, and soil surface irregularities. It can hinder feature extraction, so reducing it is an important first step. Savitzky-Golay smoothing is often used to lower spectral variability while keeping the main spectral peaks. Additionally, derivative transformations help reduce additive or multiplicative scatter effects and improve the comparability of soil samples. Systematic noise, like missing scan lines or stripes, is easy to identify. However, some noise may appear in images in an irregular pattern. Individual pixels with unusual DNs may be random noise if they are not systematic. Spatial filtering can help reduce this noise. Defective pixels are usually recognized by their significant DN differences when compared to nearby pixels in the same band. Noise reduction replaces these pixels with the average DN of surrounding pixels, often using 3×3 or 5×5 moving kernels. There are different types of filters used to remove the noise, such as convolution filtering, low-pass filter (smoothing filter), high-pass filter, edge detection filter, and adaptive filter.

13.5.3 Atmospheric Correction

While solar radiation is traveling toward the ground surface, it is absorbed or scattered by the atmosphere. The radiation that the target reflects or emits will be absorbed or scattered when it travels through the atmosphere prior to reaching a sensor. The ground receives direct solar radiation, plus sky light, which is radiation that has been scattered in the atmosphere. The sensor measures not only the radiation that has been reflected or emitted from the target, but it also measures the radiation that is a result of scattering associated with the target, plus the scattering from the atmosphere, which is referred to as path radiance. Since scattering is wavelength dependent, path radiance will vary from band to band.

We use atmospheric correction to eliminate these effects. We may start by describing the connection between the light leaving the Earth's surface and the radiance received at a sensor above the atmosphere (Eq. 13.1):

$$L_s = L_{tot}\rho T + L_p \tag{13.1}$$

- L_s = Energy detected by the sensor.
- L_{tot} = Total incoming radiation in a specific spectral band
- ρ = Fraction of reflected to incoming energy.
- T = Portion of radiation transmitted through the atmosphere.
- L_p = Path irradiance in a specific spectral band

Atmospheric effects on remotely sensed data are not regarded as inaccuracies but are instead recognized as an integral component of the signal received by the sensing instrument. Nevertheless, the removal of atmospheric effects is frequently essential, especially for scene matching and change detection analysis. Haze compensation techniques aim to reduce the impact of path radiance on remotely sensed data. One common approach in multispectral imagery is to measure the radiance over targets that have nearly zero reflectance. For instance, deep, clear water exhibits almost zero reflectance in the NIR region. Consequently, any signal detected over such areas corresponds to path radiance, which can then be subtracted from all pixels in that spectral band. This method is known as dark object subtraction. Approaches such as direct spectral conversion and empirical line correction are frequently used to adjust for these influences. For example, high-resolution multispectral images collected in Ohio were rectified using LiDAR and corrected with a low-pass filter to minimize variance caused by microtopography and image scanning.

13.5.4 Spectral Preprocessing

13.5.4.1 Spectral Transformation to Prepare the Data for Analysis

Various preprocessing steps are usually applied to the raw spectra. These steps are crucial because raw data contains lots of irrelevant noise that can negatively affect the accuracy of prediction models. Preprocessing helps to reduce noise and highlight the spectral features that are strongly connected to the soil property of interest.

For cleaning the spectra, the logarithmic transformation is a commonly used method. It maintains a linear relationship between reflectance intensity and the concentration of soil constituents. By using this formula ($A = \log[1/R]$), reflectance (R) is converted to absorbance. Soil samples don't always scatter light the same way. To fix this scattering problem, some specialized methods, such as the Kubelka–Munk function, multiplicative scatter correction (MSC), and standard normal variate (SNV), are used. These methods make the spectra more normal and better for accurate modeling. For noise reduction, averaging of spectra, the Savitzky–Golay filter, and moving average smoothing are commonly used. To emphasize diagnostic absorption features, techniques like continuum removal, baseline correction, first- and second-derivative transformations, and wavelet analysis are recommended.

13.5.4.2 Feature Extraction

After cleaning the raw spectral data, feature extraction is the next critical step to get precise surface reflectance data. This stage transforms the spectral data into a new collection of variables, or "features." These features are more closely and directly related to the exact soil characteristics. Hyperspectral sensors provide a lot of information, but the high dimensionality and dense sampling mean that many spectral bands contain redundant information. This is known as the "curse of dimensionality." It makes the data more complex and degrades the accuracy of classification and regression models if not addressed. Therefore, feature extraction is an essential step that aims to reduce redundancy and dimensionality, enhance the most important information, and model both spectral and spatial features to improve the success of data processing. This process enhances the relevant "signal" (e.g., the depth of a clay absorption band) while suppressing irrelevant variability and noise. Features can be extracted from the spectral domain, the spatial domain, or a combination of both.

13.5.4.3 Handling Multicollinearity and Redundancy

Hyperspectral datasets generally have hundreds of adjacent, strongly connected spectral bands. While fine spectral resolution gives detailed information on soil characteristics, it introduces the issue of multicollinearity. Because the sensor bands are so narrow and adjacent, the consecutive bands often measure almost the same signal. This means a lot of information is duplicated or overlaps. This duplication can potentially worsen model performance by magnifying noise and increasing computational effort without adding more information. Redundancy makes the dataset massive and slow to process. By increasing noise, it also weakens statistical and machine learning models' stability and interpretability. Redundancy arises because adjacent bands in hyperspectral data often measure nearly the same signal due to narrow bandwidths and sensor characteristics. This redundancy increases computational load without contributing additional information and can even degrade model performance by amplifying noise.

13.5.4.4 Band Selection and Feature Ranking

There are some filtering methods, such as stepwise regression, correlation-based feature selection (CFS), and mutual information ranking, that are usually used to automatically find and remove the repetitive bands. They only retain the most informative bands. For example, clay and iron oxide indices are often much better at predicting than raw reflectance values. They cut out the redundancy without sacrificing accuracy.

13.5.4.5 Dimensionality Reduction Methods

To transform the overlapping spectral bands into new independent variables, some algorithms such as principal component analysis (PCA), independent component analysis

(ICA), and minimum noise fraction (MNF) are used. These independent variables are completely uncorrelated. While minimizing noise and redundancy, these transformations preserve the most important variety in the new variables. Finally, the smaller and cleaner dataset is especially effective in extracting soil-related signals much faster and more accurately.

13.5.4.6 Regularization in Machine Learning Models

To handle multicollinearity, some machine learning methods, such as partial least squares regression (PLSR) and Ridge or Lasso regression, are particularly used. Because they can penalize redundant features and focus on the most relevant spectral information. This information helps to predict some important soil parameters like soil organic matter (SOM), cation exchange capacity (CEC), or moisture content with higher accuracy.

13.5.4.7 Ensemble and Hybrid Approaches

A combination of feature extraction techniques (like PCA) with machine learning methods (such as random forest, PLSR, SVM) is the best strategy. It allows researchers to benefit from both simplifying the data and accurately predicting outcomes. This hybrid approach improves soil property mapping by balancing interpretability and model robustness.

By addressing multicollinearity and redundancy, these approaches improve the accuracy, stability, and computational efficiency of predictive models, ultimately enhancing the extraction of meaningful soil information such as organic carbon, mineralogy, texture, and moisture.

13.5.4.8 Dimensionality Reduction Techniques

Hyperspectral sensors generate a very large amount of data, which is often difficult for standard software and computers to process. To handle this, DR techniques are applied. These methods reduce repeated and unnecessary information while keeping the most important parts that explain the maximum variation in the data. DR can be done in two ways. The first is feature selection, where only the most useful variables are kept from the original data. The second is feature extraction, where new variables are created by combining the original ones into fewer components. Several methods are commonly used for this purpose, including PCA, linear discriminant analysis (LDA), generalized discriminant analysis (GDA), random forests (RF), factor analysis, ICA, and filtering methods such as missing value ratio (MVR), low variance filter (LVF), and high correlation filter (HCF). Other approaches, like MNF, wavelet-based dimensionality reduction (WDR), and locally linear embedding (LLE), are also widely applied. Some of these methods are described in the following section.

13.5.4.9 Principal Component Analysis (PCA)

Karl Pearson initially presented PCA in 1901, and it was later developed independently by Hotelling. In signal processing, it is also known as the Karhunen–Loeve Transform (KLT). PCA creates a new set of viewing axes called principal components to project the original data. These axes are placed strategically in such a way that the first axis captures the most variety in the data, the second captures the next highest variation, and so on. The information carried by these new axes is independent and nonoverlapping. PCA uses the eigenvalue decomposition method to analyze the data's covariance or correlation matrix. It often starts by normalizing the data first. The result of PCA is presented in two main ways: loadings, which are essentially the weights showing how much each of the original measurements contributed to the new components, and component scores, which are the actual new values for every data point on these transformed, cleaner axes. Eigenvectors are loadings that are adjusted to unit length when the data are not standardized. Overall, PCA is an effective method for shrinking data, removing redundancy, and finding hidden patterns in complicated datasets.

13.5.4.10 Minimum Noise Fraction (MNF)

The MNF transformation is carried out in two stages. In the first stage, a principal component (PC) transformation is applied using the estimated noise covariance matrix, which helps to decorrelate and rescale the noise present in the data. In the second stage, a standard PC transformation is performed on this noise-whitened data. By analyzing the resulting eigenvalues and their corresponding component images, the true dimensionality of the dataset can be identified.

13.5.4.11 Wavelet Transform

The Fourier transform gives us the Wavelet transform, a powerful time-frequency analysis method that can pull useful information from a complicated signal and give us new ways to process and study data. The analysis consists mainly of discrete wavelet transform (DWT) and continuous wavelet transform (CWT). DWT is an efficient method for processing data and is useful for remote-sensing image analysis, as it typically will reduce redundancy, but may lose weak, although potentially useful information in the process. CWT, on the other hand, is a method that separates and isolates useful information from spectral data and is often the preferred method for the analysis of detailed spectral features. WT, when decomposing soil spectra into low (approximation) and high (detail) frequency coefficients, enables enhancement of subtle absorption features linked to soil properties such as organic carbon, clay, and available water, while simultaneously reducing noise in the spectra. Selecting the most informative coefficients allows for significant reductions in the dimensionality of data without giving up important spectral information. This process is particularly useful in improving predictions of soil properties when paired with machine learning models such as PLS regression, random forests, or SVM. The spectrum data are analyzed using CWT to derive a sequence of wavelet energy coefficients across various scales, represented by Eqs. 13.2 and 13.3:

$$W_f(a, b) = \int_{-\infty}^{+\infty} f(\lambda)\psi a, b(\lambda)d\lambda \tag{13.2}$$

$$\psi a, b(\lambda) = \frac{1}{\sqrt{a}}\psi\left(\frac{\lambda - b}{a}\right) \tag{13.3}$$

In this context, $f(\lambda)$ denotes the spectral reflectance; $W_f(a, b)$ signifies the two-dimensional wavelet energy coefficient, encompassing the decomposition scales (1, 2, m) and the bands (1, 2,..., n); λ indicates the band within the range of 350–2450 nm; $\psi a,b$ represents the wavelet basis function; a is the stretching factor, while b is the translation factor.

13.6 Computational Modeling and Statistics

13.6.1 Partial Least Squares Regression (PLSR) Model

PLSR is an effective method for improving analytical instrument selectivity and a robust multivariate regression approach that allows researchers to perform a range of inquiries. The purpose of PLSR is to assess the association between a dependent response vector (Y [n × 1]) and independent predictor data (X [n × m], n observations {rows}, and m variables {columns}).

PCA breaks down X into elements that better describe X than Y. PCA cannot guarantee that the elements of X that best describe X will also accurately predict Y. In contrast, PLSR determines which elements of X best predict Y. For this reason, PLSR is frequently called "better principal component regression" since it simply performs better in terms of prediction.

To address the issues regarding overfitting, an optimum number of PLSR model components should be identified by cross validation to establish an ideal balance between explained variation in the response (r^2) and goodness of modeled prediction (Q^2). R^2 and Q^2 values are comparable, though Q^2 declines with increasing model overfitting. The root mean square error of cross-validation (RMSECV) was employed to evaluate predictive performance and to determine the optimal number of components to retain in the regression model. Selecting an appropriate number of PLSR components is essential for building a model with high predictive reliability.

13.6.2 Random Forest (RF) Categorization and Regression

Classification refers to prediction of the probability of future outcomes based on prior observations. Classification models, once established, are used to predict future events and train data set through the use of established classification functions. Random forest (RF) refers to an ensemble method whereby numerous decision trees are formed in order to improve predictive accuracy. Decision trees are built using bootstrapping technique whereby a subset of the original data set is sampled with replacement. In addition, random features are selected at each decision node. The whole process is repeated several times thereby resulting in a large number of decision trees, with each tree making a prediction. Averaging and majority voting techniques are employed to generate a single prediction from the predictions made by the decision trees.

13.6.3 Evaluation Methods

The efficacy of the categorization algorithms is assessed by looking at the expected results. Both the train/test division and 10-fold cross-validation are used in the assessment process. A machine learning model's accuracy is evaluated using a confusion matrix. This matrix is a tabular representation that summarizes the outcomes of the model's predictions. It contains both actual and predicted values. The confusion matrix values are utilized to create the accuracy score. In the present study, the 10-fold cross-validation method of assessment and validation was used to evaluate the model performance. An advantage of this validation method is that it yields consistent and unbiased results with small datasets. The 10-fold cross-validation method is composed of dividing the dataset into 10-folds, constructing the model using all values of the dataset minus onefold, and testing the model on onefold. This is completed 10 times. The cross-validation error is then computed by averaging the 10 reported errors. We assessed the model based on three indices: mean square error (MSE), error rate, and coefficient of determination (R^2).

13.6.4 Support Vector Machine (SVM)

A support vector machine (SVM) is a supervised machine learning algorithm employed for classification and regression. In order to predict continuous variables (such as soil organic carbon, nitrogen, or pH) or separate data into classes (like high vs. low fertility soils), it builds an ideal hyperplane in a high-dimensional space. SVM models learn from ground-truth laboratory measurements and imaging or spectral features (Vis–NIR, MIR, UAV, or satellite data) in soil fertility research, then generalize to forecast fertility across broader regions.

13.6.4.1 How SVM Works

SVM minimizes the model's susceptibility to noise and outliers by identifying the optimal decision boundary (hyperplane) that separates data points into classes with the maximum margin. For instance, SVM increases robustness by ensuring that the boundary between productive and infertile soils is as far away from the nearest soil samples of every group.

In regression problems, called support vector regression (SVR), SVM does not attempt to fit every soil data point exactly but instead predicts continuous fertility properties (e.g., soil organic carbon or total nitrogen) within a small tolerance for error. This insensitive approach makes SVR particularly suitable for soil fertility assessment, where measurement uncertainty and natural variability are unavoidable.

Linear models frequently fall short in capturing the nonlinear and overlapping ways that soil characteristics like moisture, clay, and organic matter affect reflectance. In order to overcome this, SVM uses kernel techniques that convert soil spectral information into a higher-dimensional feature space, making it easier to separate and describe intricate relationships. Because it effectively records complex soil–spectral relationships and has been demonstrated to outperform linear models like PLSR in both Vis–NIR and MIR spectroscopy, the radial basis function (RBF) kernel is frequently used in agricultural research.

13.6.5 Application of Imaging Technology in the Assessment of Soil Fertility on a Regional and Global Basis

Numerous regional, national, and global studies have been reported where remote sensing, proximal sensing, and GIS-based technologies, satellite imageries, and UAV-mounted hyperspectral imagery have been used for soil textural predictions and fertility assessment from their spectral signatures by utilizing several environmental covariates with the help of advanced computational predictive modeling and statistics [12, 37].

13.6.6 Proximal Sensing-based Approaches

Soil reflectance measurements are commonly conducted at a point scale following standardized procedures, typically referred to as laboratory spectroscopy, where samples are analyzed from close range under controlled settings. Alternatively, reflectance data can be collected in situ at short distances using a portable spectroradiometer; however, these field-based measurements occur without strict environmental control. Laboratory spectroscopy is generally considered to provide more precise and reliable results compared to field spectroscopy. Diffuse reflectance spectroscopy, X-ray fluorescence spectrometry, visible and near-infrared (VIS-NIR) spectroscopy, shortwave and mid-wave spectroscopy (SWIR-MIR), laser-induced breakdown spectroscopy, etc., are some proximal sensing approaches based on which several physico-chemical properties of soil can be predicted nondestructively [23]. Hyperspectral imagery, especially within the SWIR range, is effective for evaluating the organic matter content in soil [17]. These approaches eliminate the requirement for traditional soil sampling and laboratory analysis and can produce precise soil texture distribution maps [38]. In a spectral study carried out by Vestergaard et al., it was reported that the absorbance spectra ranging between VNIR region (343–2200 nm) may be efficiently utilized in prediction of SOM, pH, silt, sand, and coarse sand percentage of soil with help of four separate machine learning approaches e.g., PLSR, RF, cubist, and extreme learning machine model and in each case the predictive accuracy (R^2) is more than 0.60 [39]. PXRF-spectroscopy has successfully been implemented to predict the available P, exchangeable Ca^{2+} content, and available micronutrient content in some tropical soils of Brazil, percentage base saturation, SOM content, textural fractions, salinity extent, nature of the soil parent material, secondary carbonate salt content, and classification [11]. Dasgupta et al. studied 561 soil samples under rice based cropping system collected from six different agro-climatic zones of West Bengal, India where PXRFS combining with auxiliary variables (e.g., pH, EC, SOC, and nature of parent material) have been implemented to predict the available nutrient status (e.g., Ca^{2+}, Mg^{2+}, K^{+}, Zn^{2+}, Cu^{2+}, and Mn^{2+}), exchangeable base percentage and sulfur availability index in soil with the help of four machine learning models viz. RF, PLSR, SVMR, stepwise multiple regression, and a combined average model with a good prediction accuracy [11]. In a study carried out by Pavão et al., the RF model has been applied to spectrally predict the pH, organic matter content, and CEC of surface and sub-surface soils collected from several forest areas of Brazil, where spectral data has been captured through two parallel approaches, VIS-NIR and PXRF spectroscopy where PXRFS data most accurately predicted soil pH at both surface and sub-surface level and CEC, SOM at surface level and the VIS-NIR spectroscopy based prediction was most accurate for CEC and SOM prediction at sub-surface level [40].

13.6.7 Smartphone-driven Approaches

The advancement of smartphone technology, particularly improvements in camera quality, has brought a significant change to soil analysis. Smartphones, besides their uses for communication, have become essential in environmental science due to their ability to collect image data directly from soil at the field level and process it through their high-resolution cameras, coupled with advanced software, leading to more precise and accurate data extraction and soil analysis. Recently, sensor-assembled smartphones (Android) have been utilized as a tool to predict nutrient content in soil as well as in water [12, 41]. Such advancement, in fact, has brought a revolutionary development in the field of soil mapping by integrating quick data acquisition, easier accessibility, cost-effectiveness, higher spatial resolution, and GPS information. For instance, in a recent study conducted by Lavanya et al., phosphate and nitrate concentrations in soil and water samples were measured using a smartphone-integrated sensing device, and the values were predicted through the HSV color space model with a good accuracy of less than 1% bias and 2% RSD (Figure 13.9) [12]. Moreover, digital images captured through smartphone-based sensors have been utilized successfully to predict the nature of soil texture, organic matter content, salinity level of water, soil pH, available P in soil, arsenite contamination in soil, fluoride level in drinking water, nitrate content in water, based on colorimetry or spectrometry. Some of the case studies on such smartphone-sensor-based predictions of several soil properties throughout the world have been summarized in Table 13.3.

13.6.8 Remote Sensing, Satellite Imagery, and UAV/Aircraft-based Approaches

Satellite systems, such as IKONOS, QuickBird, Hyperion, AVIRIS, Sentinel, and WorldView, offered higher spatial and spectral resolutions, improving the accuracy of soil

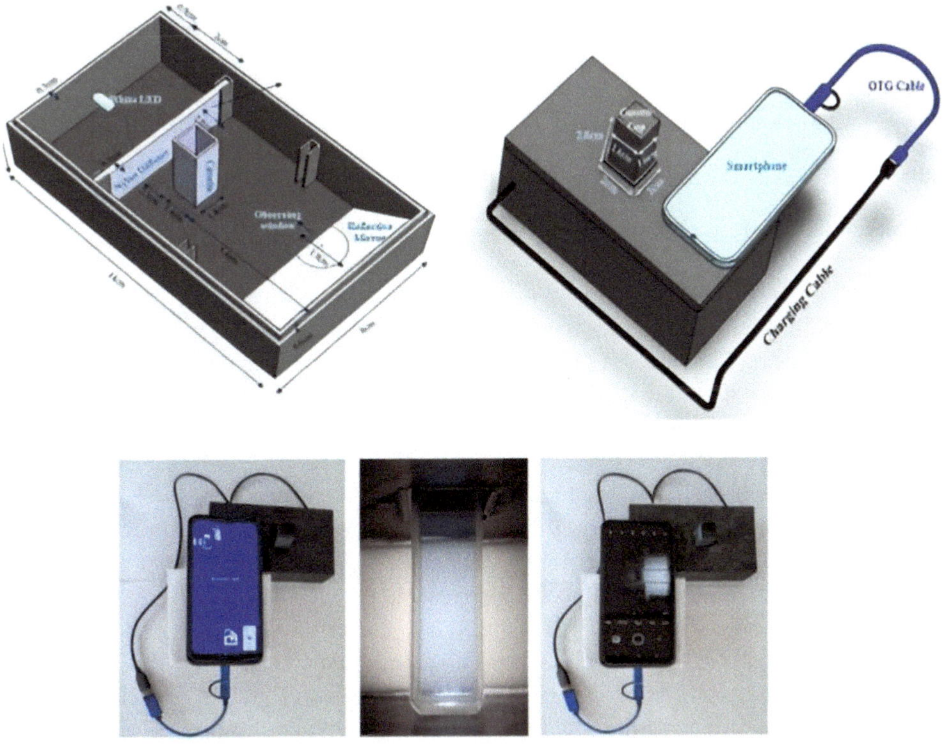

Figure 13.9 Smartphone-integrated spectroscopic device-based image acquisition and sensing. *Source:* with permission [12] 2025, Elsevier.

Table 13.3 Some case studies based on smartphone-integrated device-based soil sensing and image analysis.

Sl. No.	Physico-chemical properties of soil	Soil samples analyzed	Smartphone-based device/ sensors used	Predictive models	Accuracy of prediction			References
					R^2	RMSE	Any other measures	
1	Soil organic matter	25	Smartphone	LR[a]	0.51	14.21	–	[42]
2	Electrical conductivity	240	Smartphone	RNN[b]/ CNN[c]	0.93	–	–	[43]
3	Soil organic carbon	611	Smartphone	RF[d]	0.42–0.70	–	–	[44]
4	Soil organic matter	103	Digital camera	PLSR[e]	0.91	0.50	–	[45]
5	Iron content in soil	103	Digital camera	PLSR	0.64	0.074	–	[45]
6	Bulk Density	3	Digital camera	DNN[f]	–	1.52	–	[46]
7	Soil organic matter	371	NixPro + PXRFS[g]	RF	0.70	–	–	[47]

[a] LR – linear regression.
[b] RNN – recurrent neural network.
[c] CNN – convolutional neural network.
[d] RF – random forest.
[e] PLSR – partial least squares regression.
[f] DNN – deep neural network.
[g] PXRFS – portable X-Ray fluorescent spectroscopy; R^2 – coefficient of determination; RMSE – root mean square error.
Source: Modified from [41].

assessment and crop monitoring as an advancement critical in precision farming [3, 23]. Remote sensing using Landsat TM (Thematic Mapper) data has proven to be an effective and precise approach for estimating topsoil organic carbon (SOC) as well as blue carbon through measuring the reflection spectra from surface soil. The earliest applications of remote sensing in precision agriculture were centered on bare soil Landsat images to analyze spatial trends in SOM. Auxiliary imaged data captured through Landsat data, coupled with field data, can be utilized to assess the spatial distribution of soil phosphorus and wheat yield. Efforts to generate spatial maps of SOC concentrations have been further undertaken using higher-resolution satellite imagery such as IKONOS, SPOT data, multitemporal Landsat data, and Sentinel 1 and 2. Fathololoumi et al. introduced an improved DSM technique where multitemporal satellite data were fused with several environmental covariates and utilized to predict sand, SOC, and calcium carbonate equivalent (CCE) percentage in some soils of Iran by employing two machine learning algorithms, RF and Cubist model. Here, the cubist model outperformed the RF model with a higher R^2 value as compared to those of the RF models for both sand% ($R^2 = 0.42$), SOC% ($R^2 = 0.59$), and CCE%

($R^2 = 0.70$)[48]. Physico-chemical properties of soils, collected from marshy wetlands, have successfully been predicted with an accuracy (R^2 range 0.52–0.80) through assessing hyperspectral and multispectral images captured by Worldview-2 and QuickBird satellites [49]. Available N, P, K, and SOC content in soils collected from forest areas of Guangdong, China, have been spectrally assessed from Gaofen-1 satellite data predicted through an ANN model [50]. UAVs, i.e., drones and aircraft play pivotal roles in SSS by providing platforms for remote, high-resolution, and dynamic monitoring of soil properties across varied landscapes [5, 51].

13.7 Challenges and Limitations

It is very difficult to ensure that the collected samples truly act as a representative of spatially diverse regions. In spectral sensing, quality degradation of the spectral signals may occur due to atmospheric contamination and attenuation or due to adjacent land cover, such as vegetation and water or background elements, and irregular topography, which make calibration and corrections critical. The more advanced equipment, such as hyperspectral sensors and GIS software, isn't accessible for everyone because of their acquisition and operational costs. Large and complex datasets are generated in imaging spectroscopy, which requires extensive preprocessing, advanced computational modeling techniques, and highly skilled personnel for interpretation and information extraction. Well-equipped infrastructure is needed, which further constrains the widespread adoption. Legacy soil datasets often form the basis of digital soil databases, but these datasets are highly heterogeneous and sometimes conflicting in nature.

13.8 Future Direction and Conclusion

Spatial data integration will be further enhanced in the future with advances in GIS and remote sensing, hence enabling the incorporation of a wide range of environmental covariates such as topography, climate, and vegetation indices into soil models, improving the accuracy of the models and predictions. Deep learning and AI-enabled soil information systems facilitate early detection of soil degradation, optimized use of resources, and precision agricultural practices that enhance sustainability and productivity. Open data initiatives and collaborative platforms by government agencies and private companies should result in increased data availability and a reduction of costs associated with soil mapping. DSM tools will find broader use in urban planning and ecosystem management by helping to assess the potential for soil stability, erosion risk, and carbon storage. There is also the development of user-friendly decision-support tools using web-based (IoT – Internet of Things) platforms, imaging using smartphones, and mobile apps. Future research should focus on development of imaging based sensing technologies, along with sensor-equipped UAVs/drone systems that incorporate geospatial analysis and machine learning algorithms to facilitate efficient, accurate, and large-scale measurement of soil texture, soil fertility, landscape features, bodies of water, and vegetation.

References

1 Rossel, R.A.V., Walvoort, D.J.J., McBratney, A.B. et al. (2006). Visible, near infrared, mid infrared or combined diffuse reflectance spectroscopy for simultaneous assessment of various soil properties. *Geoderma* 131 (1–2): 59–75. https://doi.org/10.1016/j.geoderma.2005.03.007.

2 Biswas, A., Mukhopadhyay, D., Choudhury, R.K., and Biswas, A. (2025). The impact of cropping systems on soil health: a focus on the teesta-terai alluvial zone of West Bengal, India. *Journal. Indian Society of Soil Science* 73 (2): 215–233. https://doi.org/10.5958/0974-0228.2025.00020.8.

3 Choudhury, R.K., Balo, S., Ghosh, K., and Mahata, D. (2024). Paramparagat Krishi: a sustainable and holistic way of soil health management through organic and biodynamic soil inputs. *AgriGate* 4: 50–57. Accessed via: https://www.agrigatemagazine.com/_files/ugd/b546c0_59ff1a8e39ba44ab8386e5cbe8b8f2f4.pdf.

4 Bhattacharyya, R., Ghosh, B.N., Mishra, P.K. et al. (2015). Soil degradation in India: challenges and potential solutions. *Sustainability* 7: 3528–3570. https://doi.org/10.3390/su7043528.

5 Choudhury, R.K., Balo, S., Sarkar, A. et al. (2025). Concept of soil quality, its assessment and mapping through geoinformatics and GIS-based platforms. In: *Soil Degradation, Desertification and Restoration: Application of Geospatial Technology and Modeling* (ed. S.C. Pal, U. Chatterjee, and D. Ruidas). Cham: Springer Nature Switzerland AG Ch. 2. https://doi.org/10.1007/978-3-031-92277-0_2.

6 FAO (2025). World soil day – campaign materials. https://www.fao.org/world-soil-day/campaign-materials/en/ (accessed October 2025).

7 FAO (2025). World soil day – global soil partnership. https://www.fao.org/global-soil-partnership/en/ (accessed October 2025).

8 Lau, I.C., Cudahy, T.J., Heinson, G. et al. (2003). Practical applications of hyperspectral remote sensing in regolith research. *Advances in Regolith* 249–253. Accessed via: https://crcleme.org.au/Pubs/Advancesinregolith/Lau_et_al.pdf.

9 Booysen, R., Jackisch, R., Lorenz, S. et al. (2020). Detection of REEs with lightweight UAV-based hyperspectral imaging. *Scientific Reports* 10 (1): 17450. https://doi.org/10.1038/s41598-020-74422-0.

10 Malley, D.F. and Martin, P.D. (2004). Application in analysis of soils. *Near-Infrared Spectroscopy in Agriculture* 44: 729–784. https://doi.org/10.2134/agronmonogr44.c26.

11 Dasgupta, S., Chakraborty, S., Weindorf, D.C. et al. (2022). Influence of auxiliary soil variables to improve PXRF-based soil fertility evaluation in India. *Geoderma Regional* 30: e00557. https://doi.org/10.1016/j.geodrs.2022.e00557.

12 Lavanya, V., Nayak, A., Dasgupta, S. et al. (2023). A smartphone-integrated imaging device for measuring nitrate and phosphate in soil and water samples. *Microchemical Journal* 193: 109042. https://doi.org/10.1016/j.microc.2023.109042.

13 Gholizadeh, A., Saberioon, M., Rossel, R.A.V. et al. (2020). Spectroscopic measurements and imaging of soil colour for field scale estimation of soil organic carbon. *Geoderma* 357: 113972. https://doi.org/10.1016/j.geoderma.2019.113972.

14 Rossel, R.V., Brus, D.J., Lobsey, C. et al. (2016). Baseline estimates of soil organic carbon by proximal sensing: comparing design-based, model-assisted and model-based inference. *Geoderma* 265: 152–163. https://doi.org/10.1016/j.geoderma.2015.11.016.

15 Mapping Around (2025). Digital image processing in remote sensing. https://mappingaround.in/digital-image-processing-in-remote-sensing/ (accessed October 2025).

16 Humboldt-Universität zu Berlin (2025). Image properties I – introduction to remote sensing. https://eol.pages.cms.hu-berlin.de/geo_rs/S02_image-prop1.html (accessed October 2025).

17 Dasgupta, S., Debnath, S., Das, A. et al. (2023). Developing regional soil micronutrient management strategies through ensemble learning based digital soil mapping. *Geoderma* 433: 116457. https://doi.org/10.1016/j.geoderma.2023.116457.

18 Mulla, D.J. (2013). Twenty five years of remote sensing in precision agriculture: key advances and remaining knowledge gaps. *Biosystems Engineering* 114 (4): 358–371. https://doi.org/10.1016/j.biosystemseng.2012.08.009.

19 Al-lami, A.K., Abbood, R.A., Al Maliki, A.A., and Al-Ansari, N. (2021). Using vegetation indices for monitoring the spread of Nile Rose plant in the Tigris River within Wasit province, Iraq. *Remote Sensing Applications: Society and Environment* 22: 100471. https://doi.org/10.1016/j.rsase.2021.10047.

20 Drusch, M., Del Bello, U., Carlier, S. et al. (2012). Sentinel-2: ESA's optical high-resolution mission for GMES operational services. *Remote Sensing of Environment* 120: 25–36. https://doi.org/10.1016/j.rse.2011.11.026.

21 Ben-Dor, E. and Banin, A. (1995). Near-infrared analysis as a rapid method to simultaneously evaluate several soil properties. *Soil Science Society of America Journal* 59 (2): 364–372. https://doi.org/10.2136/sssaj1995.03615995005900020014x.

22 Ben-Dor, E., Irons, J.A., and Epema, A. (1999, Ch. 3). Soil reflectance. In: *Remote Sensing for the Earth Sciences: Manual of Remote Sensing* (ed. A.N. Rencz and R.A. Ryerson), 111–188. Wiley.

23 Mitran, T., Suresh, J., Sujatha, G. et al. (2024). Digital Soil Mapping: A Tool for Sustainable Soil Management. In: *Climate Change and Soil-Water-Plant Nexus* (ed. M.M. Rahman, J.C. Biswas, and R.S. Meena). Singapore: Springer Ch. 3. https://doi.org/10.1007/978-981-97-6635-2_3.

24 Ghosh, A., Mitran, T., and Mani, P.K. (2025). Estimation of soil nitrogen content influenced by different nitrogen-based management practices within rice-based cropping using diffuse reflectance spectroscopy and machine learning. *Communications in Soil Science and Plant Analysis* 56 (6): 856–878. https://doi.org/10.1080/00103624.2024.2433703.

25 Pandey, J.K., Yadav, A.K., Purushotham, M. et al. (2022). Colorimetric detection of deltamethrin pesticide from biological samples based on the peroxidase-mimic catalytic activity on MnO_2/SnO_2. *Key Engineering Materials* 928: 45–51. https://doi.org/10.4028/p-4ti1i0.

26 Demattê, J.A.M., Fongaro, C.T., Rizzo, R., and Safanelli, J.L. (2018). Geospatial soil sensing system (GEOS3): a powerful data mining procedure to retrieve soil spectral reflectance from satellite images. *Remote Sensing of Environment* 212: 161–175. https://doi.org/10.1016/j.rse.2018.04.047.

27 Lalitha, M., Dharumarajan, S., Gomez, C. et al. (2023). Spectral assessment of soil properties in semi-arid tropical regions of southern Karnataka Plateau. *Archives of Agronomy and Soil Science* 69 (11): 2065–2080. https://doi.org/10.1080/03650340.2022.2134565.

28 Ben-Dor, E., Chabrillat, S., Demattê, J.A. et al. (2009). Using imaging spectroscopy to study soil properties. *Remote Sensing of Environment* 113: 38–55. https://doi.org/10.1016/j.rse.2008.09.019.

29 Rossel, R.V. and Behrens, T. (2010). Using data mining to model and interpret soil diffuse reflectance spectra. *Geoderma* 158 (1–2): 46–54. https://doi.org/10.1016/j.geoderma.2009.12.025.

30 Malley, D.F. and Martin, P.D. (2003). The use of near-infrared spectroscopy for soil analysis. In: *Tools for Nutrient and Pollutant Management: Application to Agriculture and Environmental Quality (Eds: L. Currie, J. Hanly), Fertilizer and Lime Research Centre, New Zealand*, 371–404.

31 Rossel, R.V. and McBratney, A.B. (1998). Laboratory evaluation of a proximal sensing technique for simultaneous measurement of soil clay and water content. *Geoderma* 85 (1): 19–39. https://doi.org/10.1016/S0016-7061(98)00023-8.

32 Leone, A.P. and Sommer, S. (2000). Multivariate analysis of laboratory spectra for the assessment of soil development and soil degradation in the Southern Apennines (Italy). *Remote Sensing of Environment* 72 (3): 346–359. https://doi.org/10.1016/S0034-4257(99)00110-8.

33 Pandey, J.K., Kumar, S., Lamin, M. et al. (2022). A metaheuristic autoencoder deep learning model for an intrusion detector system. *Mathematical Problems in Engineering* 2022: 11, 3859155. https://doi.org/10.1155/2022/3859155.

34 Mitran, T., Sreenivas, K., Janakirama Suresh, K.G. et al. (2021). Spatial prediction of calcium carbonate and clay content in soils using airborne hyperspectral data. *Journal of the Indian Society of Remote Sensing* 49 (11): 2611–2622. https://doi.org/10.1007/s12524-021-01415-5.

35 Pandey, J.K. (2025). Understanding the impact of electromagnetic and optical radiation on physical and biological systems. *Journal of Nano- and Electronic Physics* 17 (5): 05023. https://doi.org/10.21272/jnep.17(5).05023.

36 Talukdar, V., Veeraiah, V., Roy, D.R. et al. (2023). An innovative approach: typicality-based collaborative filtering for recommender systems. *Proceedings of the 3rd International Conference on Advancement in Electronics & Communication Engineering (AECE)*, pp. 863–868. https://doi.org/10.1109/AECE59614.2023.10428576.

37 Beemkumar, N., Gupta, S., Bhardwaj, S. et al. (2023). Activity recognition and IoT-based analysis using time series and CNN. In: *Handbook of Research on Machine Learning-enabled IoT for Smart Applications Across Industries* (ed. N. Goel and R.K. Yadav), 350–364. IGI Global https://doi.org/10.4018/978-1-6684-8785-3.ch018.

38 Tiruneh, G.A., Meshesha, D.T., Adgo, E. et al. (2022). Use of soil spectral reflectance to estimate texture and fertility affected by land management practices in Ethiopian tropical highland. *PLoS One* 17 (7): e0270629. https://doi.org/10.1371/journal.pone.0270629.

39 Vestergaard, R.J., Vasava, H.B., Aspinall, D. et al. (2021). Evaluation of optimized preprocessing and modeling algorithms for prediction of soil properties using vis-nir spectroscopy. *Sensors* 21 (20): 6745. https://doi.org/10.3390/s21206745.

40 Pavão, S., Ribeiro, P.G., Maciel, G.P. et al. (2025). Portable XRF and Vis-NIR spectrometry for predicting chemical properties of forest soils in the Amazon: insights into sensor data dimensionality reduction. *Soil Advances* 100063. https://doi.org/10.1016/j.soilad.2025.100063.

41 Naeimi, M., Daggupati, P., and Biswas, A. (2024). Image-based soil characterization: a review on smartphone applications. *Computers and Electronics in Agriculture* 227: 109502. https://doi.org/10.1016/j.compag.2024.109502.

42 Srinivas, P., Veeraiah, V., Thejaswini, K.O. et al. (2025). The impact of electromagnetic and optical radiation on physical and biological systems. *Journal of Nano- and Electronic Physics* 17 (4): 04012. https://doi.org/10.21272/jnep.17(4).04012.

43 Gozukara, G., Anagun, Y., Isik, S. et al. (2023). Predicting soil EC using spectroscopy and smartphone-based digital images. *Catena* 231: 107319. https://doi.org/10.1016/j.catena.2023.107319.

44 Gozukara, G., Hartemink, A.E., and Zhang, Y. (2024). Illumination levels affect the prediction of soil organic carbon using smartphone-based digital images. *Computers and Electronics in Agriculture* 204: 107524.

45 Rossel, R.A.V. and McBratney, A.B. (2008). Diffuse reflectance spectroscopy as a tool for digital soil mapping. In: *Digital Soil Mapping with Limited Data (Eds: A. E. Hartemink, A. McBratney, M.d. Mendonça-Santos.* Springer, Dordrecht https://doi.org/10.1007/978-1-4020-8592-5_13.

46 Kim, D., Kim, T., Jeon, J., and Son, Y. (2023). Soil-surface-image-feature-based rapid prediction of soil water content and bulk density using a deep neural network. *Applied Sciences* 13 (7): 4430. https://doi.org/10.3390/app13074430.

47 Swetha, R.K. and Chakraborty, S. (2021). Combination of soil texture with Nix color sensor can improve soil organic carbon prediction. *Geoderma* 382: 114775. https://doi.org/10.1016/j.geoderma.2020.114775.

48 Fathololoumi, S., Vaezi, A.R., Alavipanah, S.K. et al. (2020). Improved digital soil mapping with multitemporal remotely sensed satellite data fusion: a case study in Iran. *Science of the Total Environment* 721: 137703. https://doi.org/10.1016/j.scitotenv.2020.137703.

49 Zhang, C., Mishra, D.R., and Pennings, S.C. (2019). Prediction of regional forest soil nutrients based on Gaofen-1 remote sensing data. *ISPRS Journal of Photogrammetry and Remote Sensing* 148: 221–234. https://doi.org/10.3390/f12111430.

50 Li, Y., Zhao, Z., Wei, S. et al. (2021). Prediction of regional forest soil nutrients based on Gaofen-1 remote sensing data. *Forests* 12 (11): 1430. https://doi.org/10.3390/f12111430.

51 Choudhury, R.K. and Dasgupta, S. (2026). Role of AI, remote sensing, and digital tools in regenerative farming: with a special focus on West Bengal Region of Eastern India. *SATSA Mukhapatra* 30: 359–381.

14

Consumer-Centric Applications of Visual Nutrition Analysis

Shivani Phugat and Prabudh Goel

Department of Paediatric Surgery, All India Institute of Medical Sciences, New Delhi, India

14.1 Introduction

The understanding of humans regarding food has evolved with the convergence of artificial intelligence (AI), computer vision, and nutrition science. At the vanguard of this change is the visual nutrition analysis, which presents opportunities to close the gap between intricate nutritional science and the common consumer demands. With just a photo and sophisticated AI taking care of the rest, tasks that previously required laboratory testing, expert consultations, or laborious manual tracking can now be completed.

We live in a world where the majority of the leading causes (6/10) of death are linked to diet [1], and about five billion people have micronutrient deficiencies [2], making the urgency for access to nutritional tools greater. Visual nutrition analysis applications are one such technological innovation; it is a way of democratizing nutrition knowledge, which puts advanced dietary insights into the hands of all with a smartphone. We can now move toward more proactive health management by helping people make healthy food decisions instead of just reacting to illness.

Advances in deep learning architecture, in combination with smartphones, have created an appropriate environment for this technological integration. Today, a smartphone is as technologically advanced as computers were a few years ago. This integration has also enabled the processing of complex foods in seconds, which in turn can help recognize food items, estimate portion sizes, calculate nutritional content, and also detect potential vitamin deficiencies from a picture.

14.1.1 Consumer Behavior and Digital Health Adoption

Consumers in today's world have distinct preferences in engaging with nutritional information and dietary management. Recent research indicates that people nowadays prefer smartphones for health-related activities, including dietary tracking, over traditional paper-based tracking [3]. This shift from traditional paper-based tracking to smartphone tracking has changed the expectations of consumers. The system has also changed the feedback mechanism; it has made it more personalized rather than static nutritional labels or generic dietary advice.

The research has also proven that users are more likely to log meals daily using photo-based systems rather than the manual entry methods, as it provides an improved user experience and reduces the cognitive burden [4]. This preference change, from text-heavy or complex input methods to immediate, intuitive interactions, reflects broader digital consumption patterns for visual interfaces.

The differences in the generation also play a crucial role in adoption patterns. The new generations, such as millennials and Generation Z (commonly known as Gen Z) consumers, are the primary user base for these applications, demonstrating greater comfort with AI-powered recommendations. They are also more willing to share their personal health data for personalized insights. Another interesting finding is that gamification methods are more preferred to achieve health goals [5].

The fundamental barriers to healthy eating, including the complexity of nutritional information, burden of dietary tracking, and lack of personalized nutritional guidance, are also addressed by these applications. These apps make healthy choices easier to understand, more engaging, and more achievable for people everywhere by turning good nutrition into clear visual feedback.

14.2 The Science Behind Visual Nutrition Analysis

14.2.1 Advanced Computer Vision Architectures

Modern consumer-centric visual nutrition analysis applications rely on sophisticated computer vision architectures that have evolved significantly from traditional extraction methods (Figure 14.1). The foundation of most recognition systems is convolutional neural networks (CNNs) with architectures like ResNet, VGG, Inception V3, and EfficientNet being adapted through transfer learning techniques. The models trained on ImageNet (large-scale datasets) are fine-tuned to achieve high accuracy for the identification of various food items using food-specific datasets [6–9].

To tackle the unique challenges of food analysis, specialized network designs have emerged, such as NutriNet. It is a deep CNN architecture that demonstrates the potential for a domain-specific network built specifically for food and drink image recognition [10]. Depth prediction and fusion (DPF)-nutrition addresses one of the critical challenges in visual nutrition analysis. It is an advanced approach that improves the portion estimation accuracy by integrating depth prediction modules with RGB data fusion [11]. Vision Transformers (ViTs) capture global dependencies in food images by utilizing self-attention mechanisms [12]. They are the recent innovations in the field of nutritional content detection that offer advantages over CNNs in generalizing across diverse food varieties and complex presentations of the foods. EfficientNet with Swin Transformers and Feature Pyramid Networks shows promising results in enhancing both accuracy and efficiency of food nutrient recognition [13]. Technology Assisted Dietary Assessment System (TADA) is one such example, used to measure volume, identify food type, and calculate energy and nutrient content from smartphone images. It uses advanced image

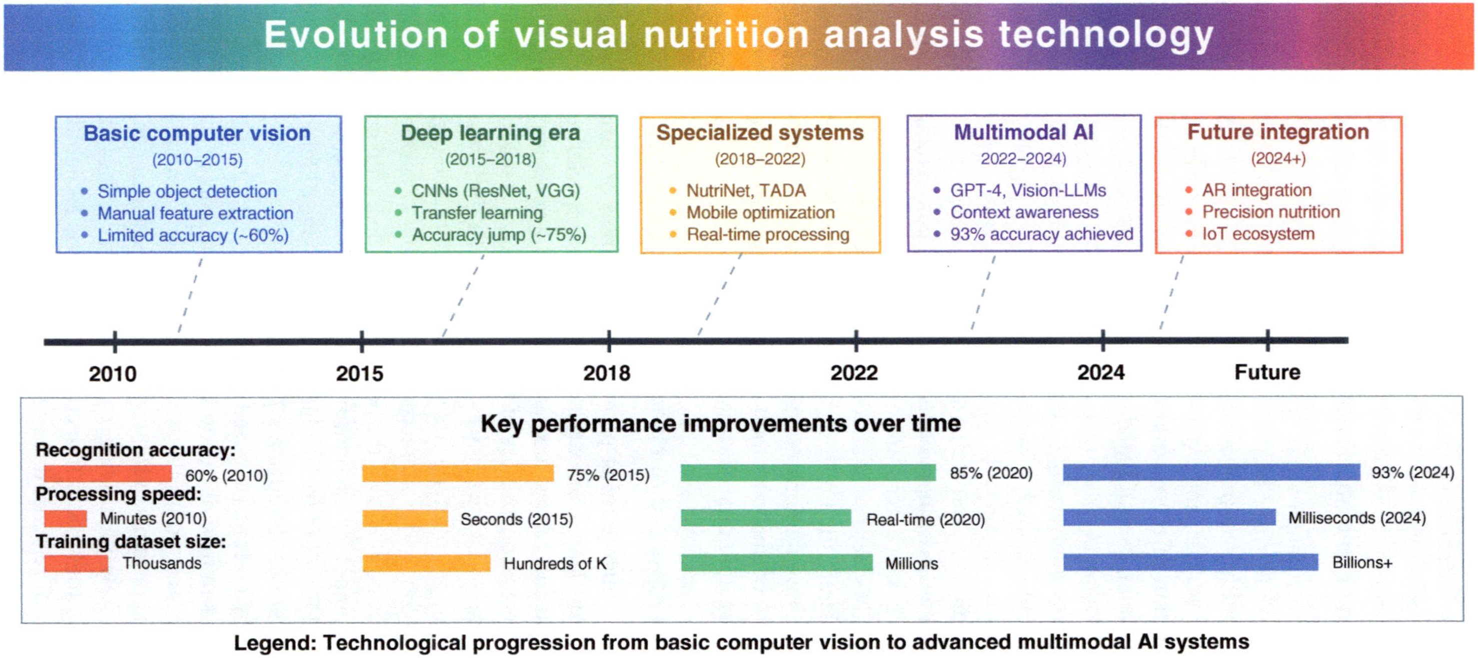

Figure 14.1 Evolution of visual nutrition analysis technology.

processing and computer vision methods [14]. It has proven both scalable and effective in real-world conditions by getting tested in over 14 user studies with over 800 participants and 60,000 food images [15].

14.2.2 Deep Learning and Multimodal Integration

The modern visual nutrition analysis systems have evolved beyond basic image classifications. Today, multitask learning has been used by learning architectures, in which a single model can predict multiple nutritional factors, including calories, macronutrients such as proteins, fats, carbohydrates, and even micronutrients (vitamins and minerals) from one image. This exploration of the natural correlation between the nutritional factors helps these systems achieve an overall higher prediction accuracy [16].

Large language models (LLMs) with computer vision have opened new avenues for a comprehensive dietary analysis by enabling systems to reason about nutritional content in ways more closely mirroring human understanding [17]. Advancements in these models have helped in the introduction of multimodal approaches that combine the visual data with textual information or even recipe databases. These systems can incorporate information about cooking methods, ingredients, and cultural context along with the image of a food item to provide more accurate and appropriate nutritional assessments. Vision-language models (VLMs) represent another particularly significant advancement and have been proven to accelerate the nutritional screening by 36.9%, setting new benchmarks in precision [18]. They have demonstrated <10% variance compared to traditional chemical analyses for over 82% of analyzed food items, enabling the systems to achieve a macro-AUCROC of 0.921 for lipid quantification.

14.2.3 GPT-4 and Advanced AI Models in Nutrition Analysis

GPT-4 has marked a huge turning point in the evolution of visual nutrition analysis. It represents a significant leap from earlier, as it is a more specialized system. It is a multimodal AI system that can process text and images and give us comprehensive nutritional insights. Research evaluating the use of GPT-4 for dietary assessment has shown both its capabilities and limitations. One study evaluating GPT-4's ability to estimate the nutritional content from photographs of meals has shown the remarkable capabilities of ChatGPT in the identification of food, with a precision rate of 93.0% in correctly identifying foods in images. Regarding the portion size, the performance varied, with good agreement for small meals and poor agreement for medium and large portions. The correlation coefficients ranged from 0.29 to 0.83 for all nutrients; however, percentage differences exceeded 10% from actual values for 13 out of 16 nutrients [19].

Another analysis by Lo et al. [20] used GPT-4 for the dietary assessment, which revealed that the system excels in food detection without any fine-tuning under challenging conditions, demonstrating exceptional adaptability. When tested with regional foods and guided with specific language prompts, it showed remarkable performance, from recognizing common foods to accurately identifying culturally specific dishes. The prompts related to African cuisine made the system transition from generic recognition of rice and bread to precise identification of regional dishes like banku and ugali.

GPT-4 addresses another challenging aspect of visual nutrition analysis, accurate portion estimation from monocular images, using contextual awareness, which is the ability to utilize its surrounding objects as references to deduce the portion sizes of the food items [20]. The system can leverage reference objects such as common utensils and plates to estimate food volume and weights, which enables more accurate calorie and nutrient calculations. In comparing the performance of GPT-4 with humans, GPT-4 yielded a similar performance as humans. For the energy, protein, and carbohydrate estimation, GPT-4 achieved an error rate of 46.3 g compared to 48.5 g of dietitians, showing a comparable accuracy in real-world dietary assessment scenarios [20].

14.2.4 Consumer Applications and Real-World Examples

There are several applications available in the market using the AI-powered food recognition technology, expanding the visual nutrition analysis. The users can now capture an image of their meals and receive immediate feedback on caloric content, macronutrient and micronutrient profiles, and portion sizes. These applications often include features for tracking daily, weekly, and monthly nutritional patterns, which enable users to identify trends and make informed dietary adjustments over time. The sophisticated user experience design principle is another component that presents complex nutritional information in digestible formats used by many mobile applications. **MyFitnessPal** [21] is one such app that tracks nutrition, leveraging machine learning algorithms to identify food items and suggest nutritional values. It has also integrated barcode scanning and photo-based food logging capabilities. **HealthifyMe** is another app that has introduced this AI-powered food recognition technology [22].

Foodvisor uses another specialized approach for food recognition and portion estimation. It utilizes deep learning models trained on millions of food images to provide nutritional analysis in an instant with a simple photo. Users have been seen to improve dietary awareness and better adherence to nutritional recommendations by using apps [23]. Photo-based food recognition has been integrated with barcode scanning by the Lose It application, which creates a hybrid approach maximizing the accuracy across different food types. The application has a feature called "Snap It," which uses CNNs to identify foods and automatically populate nutritional information. It reduces the time required by approximately 60% compared to manual entry methods [24]. This transformation from manual food logging earlier to visual-based tracking now has improved user engagement with dietary monitoring applications. The current research shows that apps that use photo-based food recognition have higher retention rates when compared to manual entry systems, mainly due to the improved user experience and reduced cognitive burden [25]. Another interesting element that has proven effective in implementing healthy behavior is gamification integrated into the applications. It includes features such as achievement badges for consistent logging, streak counters for meeting nutritional goals, and adding friends in the fitness journey. These features have also proven to improve dietary adherence [5].

Another commercially successful and scalable app utilizing these systems is **FoodAI**. It maintains high accuracy in food recognition and nutritional analysis, serving hundreds of thousands of users. The application packages computer vision into user-friendly interfaces and makes nutritional tracking accessible to diverse populations regardless of their technical expertise or nutritional knowledge [26].

Social features and community support within the applications are another crucial thing for long-term engagement and compliance. Most of the successful applications today allow users to share meals, receive feedback from nutrition professionals, and participate in challenges or educational programs. This social dimension transforms the individual dietary tracking into a community-supported wellness journey, improving adherence to healthy eating habits [27].

14.2.5 Advanced Nutritional Assessment and Personalization

AI-powered meal planning using deep generative models generates personalized weekly meal plans aligned with individual nutritional requirements, considering factors such as calorie intake, macronutrient balance, cultural preference, or any specific health conditions [28]. Users focused on weight loss receive suggestions prioritizing low-calorie, nutrient-dense meals, and people dealing with diabetes get recommendations carefully balancing carbohydrate content and glycemic index. These systems utilize the user feedback and eating patterns to evolve and refine their recommendations to better match individual preferences. This evolution pattern is also used to improve adherence to the suggested dietary plans. The recent advances in the personalized nutrition field extend to physiological responses and contextual factors such as activity levels, metabolic patterns, and health status, with the integration of wearable devices, smart scales, and continuous glucose monitors. These novel approaches represent a significant advancement over the traditional, static dietary planning methods [17].

14.2.6 Specialized Healthcare Applications

Visual nutrition analysis is particularly useful in healthcare settings, especially in conditions where precise dietary monitoring is crucial for managing specific medical conditions such as diabetes or dyslipidemia. The Connecting Health and Technology (CHAT) study has shown that image-based dietary assessment is effective in promoting healthy habits among young people over a six-month period by showing significant reductions in consumption of sugar-sweetened beverages and energy-dense, nutrient-poor foods [29]. People with diabetes can use specialized applications focusing on carbohydrate counting and glycemic management [30]. These specialized applications can identify food items and estimate their carbohydrate content accurately, essential for insulin dosing decisions. Automated carbohydrate estimation systems have been shown to achieve comparable accuracy levels as human experts [31]. These applications also address individualized challenges such as motor difficulties in Parkinson's disease by providing simplified interfaces and voice-activated controls.

14.2.7 Revolutionary Vitamin Deficiency Detection

One of the most innovative applications of visual nutrition analysis involves the detection of vitamin deficiencies through analysis of physical symptoms visible in photographs. These applications analyze images of body parts such as eyes, lips, tongue, and nails to identify visual markers associated with various vitamin deficiencies, offering a noninvasive approach that eliminates the need for blood tests while providing early warning signs of potential nutritional deficiencies [32].

The effectiveness of these applications relies on sophisticated image processing techniques and neural networks trained on datasets of symptoms associated with specific vitamin deficiencies [32]. Systems can detect angular cheilitis (cracked lips), indicating vitamin B deficiency, Beau's lines (vertical ridges in nails), suggesting various nutritional deficits, or specific tongue colorations indicating iron or vitamin B_{12} deficiencies. The accuracy of these systems has shown promising results, with some achieving over 80% accuracy in identifying specific deficiency types [33].

These applications can provide significant value in regions with limited access to healthcare. By enabling early detection of nutritional deficiencies, these tools can prompt timely interventions through dietary adjustments or supplementation, potentially preventing the progression of serious health complications related to malnutrition. The global impact of such accessible diagnostic tools could be transformative for public health initiatives in underserved communities.

14.3 Technical Innovations and Implementation Challenges

14.3.1 Real-Time Processing and Edge Computing

Sophisticated technical infrastructure, striking a balance between accuracy and performance, is necessary for the large-scale deployment of visual nutrition analysis applications (Figure 14.2). Neural networks must be optimized for mobile deployment using methods like model compression, quantization, and pruning in order to provide users with instant feedback, which requires real-time processing capabilities [34].

Contemporary applications using hybrid cloud-edge architectures use cloud computing for sophisticated nutritional analysis and database queries while performing basic food recognition locally. This method uses cloud resources to provide individualized recommendations and a thorough nutritional assessment while guaranteeing application functionality even with spotty connectivity. Truly real-time dietary analysis is now possible due to sophisticated models like RT-DETR, which have shown inference times as low as 5.7 ms while maintaining high accuracy levels [35]. The successful balancing of accuracy and computational efficiency needed for mobile applications is demonstrated by the use of lightweight models like MobileNetV2 and EfficientNet. These models make complex visual nutrition analysis possible on typical smartphone hardware by achieving competitive accuracy while preserving the processing speed required for real-time applications [36].

The accuracy and dependability of visual nutrition analysis systems are still constrained by a number of technical issues, despite notable advancements. With current systems only achieving 65–75% accuracy in estimating food volumes from monocular images, portion size estimation is still one of the most enduring problems. When food is served in bowls or containers with limited depth perception, the difficulty is especially noticeable [20].

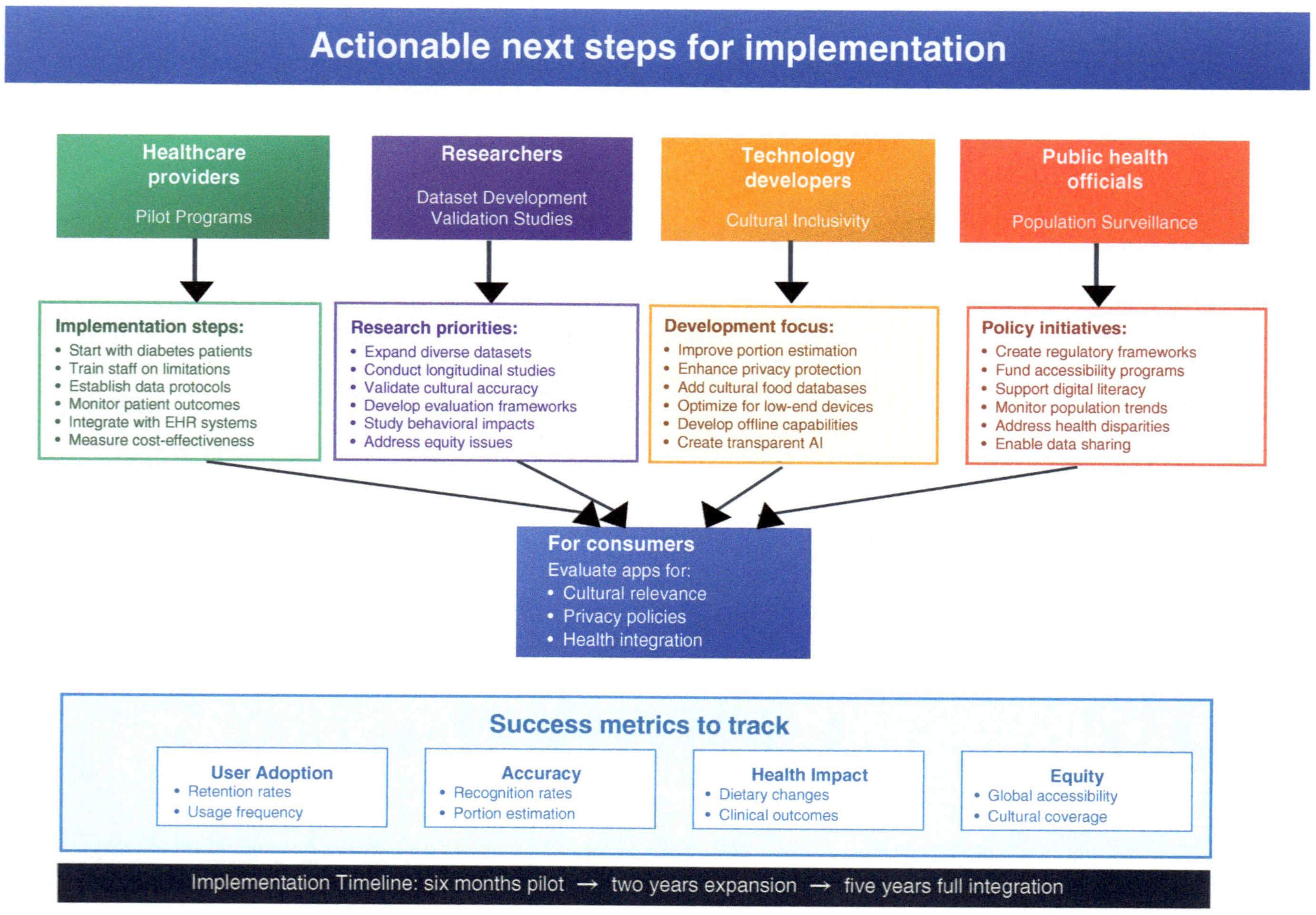

Figure 14.2 Actionable next steps for implementation.

14.3.2 Data Quality and Dataset Challenges

The performance of the visual-based nutrition analysis also depends on the quality and diversity of training datasets. The current food image datasets are extensive; however, they face the limitation of not representing the full diversity of global cuisines or even the preparation methods or food presentations [37]. This limitation can also result in reduced accuracy for foods not well-represented in training data, creating a potential for biases toward certain food types or cultural cuisines. Moreover, the labor-intensive nature of creating accurately labeled standard food datasets poses significant challenges for expanding the system capabilities to new food categories or regional cuisines [37]. The nutritional composition data is impacted by the preparation methods, ingredient sources, and regional variations in the food items, also impacting the visual labeling process.

14.3.3 Integration with Digital Health Ecosystems

Modern visual nutrition analysis applications increasingly integrate with broader digital health ecosystems, connecting with electronic health records (EHRs), telemedicine platforms, and comprehensive wellness management systems. This integration enables healthcare providers to access detailed dietary data from their patients, supporting more informed clinical decisions and personalized treatment plans. These applications serve as valuable tools for dietitians and nutritionists, who can use detailed dietary data to provide more targeted counseling and track patient progress remotely. Integration with wearable fitness trackers and health monitoring devices provides a holistic view of the relationship between diet, physical activity, and health outcomes, enabling more comprehensive wellness management. Data sharing protocols and privacy protection measures ensure that sensitive dietary information is handled securely while enabling beneficial integrations with healthcare providers and researchers. Federated learning approaches are being explored to improve model performance while preserving user privacy by learning from distributed data without centralizing sensitive information.

14.4 Challenges and Future Directions

14.4.1 Technical and Accuracy Limitations

The accuracy and dependability of visual nutrition analysis applications are limited by enduring technical issues, despite notable advancements. One of the biggest challenges is still food variability, since different cooking methods, presentation styles, cultural differences, and preparation methods can all significantly alter the appearance of the same food item. It is difficult for machine learning models to generalize across various food appearances and cultural contexts because of this variability [37]. Estimating portion size is again a significant drawback, especially when attempting to determine nutritional content from a single image without standardized reference objects. Accurately estimating food volumes from 2D images is difficult for complex meal compositions and irregular food shapes, despite the development of several methods utilizing geometric models and depth estimation [38].

Current segmentation and classification algorithms continue to face difficulties when dealing with complex meals and mixed dishes. It takes a sophisticated understanding of food composition and ingredient recognition to identify individual components in casseroles, stews, mixed salads, and fusion cuisine dishes. Overlapping foods, partially obscured objects, or meals with unclearly distinct components are common problems for current systems [39]. The creation of bigger, more varied datasets and the application of synthetic data generation techniques are recent attempts to address these issues. In order to produce more extensive training datasets that more accurately reflect the diversity and preparation variations of foods around the world, advanced data augmentation techniques and generative AI approaches are being investigated [40]. Another major challenge is mixed dish recognition, since existing CNN architectures have trouble recognizing and measuring individual ingredients in complicated, multicomponent meals.

14.4.2 Adoption and Implementation Barriers

Several practical and behavioral obstacles affect the long-term viability of consumer adoption of visual nutrition analysis applications. Despite the ease of image-based tracking, user compliance is still difficult because it takes consistent motivation and habit formation to maintain regular food logging behavior. According to studies, even though users may be excited about these apps at first, without effective engagement tactics, usage rates frequently decrease over time. System accuracy and user experience can be greatly impacted by image quality requirements. Motion blur, cluttered backgrounds, awkward camera angles, and poor lighting can all significantly lower recognition accuracy, which could irritate users and cause them to stop using the app. When eating on the go or in social dining settings, the requirement that users take crisp, well-lit pictures of their food could present challenges. Another major obstacle to widespread adoption is privacy concerns, especially since these apps gather extensive data about users' lifestyle patterns, medical conditions, and eating habits. Particularly in social settings or when handling delicate dietary information associated with medical conditions, some users might be reluctant to share pictures of food.

14.4.3 Cultural and Global Accessibility

Due to their primary training and testing on Western food datasets, current visual nutrition analysis systems frequently display cultural biases. Because of this restriction, they are less useful for users from diverse cultural backgrounds who eat foods that are underrepresented in training data. The accuracy of food recognition and the applicability of dietary recommendations are both impacted by the dearth of varied, culturally representative datasets. There are constant difficulties in standardizing nutritional data across various geographical areas and food preparation techniques. Depending on regional preparation techniques, ingredient availability, or cultural differences in recipe styles, foods with similar names may have distinct nutritional profiles. To account for these variations in nutritional analysis, sophisticated methods are needed.

To overcome these obstacles, coordinated efforts are needed to create region-specific nutritional databases, culturally sensitive user interfaces, and more inclusive datasets. Future advancements will probably concentrate on developing visual nutrition analysis systems that are more globally accessible and culturally inclusive, able to accommodate a variety of populations while honoring cultural food customs and preferences.

14.5 Future of Visual Nutrition Analysis

14.5.1 Emerging Technologies and Convergence

The intersection of several cutting-edge technologies holds the key to the future of consumer-focused visual nutrition analysis. Applications of augmented reality (AR) promise to overlay food photos with real-time nutritional data, allowing for instant dietary decision-making. By emphasizing nutritional content, recommending alternatives, or suggesting portion adjustments, such systems could help users make healthier decisions while they're still making food choices. Future applications could create genuinely customized and responsive dietary guidance systems by modifying nutritional recommendations based on stress indicators, heart rate variability, glucose levels, or sleep quality.

The accuracy of portion size estimation could be greatly increased by employing advanced computer vision techniques, such as multiview imaging or 3D food analysis capabilities using depth sensors. Current shortcomings in complex meal scenarios and mixed dish recognition should be addressed by the development of more complex object detection models and sophisticated segmentation techniques.

14.5.2 Personalization and Precision Nutrition

One of the most exciting areas of visual nutrition analysis is the shift toward more individualized nutrition. Beyond conventional one-size-fits-all nutritional recommendations, highly customized dietary guidance could be provided by sophisticated algorithms that take into account individual genetic profiles, microbiome composition, metabolic responses, and lifestyle factors [28]. More contextually relevant dietary recommendations may be possible with machine learning models that take life stage, seasonal variations, and temporal patterns into account. Depending on variables like pregnancy, aging, stages of athletic training, or recuperation from illness, these systems may modify recommendations, offering users dynamic support that adapts to their changing requirements. Long-term adherence to healthy eating practices may be enhanced by the combination of gamification strategies and behavioral psychology concepts. In order to assist users in maintaining healthy eating habits over time, future applications may make use of advanced behavior modification techniques, social support systems, and customized motivation tactics.

14.5.3 Health System Integration

An important chance to connect clinical decision-making and consumer-facing dietary tracking is the integration of visual nutrition analysis with EHRs. Data collected from consumer devices, like wearable-integrated dietary scanners or smartphone food recognition apps, could be safely sent in real time to healthcare providers in a future-focused health ecosystem. Clinicians would be able to more accurately customize interventions, identify nutritional risk factors, and track patients' eating habits over time. On a larger scale, population health analytics could make use of anonymized, aggregated dietary datasets to track nutritional trends, identify new dietary risks across various demographics or geographical areas, and inform public health surveillance. AI-driven clinical decision support systems that identify dietary deficiencies, excessive intake patterns, or disease risks associated with diet early in the care pathway may also benefit from such integration.

14.5.4 Global Impact and Democratization

Advanced visual nutrition analysis technology has global ramifications that go well beyond personal food monitoring. The development of food policies, epidemiological studies, and public health interventions could all be revolutionized by these systems (Figure 14.3). These tools could help guide focused interventions and policy decisions meant to address nutrition-related health issues by offering accurate, real-time data on dietary patterns at the population level. Accessible visual nutrition analysis applications could offer vital nutritional guidance and early deficiency detection in areas with inadequate healthcare infrastructure. In order to serve diverse global populations and address health disparities, it will be crucial to develop offline-capable applications with culturally relevant interfaces.

Closed-loop systems that maximize food production and consumption could be developed by integrating visual nutrition analysis with agricultural and food production systems. In order to enable more focused production of nutrient-dense foods, these systems may give farmers and food producers feedback regarding nutritional demand and preferences.

14.6 Conclusion

A paradigm shift in how people engage with nutritional information and oversee their dietary health can be seen in the development of consumer-centric visual nutrition analysis (Figure 14.4). These technologies, which range from basic computer vision architectures to the advanced multimodal AI systems like GPT-4, have converted abstract nutritional concepts into understandable, useful information that can be accessed through the widely used smartphone technology. Long-standing obstacles to healthy eating, such as the difficulty of tracking one's diet and the complexity of nutritional information, are addressed by the democratization of nutritional knowledge made possible by these applications. These tools enable people to make educated dietary decisions in real time by giving them immediate, visual feedback on their food choices, which may lead to better population health outcomes. However, the field still faces many obstacles, such as technical limitations in mixed dish recognition and portion estimation, as well as user experience barriers relating to adoption complexity and privacy concerns. Promising avenues to get around these restrictions and increase the clinical and public health uses of visual nutrition analysis are provided by the new developments in AR, edge computing, and health system integration. Enhancing privacy-preserving processing techniques, increasing accuracy across a variety of food cultures, and creating strong clinical validation studies to determine the long-term health effects of these technologies should be the top priorities of future research and development. To fully realize these applications' potential in promoting individual and population health, cooperation between technology developers, nutrition researchers, and healthcare providers will be crucial as they continue to develop.

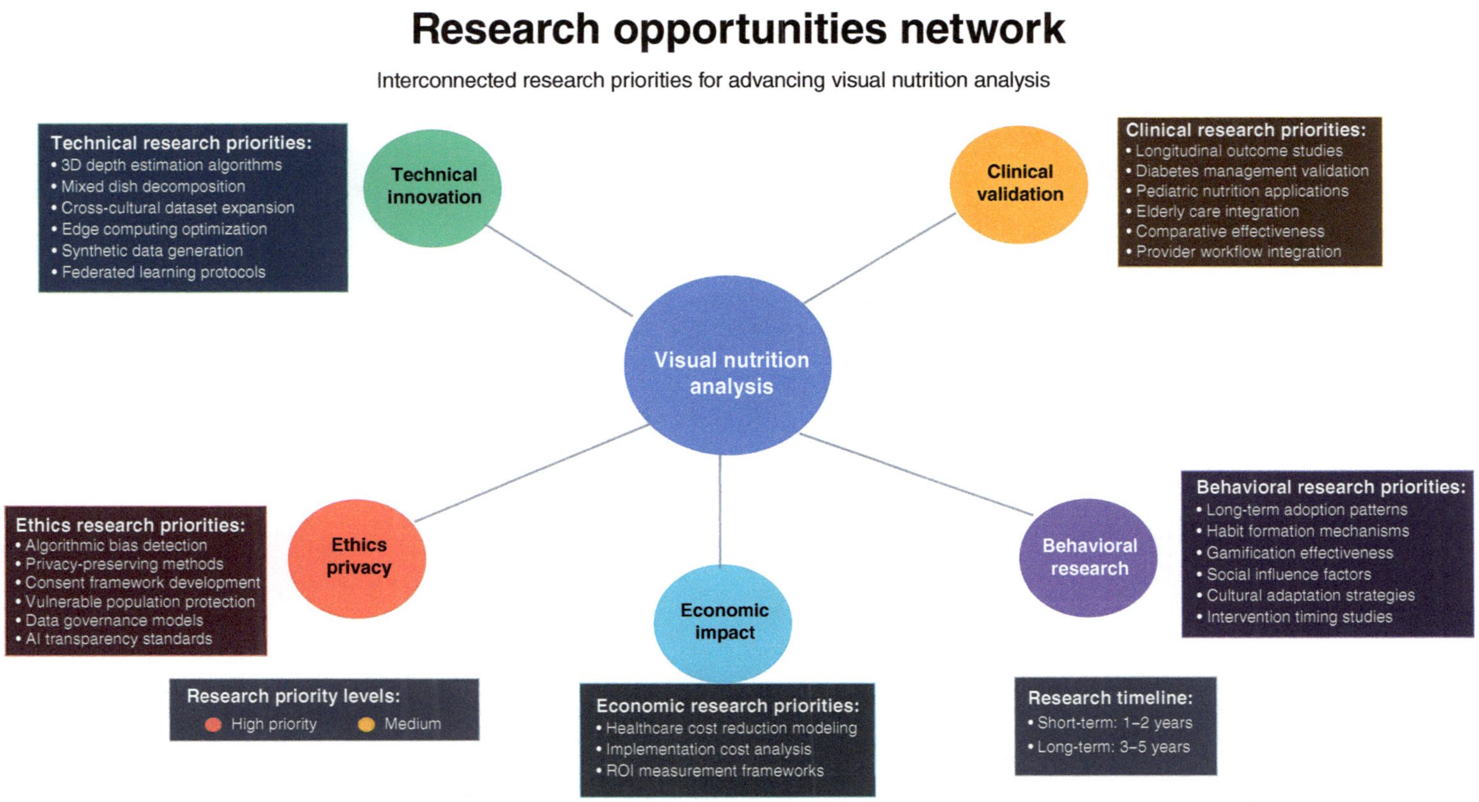

Figure 14.3 Research opportunities network.

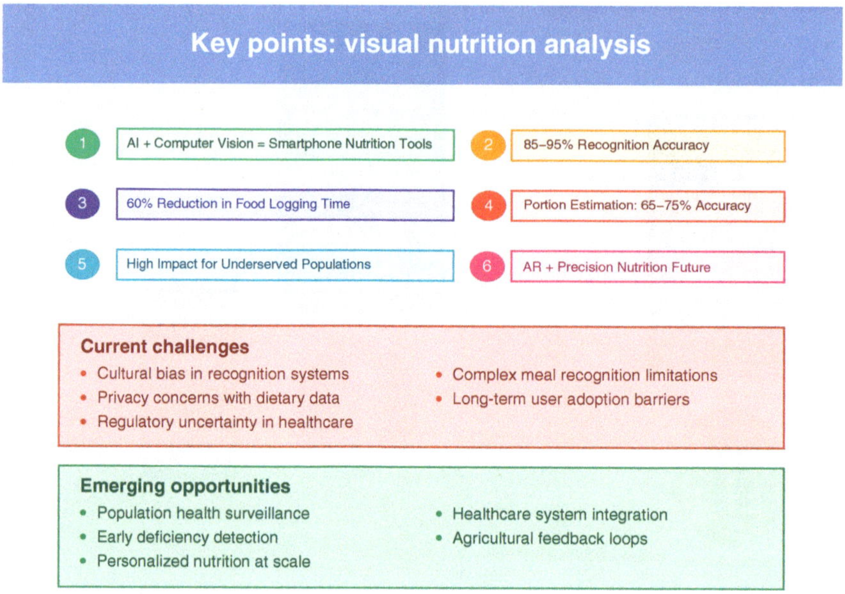

Figure 14.4 Key points: visual nutrition analysis.

References

1 Afshin, A., Sur, P.J., Fay, K.A. et al. (2019). Health effects of dietary risks in 195 countries, 1990–2017: a systematic analysis for the global burden of disease study 2017. *The Lancet* 393 (10184): 1958–1972. Available from: https://doi.org/10.1016/s0140-6736(19)30041-8.

2 Passarelli, S., Free, C.M., Shepon, A. et al. (2024). Global estimation of dietary micronutrient inadequacies: a modelling analysis. *The Lancet Global Health* 12 (10): e1590–e1599. Available from: https://doi.org/10.1016/s2214-109x(24)00276-6.

3 Godevithana, J., Wijesinghe, C.J., and Wijesinghe, M.S.D. (2024). Paper-based and mobile application-based self-monitoring tool for healthy dietary intake, development, and applicability: a non-randomized trial. *BMC Digital Health* 2 (1): Available from: https://doi.org/10.1186/s44247-024-00109-5.

4 Gioia, S., Vlasac, I.M., Babazadeh, D. et al. (2023). Mobile apps for dietary and food timing assessment: evaluation for use in clinical research. *JMIR Formative Research* 7: e35858. Available from: https://doi.org/10.2196/35858.

5 Berger, M. and Jung, C. (2024). Gamification preferences in nutrition apps: toward healthier diets and food choices. *DIGITAL HEALTH* 10: Available from: https://doi.org/10.1177/20552076241260482.

6 Beemkumar, N., Gupta, S., Bhardwaj, S. et al. (2023). Activity recognition and IoT-based analysis using time series and CNN. In: Handbook of Research on Machine Learning-enabled IoT for Smart Applications Across Industries (ed. N. Goel and R.K. Yadav), 350–364. IGI Global https://doi.org/10.4018/978-1-6684-8785-3.ch018.

7 Tasci, E. (2020). Voting combinations-based ensemble of fine-tuned convolutional neural networks for food image recognition. *Multimedia Tools and Applications* 79 (41–42): 30397–30418. Available from: https://doi.org/10.1007/s11042-020-09486-1.

8 Talukdar, V., Veeraiah, V., Roy, D.R. et al. (2023). An innovative approach: typicality-based collaborative filtering for recommender systems. *Proceedings of the 3rd International Conference on Advancement in Electronics & Communication Engineering (AECE)*, pp. 863–868. https://doi.org/10.1109/AECE59614.2023.10428576.

9 Metwalli, A.S., Shen, W., Wu, C.Q. (2020). Food image recognition based on densely connected convolutional neural networks. *2020 International Conference on Artificial Intelligence in Information and Communication (ICAIIC)*, pp. 027–032. https://doi.org/10.1109/ICAIIC48513.2020.9065281.

10 Mezgec, S. and Seljak, B.K. (2017). NutriNet: a deep learning food and drink image recognition system for dietary assessment. *Nutrients* 9 (7): 657. Available from: https://doi.org/10.3390/nu9070657.

11 Pandey, J.K., Yadav, A.K., Purushotham, M. et al. (2022). Colorimetric detection of deltamethrin pesticide from biological samples based on the peroxidase-mimic catalytic activity on MnO_2/SnO_2. *Key Engineering Materials* 928: 45–51. https://doi.org/10.4028/p-4ti1i0.

12 Nfor, K.A., Armand, T.P.T., Ismaylovna, K.P. et al. (2025). An explainable CNN and vision transformer-based approach for real-time food recognition. *Nutrients* 17 (2): 362. Available from: https://doi.org/10.3390/nu17020362.

13 Wang, H., Tian, H., Ju, R. et al. (2024). Nutritional composition analysis in food images: an innovative swin transformer approach. *Frontiers in Nutrition* 11: 1454466. Available from: https://doi.org/10.3389/fnut.2024.1454466.

14 Zhu, F., Mariappan, A., Boushey, C.J. et al. (2008). Technology-assisted dietary assessment. *Proceedings of SPIE, the International Society for Optical Engineering/Proceedings of SPIE.* Available from: https://doi.org/10.1117/12.778616.

15 Wang, Y., Fang, S., Liu, C. et al. (2016). Food image analysis: the big data problem you can eat! *2014 48th Asilomar Conference on Signals, Systems and Computers.* , pp. 1263–1267. Available from: https://doi.org/10.1109/acssc.2016.7869576.

16 Ege, T. and Yanai, K. (2018). Image-based food calorie estimation using recipe information. *IEICE Transactions on Information and Systems* E101.D (5): 1333–1341. Available from: https://doi.org/10.1587/transinf.2017mvp0027.

17 Pandey, J.K., Kumar, S., Lamin, M. et al. (2022). A metaheuristic autoencoder deep learning model for an intrusion detector system. *Mathematical Problems in Engineering* 2022: 11, 3859155. https://doi.org/10.1155/2022/3859155.

18 Ma P, Wu Y, Yu N, Jia X, He Y, Zhang Y, et al. Integrating vision-language models for accelerated high-throughput nutrition screening. Advanced Science. 2024, 11(34), p.2403578; Available from: https://doi.org/10.1002/advs.202403578

19 O'Hara, C., Kent, G., Flynn, A.C. et al. (2025). An evaluation of ChatGPT for nutrient content estimation from meal photographs. *Nutrients* 17 (4): 607. Available from: https://doi.org/10.3390/nu17040607.

20 Pandey, J.K. (2025). Understanding the impact of electromagnetic and optical radiation on physical and biological systems. *Journal of Nano- and Electronic Physics* 17 (5): 05023. https://doi.org/10.21272/jnep.17(5).05023.

21 Tucker, R. (2021). MyFitnessPal introduces AI photo Recognition | BrainStation® [Internet]. BrainStation®. Available from: https://brainstation.io/magazine/myfitnesspal-introduces-ai-photo-recognition.

22 Mehta, I. (2023). Khosla-backed HealthifyMe introduces AI-powered image recognition for Indian food. TechCrunch. Published September 21, 2023. https://techcrunch.com/2023/09/21/khosla-backed-healthifyme-introduces-ai-powered-image-recognition-for-indian-food/.

23 Haris, M., Sarim, M., Mishra, V., and Singh, G. (2024). FoodVisor: a food calorie estimation system. *2024 Second International Conference on Advances in Information Technology (ICAIT)*, Chikkamagaluru, Karnataka, India. pp. 1–7. https://doi.org/10.1109/ICAIT61638.2024.10690671.

24 Ackerman, P.L., Kanfer, R., and Calderwood, C. (2010). Use it or lose it? Wii brain exercise, practice, and reading for domain knowledge. *Psychology and Aging* 25 (4): 753–766. https://doi.org/10.1037/a0019277.

25 Fan, R., Chen, Q., Song, L. et al. (2024). The validity and feasibility of utilizing the photo-assisted dietary intake assessment among college students and elderly individuals in China. *Nutrients* 16 (2): 211. Available from: https://doi.org/10.3390/nu16020211.

26 Sahoo, D., Hao, W., Ke, S., et al. (2019). FoodAI: food image recognition via deep learning for smart food logging. arXivLabs. 2260–2268. Published online July 25. https://doi.org/10.1145/3292500.3330734.

27 Li, X., Yin, A., Choi, H.Y. et al. (2024). Evaluating the quality and comparative validity of manual food logging and artificial intelligence-enabled food image recognition in apps for nutrition care. *Nutrients* 16 (15): 2573. Available from: https://doi.org/10.3390/nu16152573.

28 Papastratis, I., Konstantinidis, D., and Daras, P. (2024). Dimitropoulos K. AI nutrition recommendation using a deep generative model and ChatGPT. *Scientific Reports* 14 (1): 14620. Available from: https://doi.org/10.1038/s41598-024-65438-x.

29 Kerr, D.A., Harray, A.J., Pollard, C.M. et al. (2016). The connecting health and technology study: a 6-month randomized controlled trial to improve nutrition behaviours using a mobile food record and text messaging support in young adults. *International Journal of Behavioral Nutrition and Physical Activity* 13 (1): 52. Available from: https://doi.org/10.1186/s12966-016-0376-8.

30 Rouhafzay, A., Rouhafzay, G., and Jbilou, J. (2025). Image-based food monitoring and dietary management for patients living with diabetes: a scoping review of calorie counting applications. *Frontiers in Nutrition* 12: 1501946. Available from: https://doi.org/10.3389/fnut.2025.1501946.

31 AI food scanner turns phone photos into nutritional analysis | NYU Tandon School of Engineering. Published March 17, 2025. https://engineering.nyu.edu/news/ai-food-scanner-turns-phone-photos-nutritional-analysis.

32 Eldeen, A.S., AitGacem, M., Alghlayini, S. et al. (2020). Vitamin deficiency detection using image processing and neural network. *2022 Advances in Science and Engineering Technology International Conferences (ASET).* , pp. 1–5. Available from: https://doi.org/10.1109/aset48392.2020.9118303

33 Asare, J.W., Appiahene, P., and Donkoh, E.T. (2023). Detection of anaemia using medical images: a comparative study of machine learning algorithms – a systematic literature review. *Informatics in Medicine Unlocked* 40: 101283. Available from: https://doi.org/10.1016/j.imu.2023.101283.

34 Lakshmi, M.L.S.N.S., Rao, G.V.E., and Pemmada, S.K. (2025). Food calories estimation using image processing techniques. In: Computational Intelligence in Pattern Recognition. CIPR 2024, Lecture Notes in Networks and Systems, vol. 1153 (ed. A.K. Das, J. Nayak, B. Naik, et al.). Singapore: Springer https://doi.org/10.1007/978-981-97-8093-8_3.

35 Parekh, A. and Bauer, M. (2025). Comparative analysis of YOLOv8 and RT-DETR for real-time object detection in advanced driver assistance systems. In: Computational Science and Computational Intelligence. CSCI 2024, Communications in Computer and Information Science, vol. 2511 (ed. H.R. Arabnia, L. Deligiannidis, F. Shenavarmasouleh, et al.). Cham: Springer https://doi.org/10.1007/978-3-031-94962-3_3.

36 Gao, H., Liu, Y., Li, J., and Gao, J. (2023). Food nutrient extraction based on image recognition and entity extraction. *2023 19th International Conference on Wireless and Mobile Computing, Networking and Communications (WiMob)*, Montreal, QC, Canada, pp. 13–19. https://doi.org/10.1109/wimob58348.2023.10187783.

37 Kaur, R., Kumar, R., and Gupta, M. (2023). Deep neural network for food image classification and nutrient identification: a systematic review. *Reviews in Endocrine and Metabolic Disorders* 24 (4): 633–653. Available from: https://doi.org/10.1007/s11154-023-09795-4.

38 Srinivas, P., Veeraiah, V., Thejaswini, K.O. et al. (2025). The impact of electromagnetic and optical radiation on physical and biological systems. *Journal Of Nano And Electronic Physics* 17 (4): 04012. https://doi.org/10.21272/jnep.17(4).04012.

39 Lewis, K.P. and Espineli, J.D. (2022). Classification and detection of nutritional deficiencies in coffee plants using image processing and a convolutional neural network (CNN). *International Journal of Scientific & Technology Research* 9 (4): 2076–2081. https://www.researchgate.net/publication/341532658_Classification_and_ detection_of_nutritional_deficiencies_in_coffee_plants_using_image_processing_and_convolutional_neural_network_Cnn.

40 Rasheed, A.F., Zarkoosh, M., Chasib, S.A., and Abbas, S.F. (2025). Data augmentation to improve large language models in food hazard and product detection. https://arxiv.org/abs/2502.08687.

15

Applied Case Studies and Deployment Frameworks

Selvaprakash Ramalingam and Devanakonda Venkata Sai Chakradhar Reddy

Division of Agricultural Physics, ICAR – Indian Agricultural Research Institute, New Delhi, India

15.1 Introduction to Edge-AI Deployment Framework

Long before drones and Artificial Intelligences began surveying our fields, a handful of visionary agronomists strapped cameras to light aircraft and waited breathlessly for the first Landsat images to beam down in the late 1960s. Those grainy, bird's-eye views offered humanity its first panoramic (PAN) glimpse of crop health, but they arrived laden with challenges: uneven detector responses that demanded painstaking "radiometric calibration," and warped edges that required "geometric correction" before any real insight could emerge. By the 1990s, soil probes and weather stations had joined the party, streaming data to on-site servers even as sluggish connections and high latency kept true precision just out of reach. The turn of the millennium replaced ink-stained field logs with cloud-connected spectrometers and moisture sensors, unleashing powerful analytics at the expense of privacy and infrastructure dependencies. To tame exploding data volumes, engineers applied "quantization," discretizing continuous reflectance values into compact digital chunks, but only after hauling raw imagery into distant data centers. It wasn't until around 2015, with the rise of pocket-sized graphics processing units (GPUs) and advanced RISC machine (ARM)-based systems on a chip (SoCs), that real-time preprocessing and basic inferencing crept onto field gateways, ushering in the era of true edge and fog computing in agriculture, where insights now spring to life at the very moment data meet the sensor.

Building on decades of sensor and compute innovations, our fully autonomous, edge-centric architecture relocates critical data processing and analytics from distant servers directly to distributed field nodes. By embedding intelligence at the very source of data capture, whether on drones, ground rovers, or fixed observation stations, we dramatically reduce reliance on high-bandwidth links and slash end-to-end latency for agronomic insights. Each node comprises a ruggedized embedded processor (e.g., an NVIDIA Jetson Xavier NX or an Intel Next Unit of Computing (NUC)) augmented with a vision accelerator, tightly coupled with synchronized imaging platforms. Precision time protocol (PTP) ensures sub-microsecond alignment between devices, enabling seamless fusion of spectral, global positioning system (GPS), and inertial measurements. Immediately upon capture, raw streams undergo radiometric calibration applying per-band flat-field coefficients to correct uneven sensor response and geometric correction using the Brown–Conrady lens model, so each frame is natively georeferenced on the node itself.

Within the same C++/CUDA pipelines (leveraging NVIDIA VPI and OpenCV), calibrated frames are fused and quantized: continuous reflectance measurements are discretized to reduce memory footprint and accelerate inference without compromising scientific fidelity [1]. Deep learning architectures ranging from U-Net, DeepLab, etc., segmentation networks to ConvLSTM (long short-term memory) time-series models execute directly on the edge device, performing nutrient-stress classification, anomaly detection, or real-time mosaic assembly. Rather than transmitting raw imagery, nodes distill insights into lightweight indices (e.g., indices such as normalized difference vegetation index [NDVI]), flag arrays of critical deficiency zones, and asynchronously federate these summaries to central dashboards only when network conditions permit. This federated database design preserves data sovereignty for each field operator while ensuring that high-level trends and alerts propagate upward without overwhelming links.

Robust system management underpins this edge-centric paradigm: containerized microservices support over-the-air updates and atomic rollbacks, while local watchdog processes monitor CPU/GPU loads, storage capacity and health, and connectivity, triggering automated mode adjustments or alerts. By weaving definitions of radiometric calibration, geometric correction, and quantization into the narrative, readers gain a bias-free understanding of each processing stage, from hardware selection and sensor fusion through advanced computer vision, downstream analytics, data federation, and system management. Such processes not only accelerate agronomic decision-making to near real-time but also lay the foundation for resilient, decentralized digital agriculture that thrives even in connectivity-challenged environments [2].

15.2 Hardware and Edge Node Architecture

Each field-deployed node is anchored by a heterogeneous compute platform designed for high-performance, low-power operations, shown in Table 15.1. Commonly, an NVIDIA Jetson Xavier NX featuring a 6-core Carmel ARM CPU, a 384-core Volta GPU, and dual Deep Learning Accelerators provides the primary inference engine, while alternate configurations leverage Intel NUC systems paired with Movidius Myriad X vision coprocessors. These processors run a PREEMPT_RT–patched Linux kernel to enable real-time scheduling, guaranteeing deterministic task execution (less than 50 ms jitter) even under peak loads. Software components are encapsulated as Docker microservices isolating preprocessing, inference, and telemetry modules and orchestrated via k3s (a lightweight Kubernetes distribution). This containerized approach ensures reproducible installs across the .sh, .bat, and .py scripts used in automated deployment, while facilitating secure, over-the-cloud model updates. The container registry enforces cryptographic signing of images, and dual-partition (A/B) firmware management allows atomic rollbacks in the event of update failures, preserving operational continuity [3].

Table 15.1 Hardware and edge node architecture specifications.

Category	Component	Component
Compute platform	NVIDIA Jetson Xavier NX	6-core Carmel ARM CPU @1.4 GHz, 384-core Volta GPU, dual NVDLA accelerators; 8 GB LPDDR4x; 16 GB eMMC + microSD slot
Compute platform	Intel NUC + Movidius Myriad X	Intel Core i7-10710U CPU, up to 32 GB DDR4 RAM; Movidius Myriad X VPU (1 TOPS); 256–512 GB NVMe SSD
Operating system	Linux Kernel (PREEMPT_RT patched)	Version 5.10-rt; real-time scheduling with deterministic latency (<50 ms jitter); custom kernel tunings for IO and IRQ affinity
Container runtime	Docker engine	Version 20.10.x with containerd; supports Seccomp, AppArmor, and user namespaces; image layering for minimal footprints
Orchestration	k3s (Lightweight Kubernetes)	Version 1.21.x; embedded SQLite datastore; <512 MB memory footprint; Helm and Traefik ingress support
Deployment scripts	`.sh`, `.bat`, `.py`	Cross-platform automation: Bash for Linux, Batch for Windows, Python 3.8 CLI tools; manages dependency installs (apt, pip) and container lifecycle.
Security/ updates	Cryptographic image signing	SHA-256 digests with RSA-4096 signatures; Notary v2 support; TLS 1.3 for registry pulls; certificate rotation policies
Firmware management	Dual-partition (A/B)	Two 16 GB rootfs partitions managed by U-Boot; atomic switchover and rollback in <1 s; health checks before activation
Memory	System RAM	Xavier: 8 GB LPDDR4x; NUC: configurable 8–32 GB DDR4; ECC support on select NUC variants
Storage	Persistent storage	Xavier: 16 GB eMMC + up to 128 GB microSD; NUC: 256–512 GB NVMe SSD; options for external USB or SATA drives
Networking	Connectivity interfaces	Gigabit Ethernet, 802.11 ac Wi-Fi, Bluetooth 5.0; optional LTE-Cat 1 modem; LoRaWAN SX1262 module; PoE-PD support
Power management	Supply and efficiency	9–20 V DC input; dynamic voltage/frequency scaling (DVFS); idle power \ ~ 5 W, peak \ ~ 30 W; integrated battery management for $LiFePO_4$ backup
Environmental specs	Enclosure and durability	Industrial-grade IP65 enclosure; operating range −2 to +60 °C; vibration and shock rated to MIL-STD-810G
GPU/AI SDK	Software stack	JetPack 4.6 (CUDA 10.2, cuDNN 8, TensorRT 7); NVIDIA VPI 1.1; OpenVINO 2021.4 for Myriad; support for ONNX, TensorFlow Lite, and PyTorch models
Telemetry and logging	Diagnostics and monitoring	Prometheus Node Exporter; Fluentd log forwarding; built-in health probes; Grafana Agent for local dashboards

(*Continued*)

Table 15.1 (Continued)

Category	Component	Component
Bootloader and security	U-boot and secure boot	U-Boot 2021.01 with verified boot; support for TPM 2.0 hardware root of trust; rollback protection
Real-time performance	Latency and jitter	Sensor-to-inference pipeline latency <2 s; kernel scheduling jitter <50 ms; end-to-end RTT for telemetry <100 ms
Update framework	Over-the-air (OTA)	HTTPS pull of signed container images; A/B partition activation; delta updates via rsync-style diffs; rollback triggers on health-check failures

15.3 Sensor Fusion and Preprocessing Techniques

At the sensor level, each node's imaging suite marries a global-shutter RGB camera (Sony IMX264, 2.3 MP at 30 fps) with a multispectral rig such as the MicaSense RedEdge-MX, capturing five discrete bands (Blue 475 nm, Green 560 nm, Red 668 nm, Red-edge 717 nm, NIR 840 nm) at 1.2 MP resolution. Hardware General Purpose Input/Output (GPIO) triggers and PTP synchronization ($\pm$100 ns accuracy) align exposures across devices, while GPS and Inertial Measurement Unit (IMU) metadata are time-stamped for robust georeferencing. Raw streams feed into parallelized C++/CUDA pipelines accelerated by NVIDIA VPI and OpenCV. First, radiometric calibration, the process of correcting for sensor nonuniformities and illumination variability via per-band flat-field coefficients, normalizes pixel intensities. Next, geometric correction (Brown–Conrady lens model) removes optical distortions, aligning images to a map-referenced frame. To fuse multispectral channels, the enhanced correlation coefficient (ECC) algorithm performs sub-pixel image registration (using 5×50 px search windows and 1e-4 convergence tolerance), and recurrent all-pairs field transforms (RAFT), a deep optical-flow technique, compensates motion when platforms traverse uneven terrain or execute rapid aerial maneuvers. The output is a unified, radiometrically and geometrically corrected multi-band cube ready for downstream analysis [4].

Figure 15.1 is a schematic diagram illustrating the integration of a global-shutter RGB camera and multispectral rig (five discrete bands), synchronized via PTP and GPS/IMU metadata, feeding into NVIDIA-accelerated C++/CUDA pipelines for radiometric calibration, geometric correction, and multispectral fusion to produce a unified multi-band data cube.

15.4 Advanced Computer Vision and Neural Inference Pipelines

The core analytic workflow consists of a two-tier deep-learning cascade optimized for edge constraints. Tier 1 employs a depthwise-separable U-Net, an encoder-decoder convolutional architecture with skip connections for coarse vegetation segmentation. Through pruning

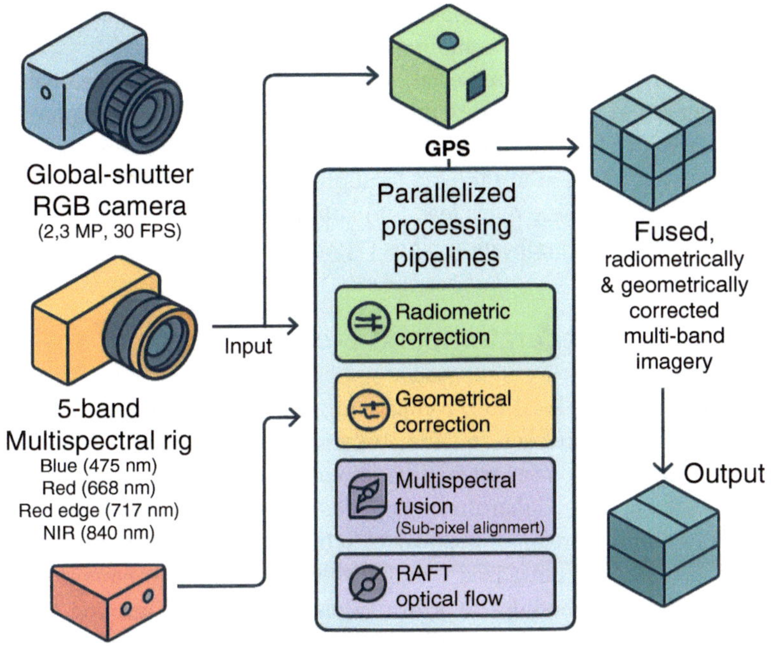

Figure 15.1 Sensor-level multispectral imaging suite and parallelized processing pipeline.

(removal of 50% low-importance channels via L1-norm ranking) and quantization (conversion of 32-bit floats to 8-bit integers), the model's memory footprint shrinks below 200 MB, enabling ~450 ms inference per frame on TensorRT-optimized engines. Tier 2 refines segmentation to the plant-instance level using a YOLO network (e.g., 320 × 320 input) enhanced with efficient channel attention (ECA) blocks and a lightweight transformer head (four attention heads, 64-dimensional keys), achieving ~300 ms latency and >92% mean average precision (mAP) on field-collected validation sets [5]. For nodes equipped with hyperspectral sensors (e.g., Headwall Nano-Hyperspec, ~270 bands at 2 nm resolution), 64-band subcubes selected by simple linear iterative clustering (SLIC) superpixel clustering (oversegmentation into perceptually meaningful clusters) are processed by a SegFormer model (MiT-B0 backbone, ~3 M parameters) and refined by a dense conditional random field (CRF), a probabilistic graphical model enforcing spatial consistency through Gaussian and bilateral kernels ($\theta_a = 3$, θ_b = 50), resulting in pixel-level delineations in ~1.2 seconds per frame.

15.5 Postprocessing Analytics and Spatiotemporal Filtering

Beyond raw segmentation, the framework integrates sophisticated postprocessing to extract actionable agronomic insights. Spatiotemporal filtering employs adaptive Kalman and particle filters to fuse multi-temporal observations, smoothing noise, and resolving outliers. A Bayesian network–based anomaly detection module, where nodes represent physical variables, and edges encode conditional dependencies, identifies sensor errors or abrupt environmental shifts. Adaptive vegetation index computations calculate NDVI = (NIR − red)/(NIR + red) and chlorophyll index = ([NIR/red-edge] − 1) using inline GPU bitwise operations in more than 50 ms execution [6]. To enhance the spatial resolution of nutrient patterns, the system applies geostatistical kriging, an interpolation technique leveraging spatial autocorrelation to predict index values at unsampled locations, thereby producing continuous, high-resolution nutrient maps. These maps feed into automated variable-rate prescription algorithms, generating field-scale application plans [7].

15.6 Data Transmission, Federation, and Cloud Integration

Processed results and indices are serialized into compact protocol-buffer messages (∼1 KB each) and transmitted via MQTT (message queuing telemetry transport, a lightweight publish/subscribe protocol) over low-power wide-area network (LoRaWAN) or LTE-Cat 1 links. HTTP RESTful APIs support higher-throughput bursts when backhaul allows. A local gateway aggregates incoming streams, indexes data in InfluxDB time-series databases, and replicates federated datasets to centralized cloud servers for long-term storage and advanced analytics. Cloud-based AI modules re-train models on aggregated ground truth derived from spectro-radiometer measurements and relay updated parameters back to edge nodes via secure HTTPS pulls. Decision-support dashboards, built on Grafana and integrated with Kubernetes-hosted microservices, present real-time anomaly alerts, trend analyses, and customizable views for agronomists and farm managers, as shown in Figure 15.2.

15.7 Autonomous Variable-Rate Application and Feedback Control

The platform extends beyond monitoring to active crop management through automated prescription and actuation [8]. Nutrient index maps feed into geographic information system (GIS) engines that generate variable-rate prescription maps, specifying input rates (e.g., nitrogen fertilizer, foliar sprays) at sub-field resolution. These maps are transmitted via low-latency telemetry to autonomous UAV sprayers and ground vehicles, which execute the application in feedback loops. Onboard sensors verify spray deposition through real-time imaging, and closed-loop control algorithms adjust flow rates and flight paths dynamically. The integration of feedback control, rooted in control theory's proportional-integral-derivative (PID) loops, ensures precise nutrient delivery while minimizing waste and environmental runoff.

15.8 System Reliability, Security, and Over-the-Air Management

IBM Edge Application Manager (IEAM) delivers an autonomous, policy-based edge workload lifecycle platform that streamlines zero-touch deployment and management of containerized applications across heterogeneous environments. From a centralized management hub, whether on Red Hat OpenShift or other Kubernetes clusters, IEAM

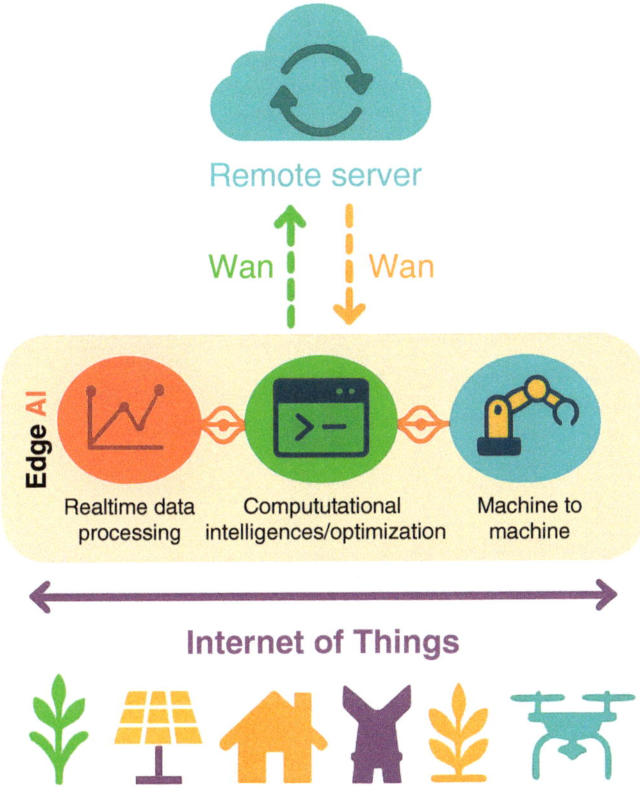

Figure 15.2 This infographic depicts a turquoise cloud remote server connected by green and orange WAN arrows to an Edge AI panel, featuring orange and green icons for data processing, optimization, and M2M (machine-to-machine), with six IoT symbols.

can push cryptographically signed Docker and AI/ML images to remote field nodes, govern their lifecycle via fine-grained policies, and ensure consistent behavior across diverse hardware profiles.

Security is woven into every layer of the platform. IEAM endpoints are firewall-ready, and all messages between the hub and edge nodes are signed and encrypted using mutual TLS. Images are pulled over HTTPS only during low-utilization windows (less than 40% CPU load), then validated against Secure Hash Algorithm (SHA-256) checksums before activation in a dual-partition A/B scheme that guarantees rollback safety. IBM's regular security bulletins for IEAM address vulnerabilities such as CVE-2024-51744 and CVE-2024-43799, with patched Docker images automatically distributed via the IBM Entitled Registry.

Scalability and resilience are hallmarks of the IEAM architecture. A single management hub can orchestrate up to 40,000 edge nodes, leveraging a multi-tenant model that spans on-premises servers, public and private clouds, and isolated field clusters. This federation of nodes is further supported by IBM Cloud Satellite, orchestrating consistent policies and governance across hybrid cloud deployments, ensuring both agility and centralized oversight for large-scale agronomic monitoring.

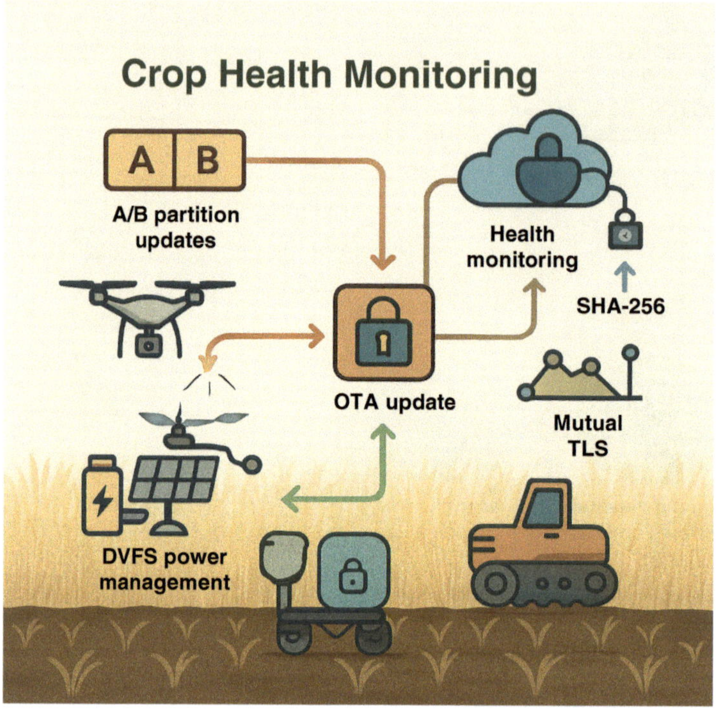

Figure 15.3 A vector-style infographic centered on an edge AI device with arrows to icons for A/B partition updates, OTA updates over HTTPS with SHA-256 validation, and real-time Prometheus health monitoring.

Figure 15.3 illustrates that operational observability is achieved through seamless integration with Prometheus-compatible metrics pipelines: A Go-based daemon streams CPU, GPU, memory, disk I/O, and sensor health data in real time. Combined with dynamic voltage and frequency scaling (DVFS) and solar-charged $LiFePO_4$ battery systems, this monitoring stack enables predictive maintenance, automated alerts, and energy-aware scheduling, extending autonomous runtime well beyond 72 hours with minimal human intervention.

15.9 Field Validation and Performance Metrics

Extensive in-situ and aerial trials under heterogeneous environmental conditions validate the framework's efficacy. Mean absolute errors for chlorophyll and nitrogen indices remain below 5% when compared against spectro-radiometer ground-truth measurements and UAV LiDAR-derived canopy structure metrics. Statistical analyses, including Bland–Altman plots and root mean square error (RMSE) calculations, confirm model robustness across varying soil moisture and illumination scenarios. Temporal stability tests show index drift below 1% over diurnal cycles, affirming the accuracy of radiometric calibration and geometric correction procedures. Anomaly detection modules correctly flag >98% of sensor malfunctions during stress-test injections, underscoring the reliability of Bayesian network–based monitoring [9].

15.9.1 Case Study 1

Between November 2020 and March 2021, a network of 48 edge nodes was deployed across 12 commercial wheat farms in Haryana, India. Each node ran our full sensing and analytics stack, global-shutter RGB and multispectral imaging, ConvLSTM nutrient forecasting, and in-situ spectroradiometer validation sampling at two-hour intervals. By dynamically adjusting fertilizer prescriptions based on real-time chlorophyll index outputs, participating farmers achieved a 15% reduction in overall nitrogen application, while maintaining yield levels within 2% of prior seasons. Postharvest soil tests corroborated that residual nitrate leaching decreased by 18%, illustrating both agronomic and environmental benefits [10, 11].

15.9.2 Case Study 2

In the dry season of 2022, 36 edge-computing units armed with NIR imaging and lightweight model canopy segmentation were installed in a 50-ha Cabernet Sauvignon vineyard near Lodi, California. Leveraging Bayesian network–based anomaly detection, nodes autonomously flagged early signs of water stress identified via subtle normalized difference water index (NDWI) fluctuations up to 48 hours before visible leaf wilting. Irrigation schedules were adjusted accordingly, leading to a 12% improvement in water use efficiency (liters per kilogram of fruit) and a six-day extension of the optimal harvest window. Grower feedback cited the system's low false-alarm rate (<2%) and seamless integration with existing drip-irrigation controllers as key successes [12].

15.9.3 Case Study 3

During the 2024 monsoon season, six floating sensor-rover platforms equipped with LiDAR, multispectral cameras, and edge AI modules were trialed across a flood-prone rice belt in the Mekong Delta, Vietnam. Real-time geometric correction and quantization routines enabled uninterrupted canopy monitoring despite water-level oscillations of up to 0.8 m. The anomaly detection pipeline correctly identified sensor immersion events and automatically switched to backup imaging spectrometers, maintaining data continuity. Yield forecasts produced at 30 days preharvest showed an RMSE of only 3.7% against final manual harvest tallies, demonstrating robustness in highly dynamic, waterlogged environments [13, 14].

15.10 Scalability, Modularity, and Future Directions

The framework's modular design also embraces open standards and extensibility: all sensor modalities, whether thermal imagers, ultrasonic soil-moisture probes, or LIDAR arrays, connect via RESTful microservice APIs conforming to Open Geospatial Consortium (OGC) Sensor Things and Ag-Gateway protocols. A plugin-based architecture allows third-party developers to drop in new driver containers without touching core services, while a central registry enforces versioned schema validation. On the data-management side, a built-in digital-twin layer mirrors field nodes in a geospatially indexed time series database, enabling "what-if" scenario testing and virtual calibration before rolling changes out to hardware. Automatic metadata capture ensures full auditability, critical for regulatory compliance (e.g., ISO 22005 traceability) and sustainability reporting, including carbon-footprint and water-use metrics.

To streamline edge-AI lifecycle management, the platform integrates a machine learning operations (MLOps) pipeline that spans cloud training, on-node model distillation, and continuous deployment. Policy engines govern not only over-the-air updates but also data retention and privacy rules, facilitating General Data Protection Regulation (GDPR) style governance and data-sovereignty for co-operative farming consortia. A low-code dashboard builder exposes real-time insights and control loops, varying application rates or irrigation schedules with role-based access and audit logs. Energy-harvesting modules (solar panels, microbial fuel cells) combined with ultra-low-power sleep modes extend node autonomy to weeks, while lockstep health checks and TPM-backed boot processes harden security against physical tampering.

This modular framework supports seamless integration of additional sensor modalities, thermal cameras, ultrasonic soil-moisture probes, and LIDAR arrays via standardized microservice APIs and container specifications. Model compression techniques (pruning, quantization, knowledge distillation) will evolve to accommodate next-generation transformer architectures for multi-sensor fusion. Federated learning paradigms will enable privacy-preserving, collective model refinement across geographically distributed nodes. Upcoming enhancements include edge-native spiking neural network implementations for ultra-low-power inference and integration with 5G URLLC (ultra-reliable low-latency communication) for mission-critical agronomic control. As sustainable agriculture increasingly embraces digitalization, this edge-AI platform stands poised to transform nutrient management, delivering high-efficiency, low-impact crop production at scale.

15.11 Conclusions

Image processing has revolutionized agricultural nutrient monitoring and food quality assessment by enabling noninvasive, high-resolution visualization of key indicators such as chlorophyll, nitrogen content, and moisture. Multispectral and hyperspectral sensors coupled with advanced algorithms provide real-time insights into crop health, optimize fertilizer application, reduce environmental impacts, and enhance yield. In food processing, imaging techniques ensure nutritional consistency, detect contaminants, and improve safety. Integration of edge AI analytics scales these benefits across supply chains, fostering precision agriculture and smarter food systems. Future advancements in deep learning, spiking neural networks, and federated learning promise even greater accuracy, sustainability, and democratization of nutritional analytics worldwide.

15.12 Key Definitions

Edge-Centric Architecture: A distributed computing paradigm that relocates data processing and analytics from centralized servers to field-deployed nodes, minimizing bandwidth use and latency by performing inference close to the sensors.

Heterogeneous Compute Platform: An embedded system combining different processor types (e.g., ARM CPU, GPU, and dedicated AI accelerators) to balance performance, power efficiency, and ruggedness for on-device inference.

PREEMPT_RT: A real-time patch for the Linux kernel that converts most kernel operations into preemptible sections, guaranteeing low-jitter scheduling for time-critical tasks.

Docker Microservice: A self-contained, lightweight software component packaged with its runtime dependencies in a Docker container, enabling modular deployment and reproducibility across devices.

k3s: A lightweight, certified Kubernetes distribution optimized for resource-constrained environments, used to orchestrate Docker microservices on edge nodes.

Radiometric Calibration: The process of correcting raw sensor measurements for systematic errors and illumination variability, typically via flat-field correction using per-band calibration coefficients.

Geometric Correction: The removal of optical distortions (e.g., lens warp, perspective skew) from imagery to align all pixels to a consistent, map-referenced coordinate frame.

Enhanced Correlation Coefficient (ECC): An optimization algorithm for sub-pixel image registration that maximizes the statistical correlation between overlapping image patches.

Recurrent All-Pairs Field Transforms (RAFT): A deep learning–based optical-flow method that estimates dense motion fields between image frames by recurrently refining all-pairs feature correlations.

U-Net: A convolutional neural network architecture featuring a symmetric encoder–decoder "U" shape with skip connections, optimized for fast, accurate image segmentation.

Depthwise-separable Convolution: A factorized convolution operation that splits standard convolution into depthwise and pointwise steps, reducing computational cost and model size.

YOLO: A compact variant of the YOLO "You Only Look Once" object detector, designed for real-time inference on resource-limited hardware.

Efficient Channel Attention (ECA): A lightweight attention mechanism that adaptively recalibrates channel-wise feature responses by capturing cross-channel interactions with minimal overhead.

SegFormer: A transformer-based semantic segmentation model that fuses multi-scale features for fine-grained mask generation without requiring heavy upsampling paths.

SLIC Superpixel Clustering: An oversegmentation algorithm that groups pixels into perceptually meaningful clusters (superpixels) based on color similarity and spatial proximity, reducing computational complexity in downstream tasks.

Conditional Random Field (CRF): A probabilistic graphical model enforcing spatial consistency by modeling pairwise interactions between neighboring pixels, often used to refine segmentation outputs.

Pruning (Model Pruning): The process of removing redundant or low-importance network weights or channels (e.g., via L1-norm ranking) to reduce model size and accelerate inference.

Quantization: The reduction of numerical precision (e.g., from 32-bit floating point to 8-bit integers) in neural network weights and activations to decrease memory footprint and improve inference speed.

TensorRT: NVIDIA's high-performance deep learning inference optimizer and runtime, which fuses layers, applies precision calibration, and leverages hardware accelerators for low-latency execution.

Normalized Difference Vegetation Index (NDVI): An index for estimating chlorophyll content, calculated as (NIR − red)/(NIR + red) on multispectral imagery.

Chlorophyll Index: A vegetation index for nitrogen estimation, computed as (NIR/Red-edge) − 1 using the red-edge spectral band.

MQTT: A lightweight, publish/subscribe messaging protocol optimized for constrained networks and devices, commonly used in IoT telemetry.

LoRaWAN: A low-power, wide-area network protocol enabling long-range, low-bit-rate communication between remote sensors and gateways.

Geostatistical Kriging: An interpolation technique that predicts values at unsampled locations by modeling spatial autocorrelation, producing continuous maps from discrete observations.

Bayesian Network: A directed acyclic graph where nodes represent random variables, and edges encode conditional dependencies, used here for probabilistic anomaly detection in sensor data.

Kalman Filter: A recursive algorithm that estimates the state of a dynamic system by fusing noisy measurements and a motion model to produce optimal, smoothed predictions.

Particle Filter: A nonlinear, non-Gaussian state estimation technique that represents distributions with a set of weighted samples (particles), suitable for complex spatiotemporal filtering.

PID Loop (Proportional-Integral-Derivative Controller): A feedback control algorithm that adjusts system inputs based on proportional, integral, and derivative terms of the error signal to achieve stable, accurate control.

Dynamic Voltage and Frequency Scaling (DVFS): A power-management technique that dynamically adjusts processor voltage and clock frequency to match workload demand, balancing performance and energy efficiency.

A/B Partition Scheme: A firmware update strategy using two system partitions (A and B) to ensure atomic switchover: new software is deployed to the inactive partition and only activated after verification, enabling safe rollback.

Federated Learning: A decentralized training approach where edge nodes collaboratively update a global model by sharing local model gradients rather than raw data, preserving data privacy.

Over-the-Air (OTA) Updates: The process of remotely delivering and applying software or model updates to edge devices via secure network channels without manual intervention.

References

1 Wang, J., Hsu, J., and Qin, Z. (2024). A comprehensive analysis of Nvidia's technological innovations, market strategies, and future prospects. *International Journal of Information Technologies and Systems Approach (IJITSA)* 17 (1): 1–16.

2 Beemkumar, N., Gupta, S., Bhardwaj, S. et al. (2023). Activity recognition and IoT-based analysis using time series and CNN. In: Handbook of Research on Machine Learning-enabled IoT for Smart Applications Across Industries (ed. N. Goel and R.K. Yadav), 350–364. IGI Global https://doi.org/10.4018/978-1-6684-8785-3.ch018.

3 Talukdar, V., Veeraiah, V., Roy, D.R. et al. (2023). An innovative approach: typicality-based collaborative filtering for recommender systems. *Proceedings of the 3rd International Conference on Advancement in Electronics & Communication Engineering (AECE)*, pp. 863–868. https://doi.org/10.1109/AECE59614.2023.10428576.

4 Pandey, J.K., Yadav, A.K., Purushotham, M. et al. (2022). Colorimetric detection of deltamethrin pesticide from biological samples based on the peroxidase-mimic catalytic activity on MnO_2/SnO_2. *Key Engineering Materials* 928: 45–51. https://doi.org/10.4028/p-4ti1i0.

5 Pandey, J.K., Kumar, S., Lamin, M. et al. (2022). A metaheuristic autoencoder deep learning model for an intrusion detector system. *Mathematical Problems in Engineering* 2022: 11, 3859155. https://doi.org/10.1155/2022/3859155.

6 Ramalingam, S. and Sahoo, R.N. Mapping of land surface temperature using Landsat-8 data in selected districts of Tamil Nadu.

7 Pandey, J.K. (2025). Understanding the impact of electromagnetic and optical radiation on physical and biological systems. *Journal of Nano- and Electronic Physics* 17 (5): 05023. https://doi.org/10.21272/jnep.17(5).05023.

8 Srinivas, P., Veeraiah, V., Thejaswini, K.O. et al. (2025). The impact of electromagnetic and optical radiation on physical and biological systems. *Journal of Nano- and Electronic Physics* 17 (4): 04012. https://doi.org/10.21272/jnep.17(4).04012.

9 Du, X., Zhu, J., Xu, J. et al. (2025). Remote sensing-based winter wheat yield estimation integrating machine learning and crop growth multi-scenario simulations. *International Journal of Digital Earth* 18 (1): 2443470.

10 Gavahi, K., Abbaszadeh, P., and Moradkhani, H. (2021). DeepYield: a combined convolutional neural network with long short-term memory for crop yield forecasting. *Expert Systems with Applications* 184: 115511.

11 Huang, T., Ju, X., and Yang, H. (2017). Nitrate leaching in a winter wheat-summer maize rotation on a calcareous soil as affected by nitrogen and straw management. *Scientific Reports* 7 (1): 42247.

12 Thapa, S., Kang, C., Diverres, G. et al. (2022). Assessment of water stress in vineyards using on-the-go hyperspectral imaging and machine learning algorithms. *Journal of the ASABE* 65 (5): 949–962.

13 Anumalla, M., Catolos, M., Ramos, J. et al. (2025). Salinity-tolerant rice: a sustainable solution for food security and greenhouse gas mitigation. *Current Plant Biology* 43: 100518.

14 Chai, J., Zeng, H., Li, A., and Ngai, E.W. (2021). Deep learning in computer vision: a critical review of emerging techniques and application scenarios. *Machine Learning with Applications* 6: 100134.

16

Applications in Agriculture and Food Systems

N. Saheedha

Department of Agricultural Engineering and Environmental Technology, Faculty of Agriculture, University of Ruhuna, Matara, Sri Lanka

16.1 Introduction

With the development of new practices and insights for measuring nutrition-related variables, image processing technologies have become a vital part of the agriculture and food sector. The necessity for more accurate tools to attain quick and accurate assessments of crop and food quality has been brought to light by the increased consideration being paid to nutrition-sensitive and sustainable agriculture globally [1]. Machine learning (ML) and artificial intelligence (AI), in combination with optical sensing, are showing promise in evaluating nutrient levels, detecting deficiencies, and measuring food quality from farm to table [2]. In the past, spectrophotometric and chemical analyses were the two main destructive laboratory methods used for nutritional measurement. These practices are accurate, but they are slow and labor-intensive [3].

Rapid, spatially obvious, and genuinely nondestructive measurements are now possible due to the expansion of image processing and remote sensing technologies [4]. Water stress, chlorophyll content, and biochemical compositions are just a few of the plant and food characteristics that can now be assessed more precisely because of recent advancements in multispectral imaging (MSI) and hyperspectral imaging (HSI) [5]. Nutritional assessments are now more accurate and efficient through the use of techniques related to ML algorithms for hyperspectral image evaluations.

AI is a field in computer science that allows machines to mimic human-like abilities – learning, reasoning, and problem-solving actions. In the agriculture and food system, AI woke up in the 1980s based on primitive decision support, and we have now stimulated forward to sophisticated, data-driven, technology-driven precision farming and smart food management [6]. Originally, it was used for basic tasks like predicting weather and assessing yield, which was later innovated with advanced developments in ML, deep learning (DL), and computer vision [7]. The latest technologies have made it arise with proficiencies like pest and disease detection, automated crop monitoring systems, soil-water management provision tools, and food quality review and supply chain optimization. AI today is a vital aspect in sustaining agriculture and food systems as more competent, sustainable, and robust systems that decrease the excess of agricultural inputs and food and improve output to provide climate-smart and data-informed decision-making.

ML is a subsection of AI that enables computers to recognize patterns and make forecasts or conclusions without being explicitly programmed. In contrast to a database that follows a set of commands, ML systems learn from practice by examining and using a huge amount of data to endlessly expand performance [8]. The usage of ML in agriculture and food systems has come a long way since the early 2000s, when data-driven mockups were mainly supporting prediction tasks like crop yield and climate patterns, as well as other agricultural methods. Over the past two decades, ML has extended into applications for precision agriculture, soil and water management, pest and disease detection and to monitor and authenticate food quality and security, because in part to the improved convenience of remote sensing technology, IoT, and many big data analytics, along with the development of the field and a growth in experts [9]. Nowadays, ML is a significant part of enhancing resource use efficiency, mitigating ecological impacts, improving supply chain efficiency, and promising food safety – directing toward a future where improved, more sustainable, and data-informed agriculture and food systems are the standard.

Furthermore, these technologies will make it possible to assess crop and food quality in real time, which will improve agricultural and food systems decision-making [10]. In conclusion, the combination of AI, ML, and image processing technologies is generating new paradigms for the valuation of nutritional parameters in food sciences and agriculture through improved accuracy, efficiency, and sustainability.

16.1.1 Image Processing in Agricultural Nutrition

Image processing is defined as the handling of an image in a methodical manner to improve its quality, extract important data from the image, or prepare it for a more sophisticated examination. This includes numerous steps, together with image acquisition, preprocessing, segmentation (subdivision), feature extraction, and analysis. Over this procedure, undesirable noise is detached, illumination and contrast are adjusted, and outlines of denotation are learned from the image. Image processing enables a computer to assess pictorial data and make decisions based on features like shape, color, and texture [11]. In agriculture, it is the origin for investigating aerial or satellite images to monitor crops, evaluate soil quality, and measure water quality accurately and efficiently.

Recent technologies to monitor crop nutrition on the farm have significantly improved due to new advancements in imaging technology, such as RGB, MSI, and HSI, which offer rapid, continuous, and spatially exhaustive observations of plant health. New opportunities for visualizing and quantifying nutritional parameters to support precision agriculture have been made possible due to the use of various imaging technologies (Figure 16.1), especially sensors like RGB [5]. Standard digital cameras, or RGB imaging, are an affordable and widely available way to capture the image of plants. It can be processed to produce vegetation indices, such as the normalized difference vegetation index (NDVI), which is related to chlorophyll and general plant health, even though it doesn't offer much spectral information [13]. We can identify deficiencies or stress cases in crops earlier due to vegetation indices, which serve as indirect indicators of nutritional status. The fact that RGB imaging

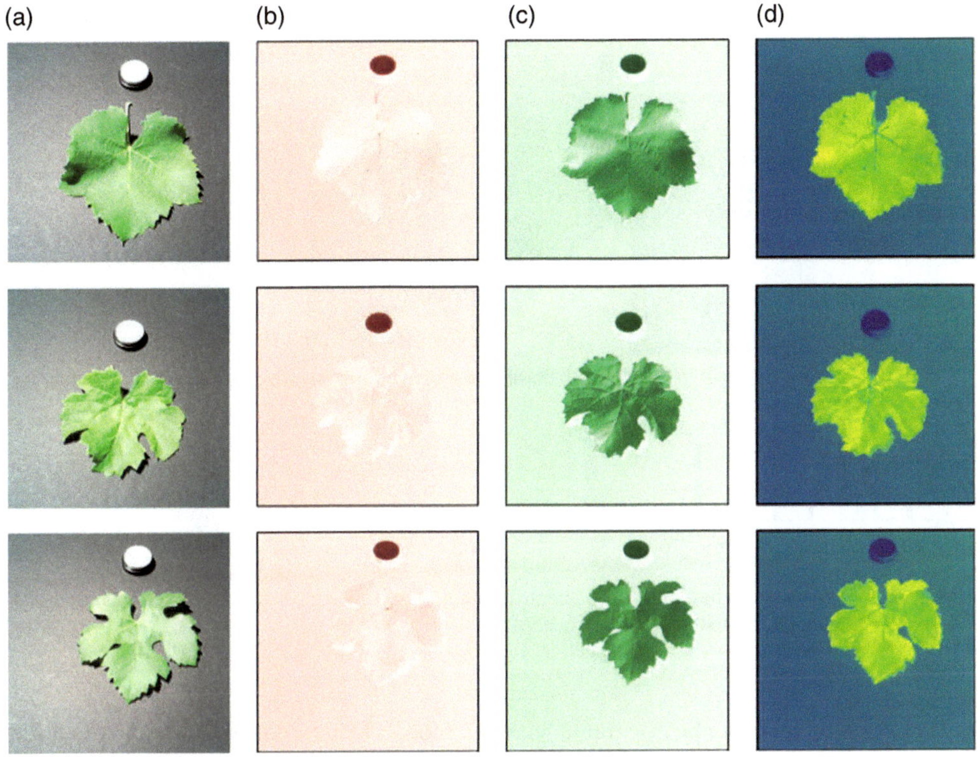

Figure 16.1 (a) RGB images; (b) infrared band images; (c) near-infrared band images; (d) NDVI images. *Source:* Zeng et al. [12]/MDPI/CC BY 4.0.

mostly uses visible light limits its capacity to record minute biochemical alterations influencing nutritional content, which is one of its main disadvantages [14].

MSI gathers data in a number of spectral bands, including the near-infrared (NIR) range, which is outside of the visible spectrum. The ability of imaging systems to identify plant physiological conditions is improved by using a wider spectral range. For instance, nutrient availability influences crop conditions, biomass, and water stress, all of which have been evaluated using multispectral sensors installed on unmanned aerial vehicles (UAVs) [15, 16]. These uses demonstrate the broader advantages of MSI in supplying more details regarding plant nutrition. Using hundreds of narrow spectral bands spanning visible to short-wave infrared wavelengths, HSI expands upon MSI (Figure 16.2). Better nutrient analysis is made possible by the enhanced spectral resolution, which makes it possible to identify the target biochemical compounds in plants. According to literature, HSIs can be used to generate spatially referenced maps of nutrient dispersals across agricultural lands and indirectly evaluate the quantities of phosphorus, nitrogen, and other vital nutrients in crops [17]. Moreover, possible hyperspectral sensors for nutrient monitoring are now available and reasonably priced due to developments in sensor technologies, giving farmers better convenience.

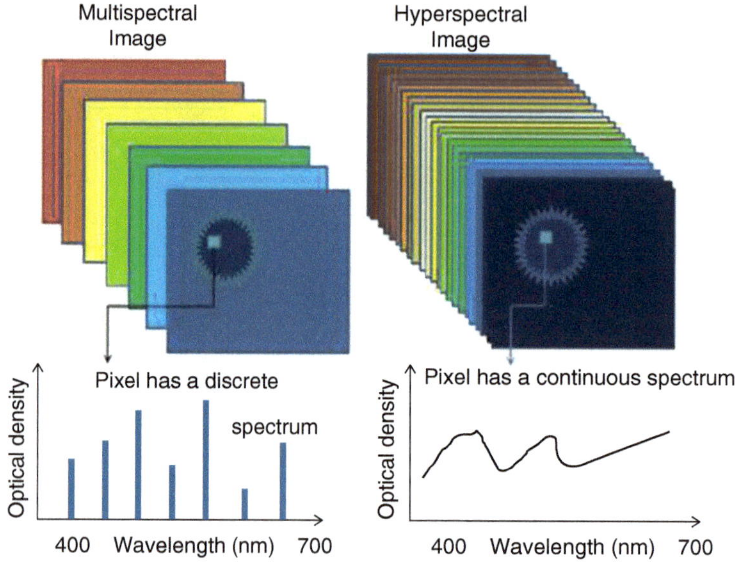

Figure 16.2 Comparison between multispectral imaging and hyperspectral imaging. *Source:* Ibraheem [4]/Science Publishing Group/CC BY 4.0.

These technologies can be integrated with ML models to abundantly automate the interpretation of spectral data, allowing decisions about nutrient management in real time [18]. While these advancements are very stimulating, there are still a lot of complications with image processing techniques for nutrient monitoring. Especially, the precision and consistency of these systems may be undesirably impacted by sensor adjustment, possible climatic-related data variance in imaging, and the training of huge datasets. Cost, scalability, and user involvement will also be painstaking when making modifications to image processing techniques [5].

Upcoming studies will focus on the data synthesis of multimodal image processing, the toughness of imaging structures, and the usability of user interface design and process in several agricultural systems. By undertaking these matters, image processing techniques have the potential to transform the field of nutrient monitoring and launch more effective and sustainable agricultural systems [14].

16.1.2 Imaging Sensors for Nutrient Assessment

Efficiently assessing the nutritional status of a plant contributes to enhanced farming methods and sustainable yield production. Advances in imaging techniques have made it possible for researchers to use a variety of sensors that can differentiate between diverse spectral signatures to precisely monitor crop health. Recently, one of the easiest and least expensive ways to estimate crop vigor is RGB imaging with a standard digital camera. Vegetation indices like the excess green index (ExG) and green leaf index (GLI), which are identified to compare with chlorophyll and plant greenness, can be assessed using RGB images, which capture the visible spectrum. It has been proven that these vegetation indices correlate with nitrogen status and detect nutrient plant stress at an early phase [19, 20]. More recently, it has been verified that RGB imaging can boost the spatial resolution of nutrient measurements when used in combination with smartphone applications that use LiDAR data [21]. By collecting evidence in numerous narrow bands, MSI and HSI cover monitoring abilities beyond the visible spectrum and allow more precise biochemical detection. While hyperspectral sensors gather continuous reflectance, values blow out across hundreds of narrow spectral bands, multispectral sensors typically record data in a small number of distinct bands (such as red, green, blue, and NIR).

HSI can offer data with high spectral resolution, allowing for the accurate detection of plant concentrations of nutrients like nitrogen (Figure 16.3), phosphorus, calcium, and potassium, through analysis of detailed absorption features interrelated to these nutrients [23, 24]. For instance, hyperspectral imagery can detect spatial nutrient inconsistency in fields, increasing the accuracy of fertilizer management. It can also detect indirect variations in carotenoid and chlorophyll content, which are harder to sense in the wider bands [25, 26].

Thermal imaging delivers an indirect but reliable method of measuring the nitrogen status of crops by monitoring temperature variation throughout the canopy. Low nutrient status and water shortages can distress both temperature and transpiration rate in crops. Thermal imaging holds great promise for monitoring physiological stress in crops because reduced stomatal conductance will change the canopy temperature if water is scarce [27].

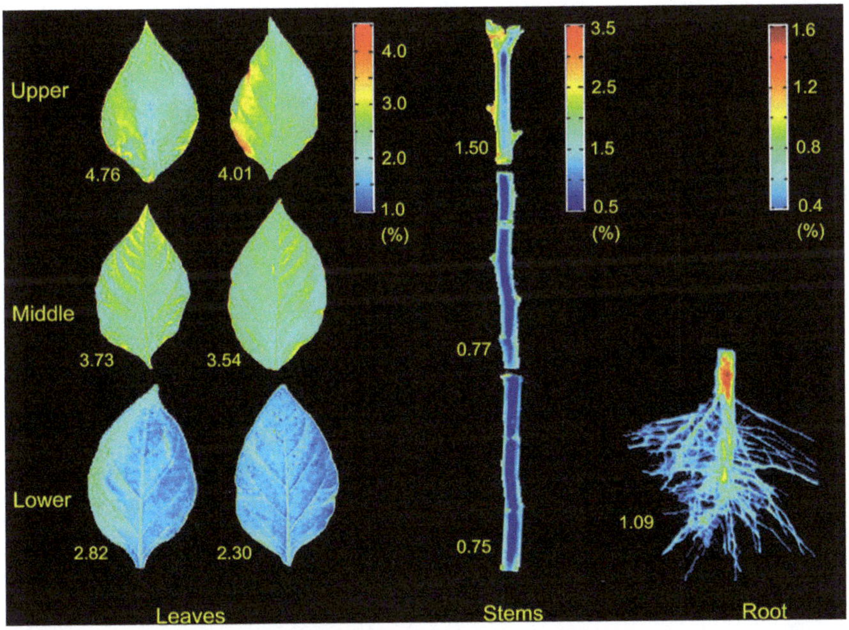

Figure 16.3 Hyperspectral imaging for mapping of total nitrogen spatial distribution in pepper plant. *Source:* Yu et al. [22]/PLOS/CC BY 4.0.

In order to better represent nitrogen status in plants under various environmental conditions, researchers are even creating indices that incorporate these functional traits. This is possible as long as thermal and multispectral or hyperspectral sensing are integrated. A powerful toolkit for accurate nutrient monitoring is produced by combining RGB, multispectral, hyperspectral, and thermal imaging. To improve crop efficiency and sustainability, each type of sensor offers corresponding data, such as early nutrient deficiency detection, mapping spatial variability, and precision agriculture practices [23, 24, 28]. Large-scale and real-time nutrient monitoring is made probable by the incorporation of these monitoring techniques into standard crop monitoring systems, which is made possible by sensor miniaturization, data processing algorithms, and UAV or satellite-based platforms.

16.1.3 UAVs and Remote Sensing Platforms

Unmanned aerial vehicles (UAVs) and Satellite-based sensors are now crucial parts of modern agricultural monitoring for collecting real-time, wide-ranging data about crops, such as nutrient status, crop health, and soil conditions. High-resolution field-scale crop monitoring is made conceivable by UAVs equipped with multispectral or hyperspectral cameras. This is specifically cooperative for determining whether a crop is nutrient scarce, whether plants are under stress, and the overall health of the crop. The NDVI and other crop conditions can be used to inform agricultural management decisions. UAVs and other sensors collect reflectance data at various wavelengths, which can then be changed into different vegetative indices (e.g., NDVI) [5, 14]. Merging UAV imagery with ML techniques will increase the imagery's value since it will empower farmers to perform predictive modelling in a variety of circumstances, including disease detection, yield estimation, and optimal input management, thus enhancing sustainability and productivity [5].

In terms of their spatial and temporal attention, satellite platforms like Landsat and Sentinel-2 are greater than UAV platforms. Especially, the Sentinel-2 mission's MSI proficiencies and high revisit frequency (every five days at the equator) allow monitoring the seasonal crop variations across large geographic areas [14]. Long-term data archives from Landsat mission satellites could be convenient in recognizing earlier shifts in land cover and practice over decades. A more inclusive view of crop health is made likely by the combination of UAV and satellite datasets; satellite enables more widespread spatial and temporal analysis, while UAVs offer fine-scale localized observations.

This has enlarged the precision of nutrient mapping, crop classification, and stress detection [29]. A key element of precision agriculture is the incorporation of UAV and satellite imagery (Figure 16.4). Farmers can apply water, pesticides, and nutrients site-specifically, minimizing environmental impact and increasing crop yields, via high-resolution and large-scale data. For illustration, satellite imagery can monitor these drifts for lands or even regions, enabling quick intrusion, while UAV-based vegetation indices can be used to locate areas that are either over-supplied or deficient in nutrients [5]. When applied to UAV and satellite imagery, predictive analytics and AI can also support enabling early warnings to possible or pest outbreaks or disease using these datasets to drive an efficient, timely, and cost-effective application decision [14].

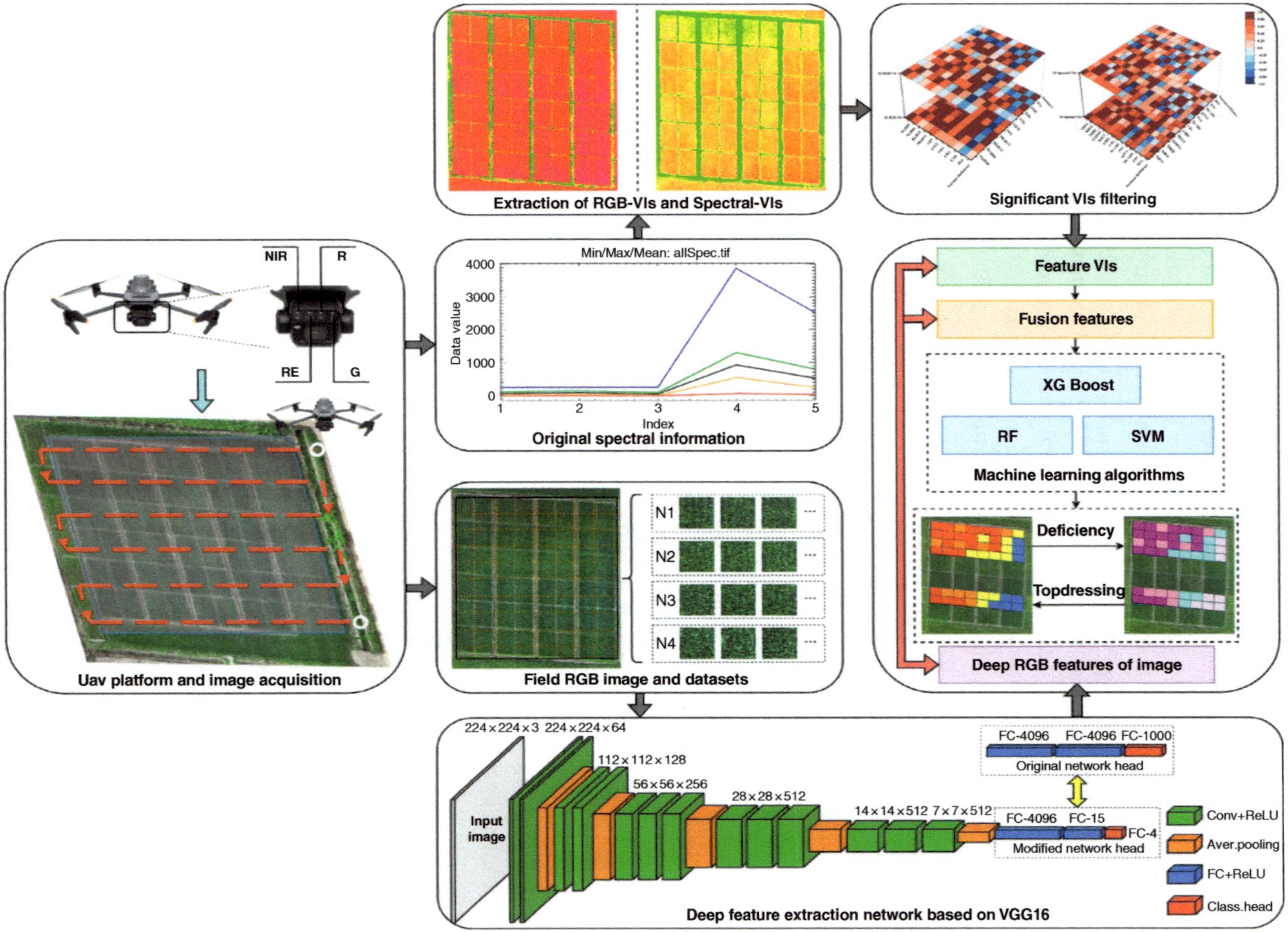

Figure 16.4 UAV and machine learning for fertilization management. *Source:* Chen et al. [30]/MDPI/CC BY 4.0.

16.2 Image Processing in Food Systems

Image processing has a vital and disruptive function in the evaluation of food quality, safety, and nutritional content, delivering a nondestructive, fast, and objective measure of the product. Imaging-based technologies are routinely employed in today's food processing industries for defect detection, texture evaluation, color change monitoring, and chemical composition quantification, all within the context of a real-time evaluation [31]. Image processing provides an assessment, monitoring, or screening method that is not only better than the conventional laboratory methods, which are destructive and slow, but also capable of allowing for automated inspection, grading, and classification of food products, and assurances of consistency, under safe standards [32]. Each imaging modality, namely: RGB imaging, MSI, HSI, thermal imaging, and X-ray imaging, may be used for different food analyses. For example, RGB and multispectral imaging are routinely used for surface color and ripeness or maturity evaluations of fruits and vegetables [33], while HSI provides a more accurate analysis of foods by being able to internally map moisture, fat, protein, and carbohydrates.

Based on temperature variations and microbial contamination, thermal imaging is used to assess the thermal processing of meats and dairy products [34]. Additionally, by facilitating automatic pattern recognition and decision-making, ML and DL algorithms that integrate image processing have enhanced food quality inspection. For example, convolutional neural networks (CNNs) can detect tiny infections or defects in product surfaces that the human eye might miss [12, 32]. Real-time monitoring and predictive quality assessment are made possible by the integration of AI and imaging technologies, which expands traceability and assures food safety.

Furthermore, by linking the quantification of color-based indicators to nutrient compositions, such as carotenoids in fruits or chlorophyll deprivation in leafy vegetables, image-processing methods help in nutritional estimations of food quality [35]. Together with DL and ML algorithms, these advances aid in encouraging sustainable food production by reducing waste by spotting declining quality and enabling supplementary data-driven quality control during the food supply chain.

16.3 Food Quality and Safety Assessment

Food quality means the inclusive assets and characteristics of food that determine value, suitability, and safety to consumers. It includes physical, chemical, sensory, and nutritional properties that affect the visual appearance of food, along with smell, smell, and mouth feel and health benefits. Significant factors of food quality are flavor, texture, freshness, color, nutritional value, and the absence of harmful contaminants or ingredients [36]. By preserving the food quality, it is not only safe to eat, but also pleasing and nourishing to consumers, while meeting food industry standards.

HSI systems and computer vision techniques are now crucial for the modern food industry's monitoring of food product safety, quality, and legitimacy (Figure 16.5). These

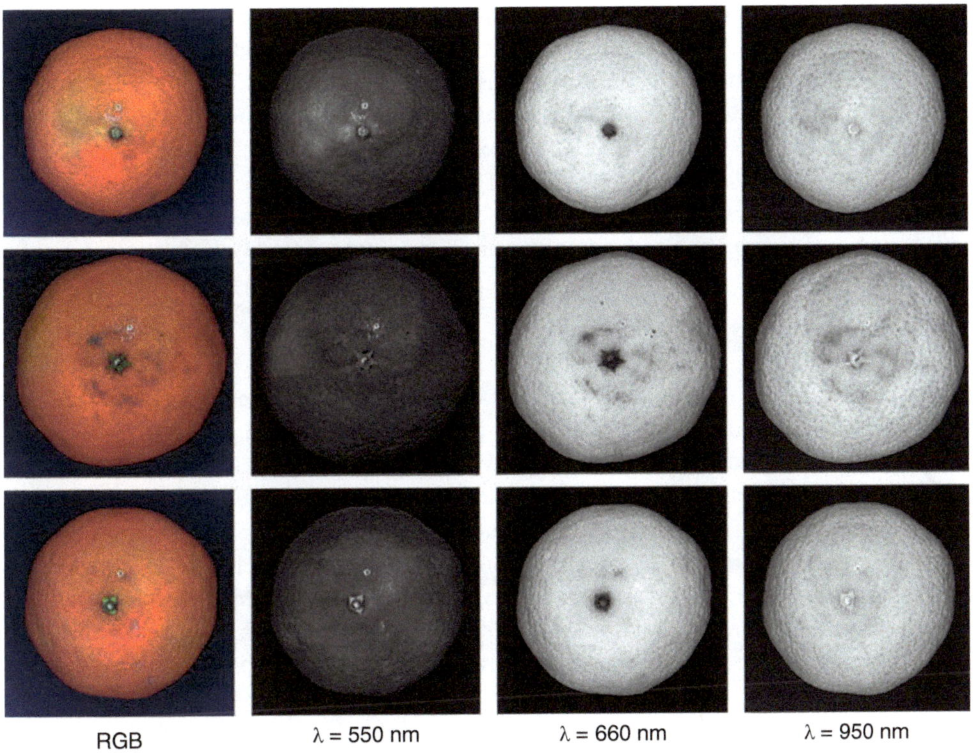

Figure 16.5 Hyperspectral imaging for food quality analysis and control. *Source:* Moltó et al. [37] / with permission of Elsevier.

technologies allow high-throughput, nondestructive, and objective inspection practices that can detect food product contamination and adulteration or spoilage with high accuracy. HSI captures both spectral and spatial data about food samples, enabling perceptions to grow in their physical appearance and how they may be recognized chemically, and precisely distinguishing between pure or fake, or contaminated food samples [10, 37].

HSI has shown early success in the identification of nondairy adulterants in milk, the identification of counterfeit honey, and even the detection of high-value ingredients that may be diluted with a lower-cost food item. In the quality control of dairy products, for example, HSI can accurately distinguish between pure and adulterated milk through examining spectral signatures that change due to protein and fat content [38]. Correspondingly, in the grain and spice industries, imaging systems can assist in confirming or identifying the use of adulterants such as synthetic dyes or unknown starches to replace or dilute a food product that may be a public health hazard [39].

Additionally, these technology applications are also essential for food safety monitoring in the industrial food processing setting, especially on automated production and packaging lines, where computer vision systems can identify and detect foreign materials, including small fragments of plastic, glass, or metal present within food products in real time. While observing global food safety protocols, this makes the consumer environment safer [40]. Also, by using spatial and spectral features to sort contaminants and variances without the need for human intrusion, DL algorithms can enhance detection abilities in imaging [30, 41]. All things well-thought-out, the use of computer vision and HSI has reformed food inspection techniques to diminish reliance on human and labor error and improve the capacity for ongoing food quality reviews. These new technologies have also facilitated traceability, transparency, and safety throughout the food supply chain, and are congruent with a sustainable approach to food production for consumers today.

16.3.1 Nutrient Composition and Quality Analysis

In reference to food and agricultural products, nutrient composition means the types and amounts of crucial nutrients they contain for growth, human health, and energy. Nutrient composition includes carbohydrates, fats, proteins, vitamins, minerals, and water. Carbohydrates offer a primary source of energy and exist in wheat, rice, and fruits. Proteins are used for growth and repair of the body, and can also be found in various pulses, milk, meat, and eggs. Fats are also concentrated sources of energy and facilitate fat-soluble vitamin absorption and are found in butter, nuts, oils, and seeds. Vitamins and minerals are required in small quantities and are used for several functions in the body, including bone health, metabolism, and blood formation. Furthermore, water is also an important constituent [42]. It provisions ingestion and controls temperature, as well as conveying nutrients. Nutrients work together to offer the nutritional value of the food and the overall quality of food and agricultural products.

HSI and MSI technologies have proven effective alternatives to traditional analytical methods used for specifically monitoring and determining internal nutrient composition in food products, including fruits, grains, and processed foods. Traditional chemical assays are frequently destructive, labor-intensive, and time-consuming. HSI and MSI are quicker, more accurate, and provide valuable spatially resolved (spatially valuable) information that can be used to monitor the nutritional composition of food products, in-line during industrial production lines [43, 44].

In crops, HSI and MSI have been touted for applications to predict protein, fat, oil, and carbohydrates (Figure 16.6). For example, HSI of soybean seeds has been fit to predict the concentrations of protein and oil with reasonable accuracy. Predictive models using spectral analysis of not only soybean seeds but also grains, such as wheat and rice, would rely on the spectral reflectance patterns to assess moisture content, starch, and other metrics valuable for biochemical status of grains, useful for support decisions on harvest and storage [45]. In the quality evaluation of fruits and vegetables, imaging systems make it possible to estimate moisture content, sugar content, and soluble solids content, which are important indicators of ripeness, taste, and shelf-life. Imaging techniques can identify internal quality attributes through subtle changes in spectral signatures due to chemical differences, which help prevent postharvesting losses and maintain the quality of the product within and between shipments. For processed foods, MSI and HSI also empower the industry to begin nutrient retention and uniformity during processing, along with fulfilling nutritional labeling tasks and regulations [46].

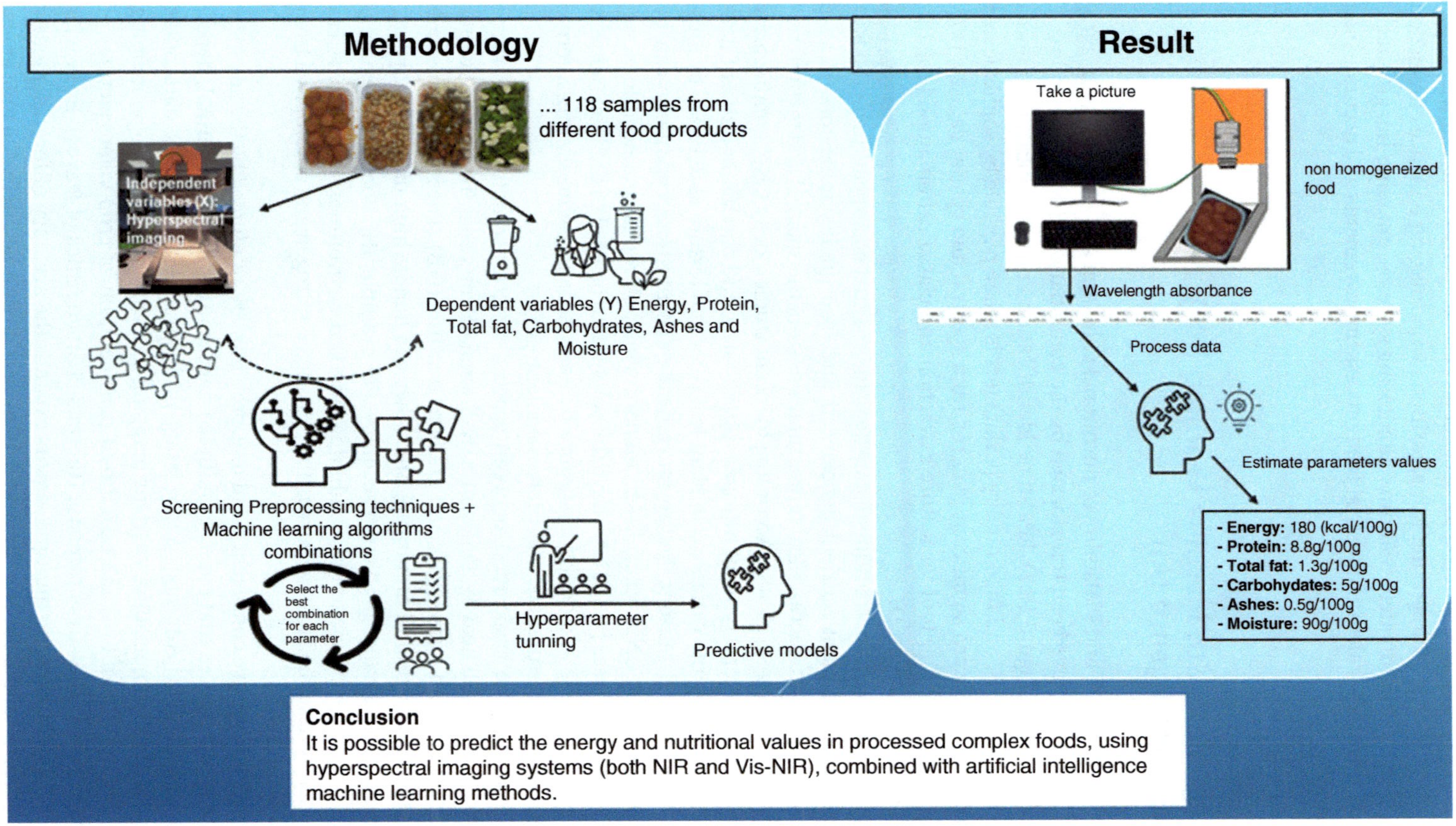

Figure 16.6 Predicting energy and nutritional values using hyperspectral imaging systems, AI, and machine learning models. *Source:* Marín-Méndez et al. [43]/with permission of Elsevier.

As a summary, MSI and HSI offer rapid, accessible, and inexpensive alternatives to traditional laboratory analysis. Integrating ML algorithms with imaging enhances prediction accuracy and generates automated, real-time monitoring systems, enabling imaging technologies to play important roles in industrial food process improvements and accurate food quality assessments.

16.4 Integration with Machine Learning (ML) and Artificial Intelligence (AI)

The accuracy, automation, and scalability of nutritional valuation and monitoring in agronomic and food systems have been transformed by the combination of ML, AI, and image processing technologies (Figure 16.7). Real-time identification and quantification of nutrients, plant stress indicators, and food quality parameters is made possible by sophisticated AI algorithms that can discern significant patterns from spectral and spatial datasets [47]. CNNs have become a popular tool for feature extraction and classification among the different AI models, especially in DL.

CNNs are more effective than traditional statistical and manual classification methods at classifying and detecting plant nutrient deficiencies and food quality indicators because they can automatically learn hierarchical representations from the spectral and textural features of crop and food images [22, 48]. According to recent studies, deep CNN architectures can accurately identify several types of nitrogen, phosphorus, and potassium nutrient deficiencies by identifying minute variations in plant leaf color and texture as detected by multispectral or hyperspectral sensors. For instance, leaf nutrient stress conditions can be accurately classified using CNN image-based models, which eliminate the need for time-consuming chemical analyses [49]. CNNs have the potential to be used in the food industry for automated food grading and quality assurance by nondestructively measuring internal quality attributes like moisture, fat, and protein content [50].

Additionally, in DL, the association between spectral reflectance data and measured nutrient concentrations still heavily depends on traditional ML methods like support vector machines (SVMs), random forest (RF), and gradient boosting (GB). These techniques are useful for large-scale agricultural monitoring since they can handle high-dimensional data, reduce overfitting, and enhance prediction accuracy. For example, RF models are regularly used for multi-parameter prediction by combining crop and environmental data, whereas SVMs are frequently described to have successfully predicted nitrogen and chlorophyll from hyperspectral imagery [51].

Moreover, model detective consistency and interpretation are reinforced by the advancement of data fusion practices that control multiple image sources of RGB, thermal, hyperspectral, and LiDAR data. Although assimilating multispectral and LiDAR data offers opportunities for three-dimensional nutrient mapping and canopy morphology, merging spectral and thermal data allows the concurrent consideration of crop vigor (pictures of leaves with reflectance) and stress response (thermal characteristics of leaves). Even with being relatively new, integrating multiple sensors expands the accuracy of forecast models while offering circumstantial data on nutrient dynamics and plant health in dynamic environments [52].

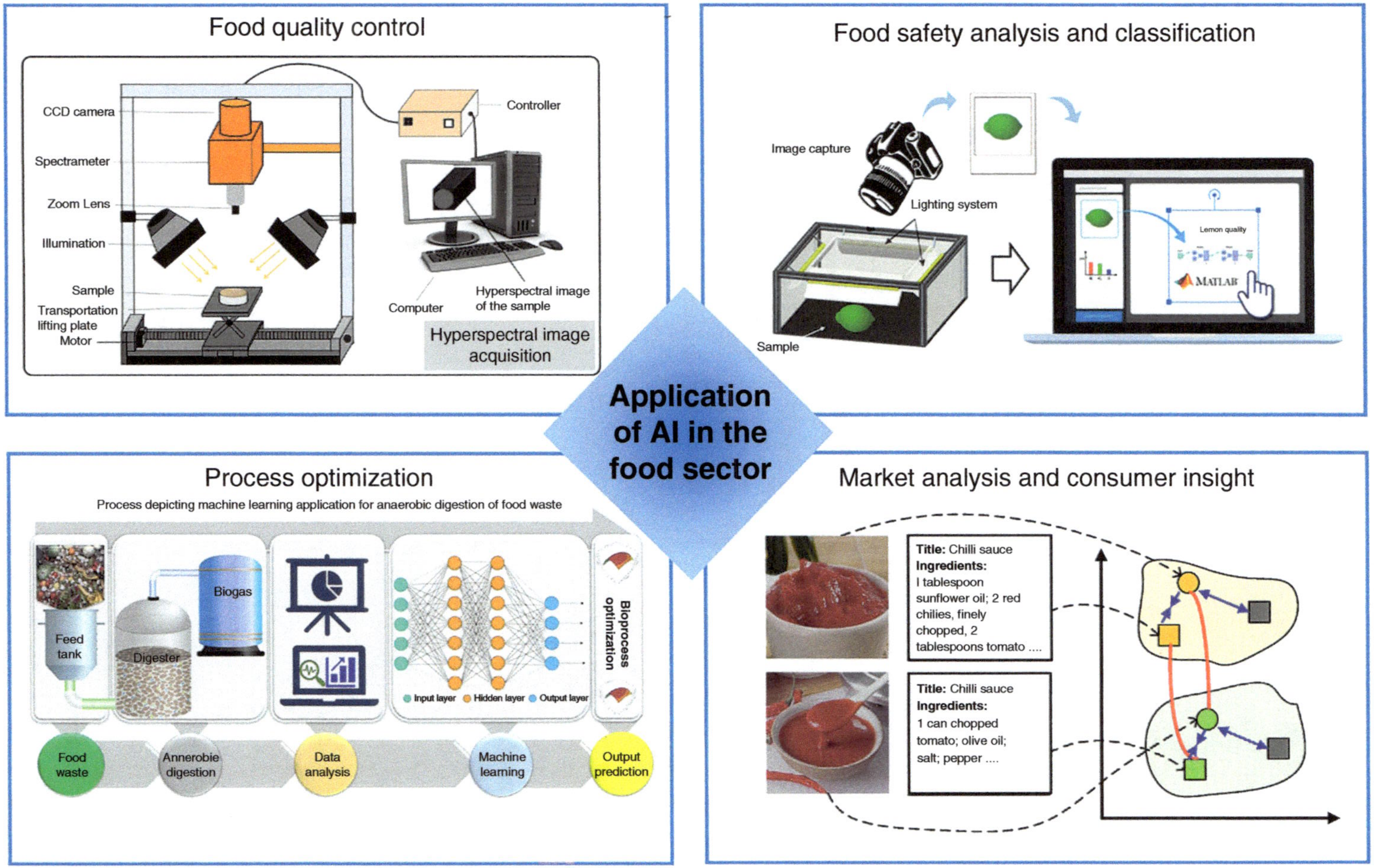

Figure 16.7 Applications of AI in the food industry. *Source:* Marín-Méndez et al. [43]/with permission of Elsevier.

Together, these AI-powered image processing tools enhance food quality control through the production chain, allow predictive yield modelling, and offer automated, real-time nutrient diagnostics. ML and AI-based image contexts will be crucial to digital food monitoring systems and precision agriculture by means of computing power and sensor developments, enabling more efficient and sustainable resource management [53].

16.5 Challenges and Limitations

While the use of image processing techniques for assessing the nutritional value of food has innovative suggestions, there are still a number of issues with their scalability or availability in diverse geographical areas. One major problem is that sensors need to be calibrated, and there is presently a lack of standardized, usable datasets in the related database, which has a negative effect on the accuracy and repeatability of results. Obviously, calibration is essential for imaging systems, together with hyperspectral and multispectral cameras, because of the variations in brightness, observing geometry, and sensor sensitivity. If calibration is not reliable across devices and/or the environment, spectral measurement will be unreliable, which will result in untrustworthy nutrient assessment [54].

Another important complication to interoperability amid imaging-based nutrition valuation systems may be data preprocessing and standardization. The dataset regularly has high heterogeneity, making cross-comparison and model generalizability across the countless datasets challenging [55]. This is because diverse studies use dissimilar imaging platforms, analytical algorithms, and spectral ranges. Also, the high dimensionality of hyperspectral images requires multifaceted processing and sophisticated computation. The execution of AI-powered analysis pipelines may be hindered by this complication, predominantly in agricultural settings with limited resources and a necessity for suitable computational infrastructure.

Another main problem is the cost of hyperspectral sensors and related hardware, which controls access to only Agri-tech and large research institutes. Accurate nutrition monitoring has a technological gap; meanwhile, smallholder farmers and other stakeholders in unindustrialized nations are not likely to be able to afford such apparatus. Apart from financial thoughts, environmental inconsistencies similar to atmospheric disturbances, shifting sunlight, and canopy geometry can affect image quality or spectral uniformity. These aspects also possibly introduce noise into the models and lower their fairness when used on real-world fields [56].

Besides, there is presently no agreement on uniform standards for the assembly, preprocessing, and interpretation of images in diverse studies and application domains. The absence of broadly recognized protocols restricts reproducibility and delays the development of standard datasets, which are crucial for model justification. In order to overcome the above-mentioned challenges, researchers, policymakers, and industry must work together to develop reasonably priced sensor technologies, open-access datasets, and scalable analysis frameworks. By taking an integrated tactic, entry barriers for imaging-based nutrition estimation tools would be minimized, and global sustainability and impartiality in digital agriculture would be heightened.

16.6 Future Perspectives and Emerging Trends

The integration of cloud-connected, portable, and AI-driven technologies is anticipated to advance image-based nutrition visualization in the future. Consequently, food and agricultural systems will turn out to be more available and effective. For on-site nutrient monitoring and food quality valuation, more farmers and food producers can use smartphone imaging platforms. These platforms make precision agriculture tools open to all, doing away with the requirement for complex equipment. These mobile applications can quickly analyze photos to detect nutrient deficiencies, estimate protein or moisture content, and deliver real-time, actionable advice when integrated with ML algorithms [57].

These systems are boosted by the usage of edge computing and Internet of Things (IoT) devices, which allow real-time data analysis directly on fixed sensors or transferable devices. This lessens delays and reduces the necessity for cloud computing. These systems enable ongoing crop and food product monitoring. They make it conceivable to identify issues quickly and produce results efficiently. Still, analysis capabilities are lengthened through new imaging methods such as 3D imaging and hyperspectral data fusion. These technologies make it possible to weigh the physical and compositional potentials of plants and foods at the same time.

For example, combining spectral data from hyperspectral sensors with spatial data from 3D models permits precise evaluation of fruit ripeness, grain quality, and leaf nutrient distribution [58]. As well as analytical frameworks for the management of nutrients in fluctuating weather conditions are being formed by combining remote sensing and crop modelling. These combined technologies use real-time imaging data to model nutrient uptake, crop growth, and yield predictions. This inspires flexible management techniques and sustainable farming approaches [43].

Such machineries could minimize postharvest losses, maximize the use of water and fertilizer resources, and increase food security by offering detailed and timely insights. These returns will contribute to the growth of stronger and sustainable food systems. Precision nutrition monitoring will perhaps become more viable and widely used in a diversity of agricultural and food production settings as sensor expenses weaken and AI models expand.

16.7 Conclusion

The technique we imagine and measure nutrition in agriculture and food systems has been pointedly changed by image processing technologies. These techniques disregard the necessity for long-time and labor-intensive traditional laboratory trials by empowering rapid, nondestructive, and accurate estimation of plant nutrient levels and food quality. Also, Image-based systems can more precisely and reliably predict yields, detect nutrient shortages, and evaluate food quality when paired with ML and AI. Precision agriculture is reinforced by the practice of these technologies. They encourage sustainability ideas, minimize negative impacts on the environment, and maximize resource use efficiency.

The applications of these tools are extended by recent developments like smartphone imaging, IoT integration, edge computing, and multi-sensor data synthesis, which make them more beneficial and accessible for a series of users, including smallholder growers. Even with these advances, there are still a number of hitches. Extensive use is hindered by complications like sensor calibration, high computational requirements, data standardization, costly equipment, and environmental variations. Legislators, researchers, and business administrators must work together to develop open-access datasets, sensibly priced sensors, and uniform analysis techniques to address these challenges.

In brief, AI-driven, accessible, and user-friendly systems are vital for the future of nutrition monitoring through image processing. These systems have to offer rapid, useful statistics for safeguarding food quality and handling agriculture sustainably. For these potent tools to excellently support resource efficiency, environmental sustainability, and global food security, continuing research, technological expansions, and wide-ranging distribution approaches are essential.

References

1 Sishodia, R.P., Ray, R.L., and Singh, S.K. (2020). Applications of remote sensing in precision agriculture: a review. *Remote Sensing* 12 (19): 3136. https://doi.org/10.3390/rs12193136.

2 García-Vera, Y.E., Polochè-Arango, A., Mendivelso-Fajardo, C.A., and Gutiérrez-Bernal, F.J. (2024). Hyperspectral image analysis and machine learning techniques for crop disease detection and identification: a review. *Sustainability* 16 (14): 6064. https://doi.org/10.3390/su16146064.

3 Saha, D. and Manickavasagan, A. (2021). Machine learning techniques for analysis of hyperspectral images to determine quality of food products: a review. *Current Research in Food Science* 4: 28–44. https://doi.org/10.1016/j.crfs.2021.01.002.

4 Ibraheem, I. (2015). Early detection of melanoma using multispectral imaging and artificial intelligence techniques. *American Journal of Biomedical and Life Sciences* 3 (2-3): 29–33. 3 (Special Issue: Spectral Imaging for Medical Diagnosis "Modern Tool for Molecular Imaging"), 29–33. https://doi.org/10.11648/j.ajbls.s.2015030203.16.

5 Ram, B.G., Oduor, P., Igathinathane, C. et al. (2024). A systematic review of hyperspectral imaging in precision agriculture: analysis of its current state and prospects. *Computers and Electronics in Agriculture* 222: 109037. https://doi.org/10.1016/j.compag.2024.109037.

6 Kacho, N.F., Hussain, M.Y., and Banday, S. (2024). Artificial intelligence in agriculture. In: Futuristic Trends in Agriculture Engineering & Food Sciences V3B1, 652–658. IIP Series https://doi.org/10.58532/v3bcagp1ch51.

7 Dubey, S. (2025). AI-based weather prediction. *Journal of Applied Science and Education (JASE)* 5 (2): 1–10. https://doi.org/10.54060/a2zjournals.jase.104.

8 Beemkumar, N., Gupta, S., Bhardwaj, S. et al. (2023). Activity recognition and IoT-based analysis using time series and CNN. In: Handbook of Research on Machine Learning-enabled IoT for Smart Applications Across Industries (ed. N. Goel and R.K. Yadav), 350–364. IGI Global https://doi.org/10.4018/978-1-6684-8785-3.ch018.

9 Srivastav, A.K. and Das, P. (2025). Data analytics and machine learning for crop management. In: Biotechnology and IoT in Agriculture and Food Production. Berkeley, CA: Apress https://doi.org/10.1007/979-8-8688-1469-3_23.

10 Ahmed, M.T., Monjur, O., Khaliduzzaman, A. et al. (2025). A comprehensive review of deep learning-based hyperspectral image reconstruction for agri-food quality appraisal. *Artificial Intelligence Review* 58: 96. https://doi.org/10.1007/s10462-024-11090-w.

11 Talukdar, V., Veeraiah, V., Roy, D.R. et al. (2023). An innovative approach: typicality-based collaborative filtering for recommender systems. *Proceedings of the 3rd International Conference on Advancement in Electronics & Communication Engineering (AECE)*, pp. 863–868. https://doi.org/10.1109/AECE59614.2023.10428576

12 Zeng, M., Zhu, X., Wan, L. et al. (2025). Data-driven prediction of grape leaf chlorophyll content using hyperspectral imaging and convolutional neural networks. *Applied Sciences* 15 (10): 5696. https://doi.org/10.3390/app15105696.

13 Lazarević, B., Gunjaca, J., Safner, T. et al. (2024). Multispectral and chlorophyll fluorescence imaging for the detection of nutrient deficiency symptoms in common bean. *Journal of Central European Agriculture* 25 (2): 416–432. https://doi.org/10.5513/JCEA01/25.2.4305.

14 Wei, Z. and Fang, W. (2024). UV-NDVI for real-time crop health monitoring in vertical farms. *Smart Agricultural Technology* 8: 100462. https://doi.org/10.1016/j.atech.2024.100462.

15 Grbović, Ž., Ivošević, B., Budjen, M. et al. (2025). Integrating UAV multispectral imaging and proximal sensing for high-precision cereal crop monitoring. *PLoS One* 20 (5): e0322712. https://doi.org/10.1371/journal.pone.0322712.

16 Wang, D., Struik, P.C., and Liang, L. (2024). Developing remote sensing- and crop model-based methods to optimize nitrogen management in rice fields. *Computers and Electronics in Agriculture* 220: 108899. https://doi.org/10.1016/j.compag.2024.108899.

17 Bourriz, M., Hajji, H., Laamrani, A. et al. (2025). Integration of hyperspectral imaging and AI techniques for crop type mapping: present status, trends, and challenges. *Remote Sensing* 17 (9): 1574. https://doi.org/10.3390/rs17091574.

18 Kharel, T., Tyler, H., Mubvumba, P. et al. (2025). Machine learning on multi-spectral imagery to estimate nutrient yield of mixed-species cover crops. *Agricultural & Environmental Letters* 10: 10. https://doi.org/10.1002/ael2.70009.

19 Berger, K., Machwitz, M., Kycko, M. et al. (2022). Multi-sensor spectral synergies for crop stress detection and monitoring in the optical domain: a review. *Remote Sensing of Environment* 280: 113198. https://doi.org/10.1016/j.rse.2022.113198.

20 Gitelson, A.A., Kaufman, Y.J., Stark, R., and Rundquist, D. (2002). Novel algorithms for remote estimation of vegetation fraction. *Remote Sensing of Environment* 80 (1): 76–87. https://doi.org/10.1016/S0034-4257(01)00289-9.

21 Harikumar, A., Shenhar, I., Pebes-Trujillo, M.R. et al. (2025). Harnessing smartphone RGB imagery and LiDAR point cloud for enhanced leaf nitrogen and shoot biomass assessment: Chinese spinach as a case study. *Frontiers in Plant Science* 16: 1592329. https://doi.org/10.3389/fpls.2025.1592329.

22 Yu, K.-Q., Zhao, Y.-R., Li, X.-L. et al. (2014). Hyperspectral imaging for mapping of total nitrogen spatial distribution in pepper plant. *PLoS One* 9 (12): e116205. https://doi.org/10.1371/journal.pone.0116205.

23 Dung, C.D., Trueman, S.J., Wallace, H.M. et al. (2023). Hyperspectral imaging for estimating leaf, flower, and fruit macronutrient concentrations and predicting strawberry yields. *Environmental Science and Pollution Research International* 30 (53): 114166–114182. https://doi.org/10.1007/s11356-023-30344-8.

24 Yu, H., Kong, B., Hou, Y. et al. (2022). A critical review of applications of hyperspectral remote sensing in crop monitoring. *Experimental Agriculture* 58: e26. https://doi.org/10.1017/S0014479722000278.

25 Misbah, K., Laamrani, A., Khechba, K. et al. (2022). Multi-sensors remote sensing applications for assessing, monitoring, and mapping NPK content in soil and crops in African agricultural land. *Remote Sensing* 14 (1): 81. https://doi.org/10.3390/rs14010081.

26 Galieni, A., D'Ascenzo, N., Stagnari, F. et al. (2021). Past and future of plant stress detection: an overview from remote sensing to positron emission tomography. *Frontiers in Plant Science* 11: 609155. https://doi.org/10.3389/fpls.2020.609155.

27 Mertens, S., Verbraeken, L., Sprenger, H. et al. (2023). Monitoring of drought stress and transpiration rate using proximal thermal and hyperspectral imaging in an indoor automated plant phenotyping platform. *Plant Methods* 19 (1): 132. https://doi.org/10.1186/s13007-023-01102-1.

28 Almeida, T.A.B., Montenegro, A.A.A., da Silva, R.A.B. et al. (2024). Evaluating daily water stress index (DWSI) using thermal imaging of neem tree canopies under bare soil and mulching conditions. *Remote Sensing* 16 (15): 2782. https://doi.org/10.3390/rs16152782.

29 Allu, A.R. and Mesapam, S. (2024). Fusion of different multispectral band combinations of Sentinel-2A with UAV imagery for crop classification. *Journal of Applied Remote Sensing* 18 (1): 016511–016511. https://doi.org/10.1117/1.jrs.18.016511.

30 Chen, B., Su, Q., Li, Y. et al. (2025). Field rice growth monitoring and fertilization management based on UAV spectral and deep image feature fusion. *Agronomy* 15 (4): 886. https://doi.org/10.3390/agronomy15040886.

31 Pandey, J.K., Yadav, A.K., Purushotham, M. et al. (2022). Colorimetric detection of deltamethrin pesticide from biological samples based on the peroxidase-mimic catalytic activity on MnO_2/SnO_2. *Key Engineering Materials* 928: 45–51. https://doi.org/10.4028/p-4ti1i0.

32 Yahya, M.A., Yahya, M.A., Udhayanithi, D., and Venkataraman, A. (2023). Automatic fruit quality inspection system using image processing. *International Journal of Scientific Research in Science, Engineering and Technology* 10 (4): 01–08. https://doi.org/10.32628/ijsrset23103177.

33 Qin, J., Chao, K., Kim, M. et al. (2013). Hyperspectral and multispectral imaging for evaluating food safety and quality. *Journal of Food Engineering* 118 (2): 157–171. https://doi.org/10.1016/j.jfoodeng.2013.04.001.

34 Barbin, D.F., ElMasry, G., Sun, D.-W., and Allen, P. (2013). Non-destructive determination of chemical composition in intact and minced pork using near-infrared hyperspectral imaging. *Food Chemistry* 138 (2–3): 1162–1171. https://doi.org/10.1016/j.foodchem.2012.11.120.

35 Cubero, S., Aleixos, N., Moltó, E. et al. (2011). Advances in machine vision applications for automatic inspection and quality evaluation of fruits and vegetables. *Food and Bioprocess Technology* 4 (4): 487–504. https://doi.org/10.1007/s11947-010-0411-8.

36 Pandey, J.K., Kumar, S., Lamin, M. et al. (2022). A metaheuristic autoencoder deep learning model for an intrusion detector system. *Mathematical Problems in Engineering* 2022: 11, 3859155. https://doi.org/10.1155/2022/3859155.

37 Moltó, E., Blasco, J., and Gómez-Sanchís, J. (2010). Analysis of hyperspectral images of citrus fruits. In: Hyperspectral Imaging for Food Quality Analysis and Control (ed. D.-W. Sun), 321–348. Academic Press https://doi.org/10.1016/B978-0-12-374753-2.10010-3.

38 Kimbahune, S., Syed, M.G., Mithun, B.S. et al. (2016). Hyperspectral sensing-based analysis for determining milk adulteration. *Proceedings of SPIE 9860, Sensing for Agriculture and Food Quality and Safety VIII (Paper 98600E)*. https://doi.org/10.1117/12.2223439.

39 Patel, D., Bhise, S., Kapdi, S.S., and Bhatt, T. (2024). Non-destructive hyperspectral imaging technology to assess the quality and safety of food: a review. *Food Production, Processing and Nutrition* 6 (1): 69. https://doi.org/10.1186/s43014-024-00169-7.

40 Lorente, D., Aleixos, N., Gómez-Sanchis, J. et al. (2012). Recent advances and applications of hyperspectral imaging for fruit and vegetable quality assessment. *Food and Bioprocess Technology* 5 (4): 1121–1142. https://doi.org/10.1007/s11947-011-0725-1.

41 Ahmed, M., Monjur, O., Alin, K., and Kamruzzaman, M. (2024). A comprehensive review of deep learning-based hyperspectral image reconstruction for agri-food quality appraisal [preprint]. Research Square. https://doi.org/10.21203/rs.3.rs-5240049/v1.

42 Thompson, F.E. and Subar, A.F. (2013). Dietary assessment methodology. In: Nutrition in the prevention and treatment of disease, 3e (ed. A.M. Coulston, C.J. Boushey, and M.G. Ferruzzi), 5–46. Academic Press https://doi.org/10.1016/B978-0-12-391884-0.00001-9.

43 Marín-Méndez, J.J., Luri Esplandiú, P., Alonso-Santamaría, M. et al. (2024). Hyperspectral imaging as a non-destructive technique for estimating the nutritional value of food. *Current Research in Food Science* 9: 100799. https://doi.org/10.1016/j.crfs.2024.100799.

44 Reis, M. and Al-Sarayreh, M. (2024). Hyperspectral imaging combined with chemometrics. In: Advances in Food and Agricultural Analysis (ed. L.M.L. Nollet and F. Toldrá), 61–88. Elsevier https://doi.org/10.1016/B978-0-443-21493-6.00004-6.

45 Giri, A., Sagan, V., Alifu, H. et al. (2024). A wavelet decomposition method for estimating soybean seed composition with hyperspectral data. *Remote Sensing* 16 (23): 4594. https://doi.org/10.3390/rs16234594.

46 Umuhoza, A., Semyalo, D., Pahlawan, M.F.R. et al. (2025). Integration of hyperspectral imaging and chemometrics for internal quality evaluation of packaged and non-packaged fresh fruits. *Agriculture* 15 (16): 1718. https://doi.org/10.3390/agriculture15161718.

47 Srinivas, P., Veeraiah, V., Thejaswini, K.O. et al. (2025). The impact of electromagnetic and optical radiation on physical and biological systems. *Journal of Nano and Electronic Physics* 17 (4): 04012. https://doi.org/10.21272/jnep.17(4).04012.

48 Kamilaris, A. and Prenafeta-Boldú, F.X. (2018). Deep learning in agriculture: a survey. *Computers and Electronics in Agriculture* 147: 70–90. https://doi.org/10.1016/j.compag.2018.02.016.

49 Xu, Z., Guo, X., Zhu, A. et al. (2020). Using deep convolutional neural networks for image-based diagnosis of nutrient deficiencies in rice. *Computational Intelligence and Neuroscience* 2020: 1–12. https://doi.org/10.1155/2020/7307252.

50 Gul, N., Muzaffar, K., Shah, S.Z. et al. (2024). Deep learning hyperspectral imaging: a rapid and reliable alternative to conventional techniques in the testing of food quality and safety. *Quality Assurance & Safety of Crops and Food* 16 (1): 78–97. https://doi.org/10.15586/qas.v16i1.1392.

51 Okyere, F.G., Cudjoe, D., Sadeghi-Tehran, P. et al. (2023). Modeling the spatial-spectral characteristics of plants for nutrient status identification using hyperspectral data and deep learning methods. *Frontiers in Plant Science* 14: 1209500. https://doi.org/10.3389/fpls.2023.1209500.

52 Pandey, J.K. (2025). Understanding the impact of electromagnetic and optical radiation on physical and biological systems. *Journal of Nano- and Electronic Physics* 17 (5): 05023. https://doi.org/10.21272/jnep.17(5).05023.

53 Dhal, S. and Kar, D. (2025). Leveraging artificial intelligence and advanced food processing techniques for enhanced food safety, quality, and security: a comprehensive review. *Discover Applied Sciences* 7: 75. https://doi.org/10.1007/s42452-025-06472-w.

54 Balakrishnan, P., Leema, A.A., Jothiaruna, N. et al. (2025). Artificial intelligence for food safety: from predictive models to real-world safeguards. *Trends in Food Science & Technology* 163: 105153. https://doi.org/10.1016/j.tifs.2025.105153.

55 Basu, A., Warzel, D., Eftekhari, A. et al. (2019). Call for data standardization: lessons learned and recommendations in an imaging study. *JCO Clinical Cancer Informatics* 3 (3): 1–11. https://doi.org/10.1200/CCI.19.00056.

56 Njoroge, S., Mugi-Ngenga, E., Limo, B., and Fakoya, O.E. (2025). Precision agriculture in Africa: challenges and opportunities. *Growing Africa* 4 (1): 2–5. https://doi.org/10.55693/ga41.mbuf4046.

57 Abiri, R., Rizan, N., Balasundram, S.K. et al. (2023). Application of digital technologies for ensuring agricultural productivity. *Heliyon* 9 (12): e22601. https://doi.org/10.1016/j.heliyon.2023.e22601.

58 Morchid, A., El Alami, R., Raezah, A.A., and Sabbar, Y. (2024). Applications of Internet of Things (IoT) and sensor technology to increase food security and agricultural sustainability: benefits and challenges. *Ain Shams Engineering Journal* 15 (3): 102509. https://doi.org/10.1016/j.asej.2023.102509.

17

Combining AI with Automated Quality Control in the Food Sector

B. Leena[1], S. Karthikeyan[1], and Razaidi Hussin[2]

[1] *Department of ECE, KGiSL Institute of Technology, Coimbatore, Tamil Nadu, India*
[2] *Faculty of Electronic Engineering & Technology (FKTEN), Universiti Malaysia Perlis (UniMAP), Arau, Malaysia*

17.1 Introduction

Over the past 10 years, a number of food events have increased the significance of food quality management in the agri-food sector. The demand for food has shown an exceptional development trajectory due to rising living standards and an expanding worldwide population. Studies have predicted a 60–97% increase in the world's food needs by 2050. Food processing and supply chain management face previously unheard-of difficulties as a result of the rising demand for food, which has also increased expectations on agricultural output [1]. Therefore, to satisfy the growing demand for food while maintaining food safety and quality, the food sector must look for creative solutions.

In addition to altering other industries, the rise of artificial intelligence (AI) technology in recent years has had a swift impact on the food sector, bringing about dramatic developments, outlined by Addanki, Mounika, et al. [2]. As AI technology continues to advance in areas such as image recognition, machine learning (ML), and Natural Language Processing (NLP), it offers robust technical assistance for the intelligent, automated production methods used in the food business. Businesses can apply intelligent management in supply chain optimization, equipment predictive maintenance, and food quality control, among other areas, by integrating AI technology into the operation of equipment used in the food processing process. This lowers production costs and significantly increases production efficiency [3].

The food sector bears a heavy burden of meeting consumer demands and desires while ensuring food safety, security, and nutritional qualities. Given the current circumstances, automated systems are essential to the food business in order to produce food with the highest level of precision and efficiency while maintaining the necessary product quality control. Generally speaking, it refers to science and technology. In the contemporary food sector, automation methods have become more prevalent in recent years. From the selection of raw materials to servicing, several food sectors have already integrated highly technological automated systems into their production lines. The industry's automated system's main goal is to complete the procedure as accurately and efficiently as possible. For the food industry, this means that there are more benefits than using manual methods.

One of the most important objectives when thinking about the food industry is quality assurance, along with other areas. In order to succeed in the very competitive food market, automated systems can provide high-quality products. In comparison to those that produce food using manual methods, the superior product increases market share and fosters more consumer loyalty, resulting in greater profitability for the industry. For product quality assurance in the manual processing business, the human visual inspection approach is employed. For food businesses to use automated approaches for product quality assessment, automated technologies are necessary because human visual processes are laborious, time-consuming, and prone to error. As a result, increased product quality receives the greatest automation emphasis.

AI-based or autonomous models are employed in practically all technological domains. It allows people all around the world to effectively solve problems, computerize the food business, and transform its goods. The two main categories of AI's major contributions to the food industry are food security and food quality management. In Figure 17.1, the region under each class is displayed.

Quality control and safe food rely on people, and technology is depicted in Figure 17.2. Quality varies across manufacturers and between lots of produce, and there are several unique features of food items and production methods that affect both product quality and quality assurance in manufacturing processes.

The following resources can be effectively used to optimize the quality of food processing:

- Machine/computer vision
- Online sensors
- Expert systems
- Robotics technology
- Integrated manufacturing using computers
- Adaptable production methods

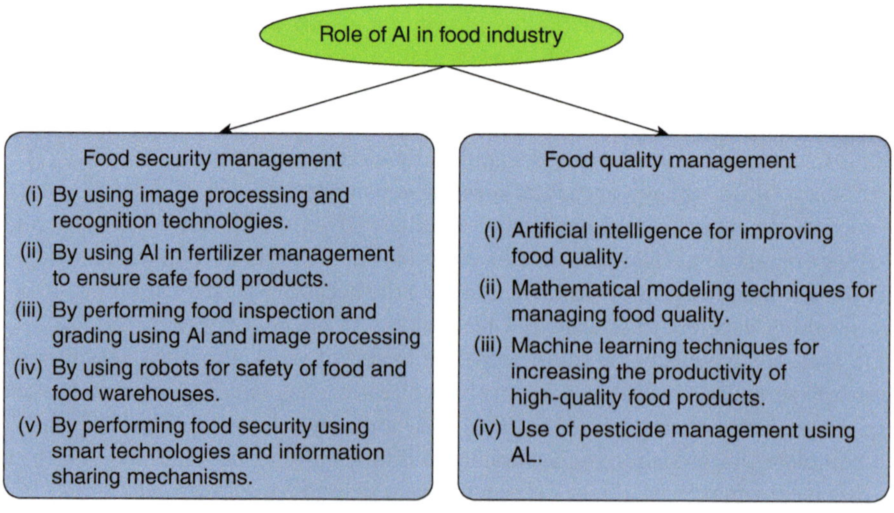

Figure 17.1 AI role in the food sector.

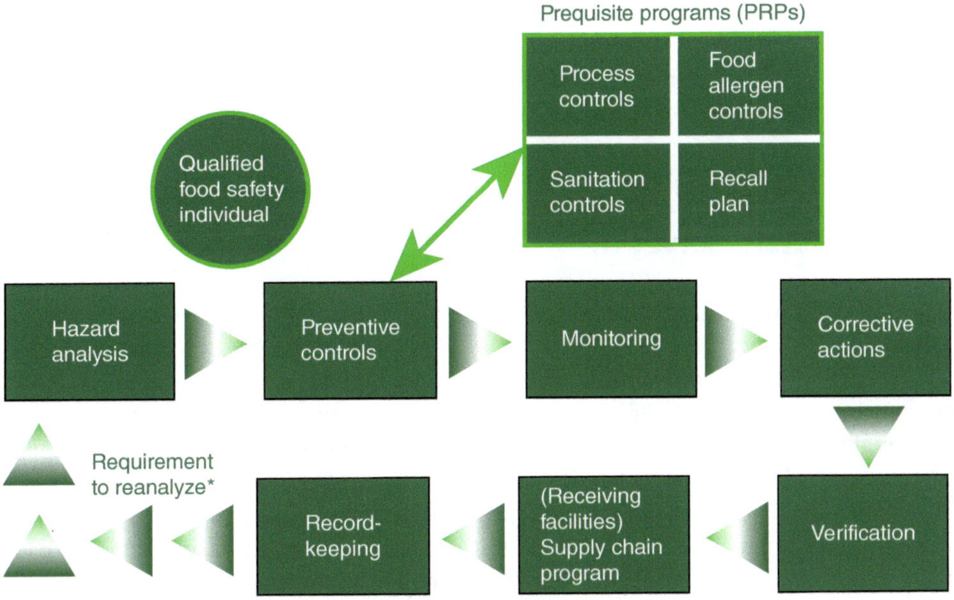

Figure 17.2 Quality control and safe food rely on people and technology.

In the food sector, convolutional neural networks (CNNs) are very valuable and promising for detecting food and quality safety because of their exceptional ability in processing image data and extracting image attributes. CNNs can automatically detect substandard items on food packing lines or quickly identify sections that have rotted by examining the texture of the fruit's surface. Additionally, CNNs and super-resolution imaging technologies can accurately evaluate minute pollutants, greatly increasing the accuracy of food quality prediction [4].

Recurrent neural networks (RNNs) are especially crucial for risk forecasting in the production and distribution of food because of their exceptional performance in processing time series data. Predictive time series modeling is a better fit for RNNs because of its ability to capture temporal correlations in dynamic processes [5]. In food manufacturing settings, for instance, RNNs can forecast possible contamination concerns by examining temporal variations in processing parameters like temperature and humidity. By employing RNNs to monitor transit conditions in real time, cold chain logistics can prevent food from rotting. An RNN version known as Long Short-Term Memory (LSTM) networks efficiently capture long-term dependencies, which makes them perfect for complicated predictive tasks for food safety data [6].

This study looks at how AI technology might be used in the food industry and quality assurance automation, with a particular focus on describing AI in the automation of different food processes, given the enormous potential that AI has shown in this area [7]. Second, it talks about the ethical and technical difficulties that AI is facing when it comes to automating the food business and suggests options for related studies. This evaluation aims to offer useful sources and perspectives on automation and innovation in the food industry.

This format will be used for the chapter. Section 17.1 provides a general overview of the research. Section 17.2 compiles the latest study findings. The general structure and internal operations of the proposed AI-based CNN and RNN system for automatic quality control in the food sector are briefly explained in Section 17.3. Section 17.4 provides a thorough explanation of the experimental methods and results. Section 17.5 contains the work's last section, which is the conclusion [8, 9].

17.2 Literature Survey

This study offers a literature review of AI in the food sector while preserving all facets of the field.

A new computerized method based on fractal theory and AI techniques was presented in this study by Castillo and Meliif [4] to operate an autonomous assessment of the manufacturing quality. This research combines the use of AI in the food sector with fractal dimension as a metric for classifying chemicals and microorganisms [10].

Automated quality control (AQC) has a good economic and environmental impact when it comes to waste reduction. Xiong, Felix, et al. suggested a comprehensive framework for quality control that assesses various aspects of visual, informational data in order to facilitate future use cases in science and industry [11]. The framework comprises a model training pipeline based on computer vision and an AQC architecture design. Following a generic design, the framework is put into practice using a real-world packaging use case. Incorporating the DCAI paradigm into the framework is an innovative method for developing quality control solutions. Finally, the tests are conducted on the use case solution using real data. Consequently, it is demonstrated that the system's use in a real-world industry use case operates smoothly and produces excellent outcomes. The great majority of artifacts that are packed are rapidly and accurately categorized. Both conventional computer vision methods and DL-based methods are combined and compared.

Song, Xiaodong, et al. explain AI technology and its possible uses in the automation of the food sector, emphasizing its use in production processing. It also illustrates the advantages of more efficiency, lower costs, and higher-quality products. However, the promise of AI in the food industry has also prompted individuals to consider the benefits and drawbacks of incorporating intelligent food automation processing systems or equipment into food business applications. In an effort to encourage and expedite the use of AI approaches in food processing and production, this work aims to provide recommendations and guidance to governments, researchers, and professionals in the food industry [12].

The application of DL algorithms to automatically identify surface damage and cracks in chicken eggs is highlighted. To improve egg quality control in the food sector, eggs with physical damage, such as cracks, fractures, or other surface faults that could affect their quality, must be precisely recognized [13]. Eight hundred photos of eggs were used in the study, divided into two groups: damaged eggs and undamaged (intact) eggs. CNN-based DL models include GoogLeNet, MobileNet-v2, Visual Geometry Group (VGG)-19, and residual network (ResNet)-50. The classification accuracy of GoogLeNet was 98.73%, VGG-19 was 97.45%, MobileNet-v2 was 97.47%, and ResNet-50 was 96.84%. According to the findings, the GoogLeNet model has the highest damage identification accuracy rate (98.73%) [10]. In order to detect egg damage automatically, this study uses DL and AI approaches. With prompt intervention, this method minimizes product losses by giving manufacturers the capacity to identify defective eggs more quickly and correctly. In addition, these technologies provide a more effective way to identify and categorize damaged eggs than conventional techniques [14].

The virtualization method or computational model suggested in this work is centered on food product quality control and is built on collaboration and data sharing among the key players in the process, such as quality managers and inspection authorities. Intelligent algorithms that forecast the sensory qualities of food are included in the ICatador cloud platform, which supports virtualization (digitalization). García-Esteban, Juan Alberto, et al. have used our virtualization method in the ripening cheese elaboration as a validation scenario. The merit figure associated with a product's sensory quality can be quickly and methodically determined by the quality manager thanks to sophisticated visualization tools. For the purpose of adapting to the product elaboration process, intelligent algorithms are enhanced by the vast amount of data that is available from the elaboration process itself. In this manner, we can continuously improve the ML method to suit the specific process and employ it efficiently to boost competitiveness [15].

A computer vision system driven by ML was used by Chen, Shumian, et al. to distinguish between rice grains that had defects, including spotting, fracturing, or chalkiness. An Support Vector Machine (SVM) and near-infrared images were used to approach employing computer intelligence and analysis. The work's findings showed that, during a record-breaking 0.15 seconds execution time, the system was highly effective, achieving 99.3% accuracy in recognizing broken kernels and 93.6% precision in detecting spotted or damaged grain locations [16].

This study combined CNN and random forest to offer a DL method for brain tumor segmentation. Preprocessing, feature extraction, and image classification are some of the processes that are included in this approach. Consequently, based on the selected parameters, the photos are categorized using random forest. The tumor in an MRI image is segmented using a kernel-based CNN approach. The experimental results showed that the suggested method accurately segments brain tumors with a 95.4% accuracy rate when compared to current algorithms [17].

Ireri, David, et al. created a computer vision system driven by ML that distinguished and ranked tomatoes according to their visual appearance (from RGB photos). Utilizing an RBF kernel-SVM classifier, the system successfully identified calyx and other features, with an accuracy of 96% for both damaged and defective tomatoes. Technology showed a lot of potential as an inline tomato sorting tool overall, able to streamline the process while guaranteeing adherence to quality requirements [18].

Wang, Jing, et al. developed a better traceability approach in which AI and ML models evaluate the quality of food. Their suggested method offers real-time assessments of food quality in the whole supply chain by combining fuzzy categorization and ANN models, which enable both forward tracking and dynamic tracing. In a case study involving a pork producer, researchers found that the method was effective in providing accurate evaluations and ensuring food quality [19].

Wonggasem, Kris, et al. created an automated quality inspection system that uses pictures of newborn corn to distinguish between "pass" and "fail" categories. DL models that are comparatively more sophisticated are contrasted with a conventional image processing technique that uses the threshold concept. Specific subtypes of DL models, specifically CNN, were used. Extensive studies comparing various network designs and associated hyperparameters were carried out. To further aid in network comprehension, a Shapley additive explanation (SHAP) framework was employed. Up to 99.06% accuracy was the best result with EfficientNetB5 networks, with comparatively higher input sizes, compared to 95.28% from conventional image processing. We primarily contribute the SHAP framework, a data application on baby corn photos, a number of model trials, and a region of interest identification [20].

Iwendi, Celestine, et al. suggested a DL remedy for medical datasets based on health data that chooses the right diet on its own and suggests foods for patients based on their individual conditions and other characteristics. The study's goal was achieved by implementing several ML and DL methods, such as logistic regression, Naive Bayes, RNN, Multilayer Perceptron (MLP), Gated Recurrent Unit (GRU), and LSTM. A total of 1000 goods, each with eight features, and information from 30 individuals with 13 attributes associated with different diseases were included in the medical dataset, which was collected via hospitals and the internet. The LSTM algorithm outperformed the traditional approaches in the parameters of accuracy (96.74%), precision, recall, and F1 score, according to an evaluation of the various ML and DL systems' performances explained in Sivanantham et al. [21].

17.3 System Design

CNNs and RNNs combined to provide a complementary DL framework for food quality assurance by using their own technological advantages in image detection and time series forecasting. Food ingredients and appearance can be effectively identified thanks to CNNs' multi-feature recognition capabilities. In the meantime, RNNs' time series prediction skills are excellent for dynamic risk warning in the food processing industry. When CNNs and RNNs work together to forecast food quality, they not only increase performance efficacy but also guarantee food safety by leveraging their individual capabilities. CNNs' capacity to extract characteristics from food images and then utilize those features in conjunction with RNNs to assess patterns over time in food processing control and production lines enables real-time food appearance identification and comprehensive composition analysis. Figure 17.3 depicts the automatic quality control system's flowchart.

17.3.1 Convolutional Neural Network

The CNN framework consists of three fundamental components: convolutional layers, max-pooling, and sparse connectivity. Neurons on neighboring layers use the pixels in the image that are spatially dependent on them to their advantage.

Most important of all, the CNN's convolution kernel is where the phrase "convolutional neural network" originates. A two-dimensional matrix contains a convolution kernel with n columns, and each point has corresponding weights. Convolution kernels, often known as a neuron's receptive field, resemble neurons in size. The CNN's receptive field is filled by the weights and pixel values of the convolution kernel at the relevant points in the picture. This function is performed by the k convolution kernels of the system. Following the counting of

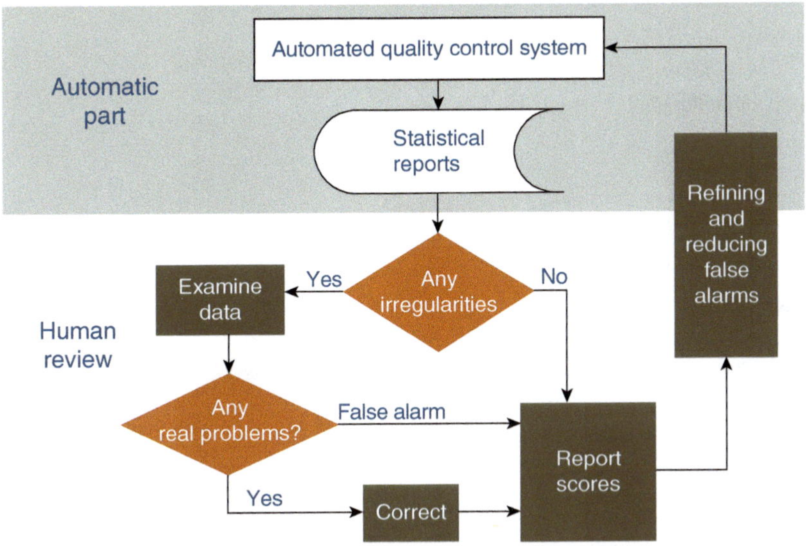

Figure 17.3 Flowchart for an automated quality control system.

all pixels in the image, go to the next location in the image based on the step size. Currently, this pixel matrix is the source image's feature map.

$$O_w = \left\lfloor \frac{i_w - n + 2_p}{s} \right\rfloor + 1 \tag{17.1}$$

$$Oh = \left\lfloor \frac{i_h - n + 2_p}{s} \right\rfloor + 1 \tag{17.2}$$

Averaging, maximal combining, and other methods are the most widely used pooling algorithms. We select the computation that corresponds to the highest pooling strategy. Using a size 2×2 filter, the region of the image where each pixel has the highest value between 2 is chosen and kept as its identifying feature. The filter creates a feature map after repeating this procedure for the subsequent range. Both the general classification of the data and the attainment of structured output by the local job have improved with end-to-end training for the convolution network. As a result, it has been widely applied in the domains of data classification and detection. Because the data dimensions are too large, CNNs will encounter issues very rapidly when utilizing a typical multi-layer perceptron, where every layer is completely connected [22, 23]. CNN doesn't have any information on space available. If we employ the pooling layer, the final efficiency can be fairly poor. CNNs lose a lot of information in the pooling layer, which reduces, because only the most active neurons can transfer information between layers, the spatial resolution [24, 25].

The pseudocode for a CNN is displayed in Figure 17.4. This will prevent CNNs from distinguishing between postural differences and other factors. Although it usually requires more time and computer power, overtraining from all angles is one method to avoid this issue. Field-of-view data with a high degree of overlap, mutual masking, and several backgrounds cannot be reliably identified by a traditional CNN [8, 9]. Here, we describe the

Input:

d: dataset,

1: actual labels within the dataset

W: Word-to-Vec grid

Output:

Score for the CNN-trained model on the test dataset

Let f represent the 3D feature matrix.

For i in dataset **do**

Allow The feature set matrix of sample I is represented by f.

For jin i **do**

Vj - vectorize(j, w)

Append vj to f

Append fi to f

Ftrain, ftest, Itrain, Itest utilizing the feature set and labels, create train and test subsets.

M-CNN (ftrain, Itrain)

Analysis of scores (i, Itest, M)

Return score

Figure 17.4 CNN pseudocode.

removal of the fully connected layers, how the suggested approach employs two convolutional layers as a skip layer, and the skip and pooling layer parameters. In particular, CNN often adds multiple fully connected layers to its tail.

17.3.2 Recurrent Neural Network (RNN)

Furthermore, RNNs have a large computing overhead, which might impact the effectiveness of real-time predictions, particularly when processing lengthy data sequences. These limitations show up in food quality control safety risk prediction when it comes to the incapacity to precisely comprehend how long-term historical data affects future hazards. RNNs could find it difficult to identify important events that took place a few steps earlier, for instance, while identifying possible risks in the food processing chain.

LSTM networks successfully solve these problems by implementing input, output, and forget gates. Because LSTM can recognize deep patterns from complex time series data and selectively forget irrelevant information while retaining helpful data, it can better capture the interaction between long-term and short-term factors when it comes to food quality management risk prediction. Examining how storage conditions, multi-step procedures, and historical monitoring information interact is more accurate with LSTM. More resilient to noise and outliers, LSTM excels at managing multivariate inputs and nonlinear changes. It can also simulate dynamic interactions between important components at various phases.

This is challenging for conventional time series analysis methods. They can only depict short-term dependence to record long-term dynamic changes in the food supply chain. On the other hand, because of their special abilities for dynamic updating, real-time prediction, and gating, LSTM-based models may effectively capture long-term dependencies in time series and automatically learn nonlinear dynamic patterns. Consequently, the use of LSTM provides greater food quality safety risk prediction accuracy and reliability. The implementation results of our newly developed system will be discussed in the upcoming section.

17.4 Result and Discussion

In the food sector, the performance evaluation metrics for automatic quality control analysis utilizing AI and DL are accuracy, sensitivity, specificity, and processing time. Based on these characteristics, the proposed CNN with RNN is compared to three traditional approaches: Deep neural networks (DNN), artificial neural networks, and Elman neural networks (ENN). The newly proposed integrated (CNN-RNN) automated food quality control and detection system using AI technology achieved an accuracy of 96.28% with a minimum processing time duration of 24.63 ms.

17.4.1 Confusion Matrix

The efficiency of our trained ML methods is predicted using the performance evaluation measures. It also lets us determine how effectively a learning approach will work on a dataset that has never been utilized before. Effective performance and output determination are achieved using a confusion matrix. Figure 17.5 provides an explanation and illustration of how the confusion matrix operates.

17.4.2 Accuracy

Dividing the total number of projections by the number of accurate forecasts determines how efficient the model is.

$$\textbf{Accuracy} = \frac{\text{TP} + \text{TN}}{\text{TP} + \text{TN} + \text{FN} + \text{FP}}$$

Figure 17.5 Confusion matrix.

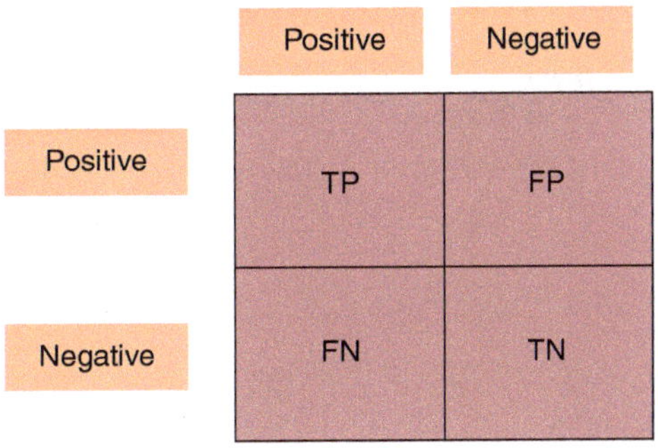

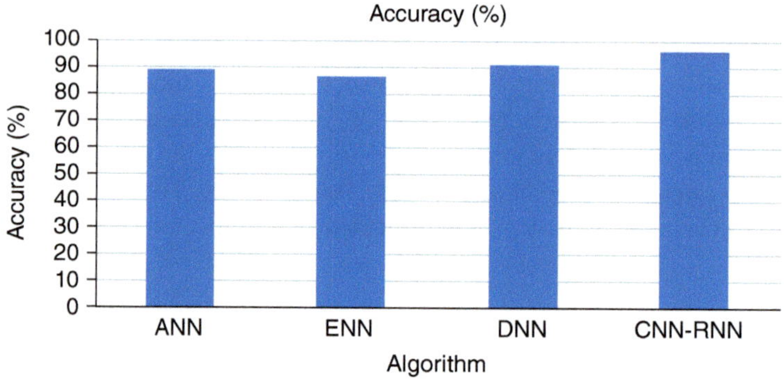

Figure 17.6 Accuracy graph.

Table 17.1 Accuracy outcome.

Algorithm	Accuracy (%)
ANN	88.64
ENN	86.43
DNN	91.15
CNN-RNN	96.28

Figure 17.6 clearly displays the graph of accuracy for the current and suggested systems from Table 17.1. Our suggested CNN-RNN system performs better than the other systems, with an accuracy of 96.28%, according to this graph.

17.4.3 Sensitivity

Sensitivity is computed as the ratio of correct positive predictions to all positive values. The formula for calculating sensitivity is as follows:

$$\textbf{Sensitivity} = \frac{\text{TP}}{\text{TP} + \text{FN}}$$

Table 17.2 displays the sensitivity output graph of the current and suggested systems in Figure 17.7. When it comes to AQC in the food sector, the recently suggested CNN-RNN algorithm outperforms all previous algorithms with a sensitivity of 93.26%.

17.4.4 Specificity

To calculate specificity, divide the number of accurate negatives by the overall number of negative forecasts. The formula that follows is used to determine the specificity.

Table 17.2 Sensitivity outcome.

Algorithm	Sensitivity (%)
ANN	86.25
ENN	84.89
DNN	87.67
CNN-RNN	93.26

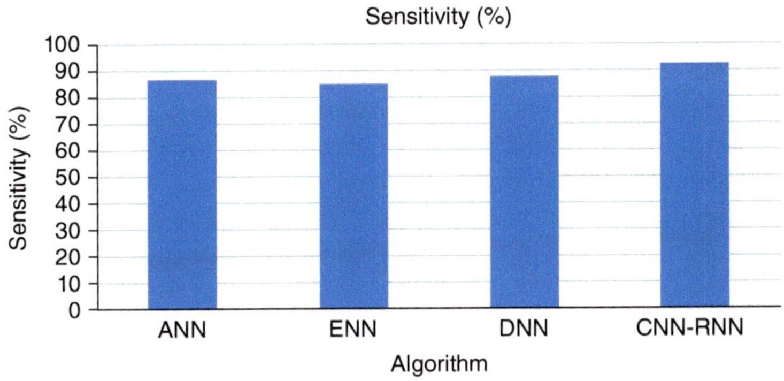

Figure 17.7 Sensitivity graph.

$$\textbf{Specificity} = \frac{\text{TN}}{(\text{TN} + \text{FP})}$$

Figure 17.8 displays the specificity output graph for the current and suggested systems from Table 17.3. When compared to other current algorithms, the CNN-RNN algorithm's innovative method yields a greater specificity result of 91.58%.

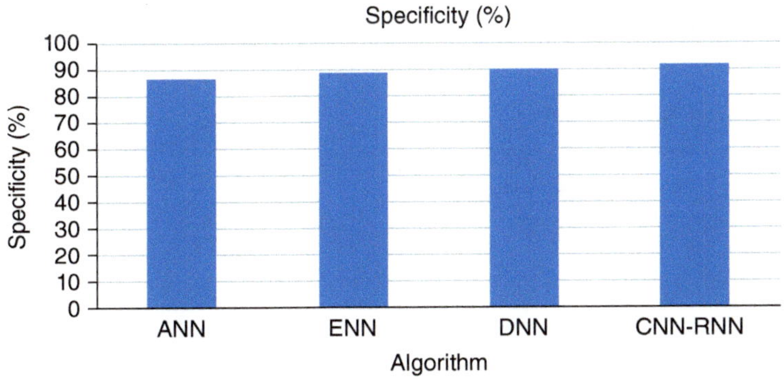

Figure 17.8 Specificity graph.

Table 17.3 Specificity outcome.

Algorithm	Specificity (%)
ANN	85.42
ENN	88.35
DNN	89.13
CNN-RNN	91.58

17.4.5 Processing Time

The entire time duration is the total amount of time something lasts from start to finish.

Figure 17.9 presents from the Table 17.4 is a comparison of the quality analysis and processing time duration for the suggested method and the current approaches. The suggested AI-based automatic quality detection and control in the food sector using a hybrid CNN-RNN method took a time duration of 24.63 ms, which took less processing time than the other traditional models.

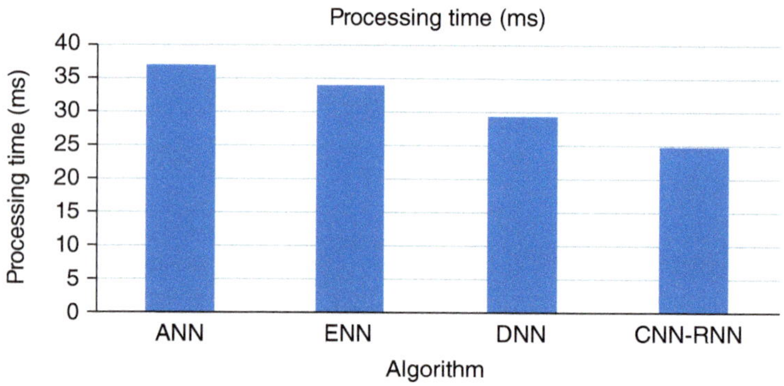

Figure 17.9 Processing time duration graph.

Table 17.4 Outcome for processing time duration.

Algorithm	Processing time (ms)
ANN	37.29
ENN	33.82
DNN	29.15
CNN-RNN	24.63

Several methods for AQC using AI approaches, ANN, ENN, and DNN, are utilized to determine the processing time duration values, accuracy, sensitivity, and specificity. Their results are computed and recorded. Based on the recorded numbers, a bar chart was created. In accordance with the output results, when compared to the current systems, our suggested hybrid CNN-RNN model produces overall better output outcomes for AQC in the food sector.

17.5 Conclusion

Food was once produced and processed by hand, but as automated equipment advanced, and the subsequent assistance of AI technology, it became easy to see how AI technology plays a revolutionary and inseparable role in this sector. The benefits of increased quality, safety, and food security have been demonstrated in every area of the food sector. As the technology advances, costs come down, and data management procedures are enhanced, the potential of AI in food sector uses will be investigated and utilized extensively, despite shortcomings in data authenticity, integrity, and implementation costs. By enhancing contaminant identification, defect detection, and product consistency, we offer an optimization framework that can be applied to neural network classification in the food industry to enhance AQC. The suggested system uses DL approaches, including RNNs for anomaly detection and predictive maintenance and CNNs for image-based quality assessment. With a minimal processing time of 24.63 ms, the recently suggested combined (CNN-RNN) automated food quality inspection and detection system leveraging AI technology achieved an accuracy of 96.28%.

References

1 Mavani, N.R., Ali, J.M., Othman, S. et al. (2022). Application of artificial intelligence in the food industry—a guideline. *Food Engineering Reviews* 14 (1): 134–175.

2 Sivanantham, K. and Kumar, R.M. (2023). Kumar Ashishand Jain, Rachna and Vairamani, Ajantha Devi and Nayyar, Anand,. Different approaches to background subtraction and object tracking in video streams: a review. Springer Nature Singapore In: Object Tracking Technology: Trends, Challenges and Applications, 23–39.

3 Addanki, M., Patra, P., and Kandra, P. (2022). Recent advances and applications of artificial intelligence and related technologies in the food industry. *Applied Food Research* 2 (2): 100126.

4 Castillo, O. and Meliif, P. (2024). Automated quality control in the food industry, combining artificial intelligence techniques with fractal theory. *WIT Transactions on Information and Communication Technologies* 10–15.

5 Xiong, F., Kühl, N., and Stauder, M. (2024). Designing a computer-vision-based artifact for automated quality control: a case study in the food industry. *Flexible Services and Manufacturing Journal* 36: 1422–1449.

6 Song, X., Zhang, X., Dong, G. et al. (2025). AI in food industry automation: applications and challenges. *Frontiers in Sustainable Food Systems* 9: 1575430.

7 Chandrasiri, G.S.M., Wijenayake, A.K.I., and Arachchige, U.S. (2022). Development of automated systems for the implementation of food processing. *Journal of Research Technology & Engineering* 3 (1): 8–18.

8 Pandey, J.K. (2025). Understanding the impact of electromagnetic and optical radiation on physical and biological systems. *Journal of Nano- and Electronic Physics* 17 (5): 05023. https://doi.org/10.21272/jnep.17(5).05023.

9 Srinivas, P., Veeraiah, V., Thejaswini, K.O. et al. (2025). The impact of electromagnetic and optical radiation on physical and biological systems. *Journal of Nano- and Electronic Physics* 17 (4): 04012. https://doi.org/10.21272/jnep.17(4).04012.

10 Cengel, T.A., Gencturk, B., Yasin, E.T. et al. (2025). Automating egg damage detection for improved quality control in the food industry using deep learning. *Journal of Food Science* 90 (1): e17553.

11 García-Esteban, J.A., Curto, B., Moreno, V. et al. (2018, July). A digitalization strategy for quality control in the food industry based on Artificial Intelligence techniques. *2018, IEEE 16th International Conference on Industrial Informatics (INDIN)*, pp. 221–226. IEEE.

12 Wonggasem, K., Chakranon, P., and Wongchaisuwat, P. (2024). Automated quality inspection of baby corn using image processing and deep learning. *Artificial Intelligence in Agriculture* 11: 61–69.

13 Leena, B. (2023). Deep learning-based convolutional neural network with random forest approach for MRI brain tumour segmentation. In: System Design for Epidemics Using Machine Learning and Deep Learning, 83–97. Cham: Springer International Publishing.

14 Chen, S., Xiong, J., Guo, W. et al. (2019). Colored rice quality inspection system using machine vision. *Journal of Cereal Science* 88: 87–95.

15 Ireri, D., Belal, E., Okinda, C. et al. (2019). A computer vision system for defect discrimination and grading in tomatoes using machine learning and image processing. *Artificial Intelligence in Agriculture* 2: 28–37.

16 Wang, J., Yue, H., and Zhou, Z. (2017). An improved traceability system for food quality assurance and evaluation based on fuzzy classification and neural networks. *Food Control* 79: 363–370.

17 Iwendi, C., Khan, S., Anajemba, J.H. et al. (2020). Realizing an efficient IoMT-assisted patient diet recommendation system through a machine learning model. *IEEE Access* 8: 28462–28474.

18 Wang, Y., Wu, J., Deng, H., and Zeng, X. Bai Yuan Ding (2021). Food image recognition and food safety detection method based on deep learning. *Computational Intelligence and Neuroscience*
2021 (1): 1268453.

19 Hu, J., Wang, X., Zhang, Y Jianru Xue. et al. (2020). Time series prediction method based on a variant LSTM recurrent neural network. *Neural Processing Letters* 52: 1485–1500.

20 Kalimuthu, S., Praveen, P.B., Mahalakshmi, J. et al. (2024). Blockchain-based IoT healthcare data using narrow elliptical curve with particle swarm optimization (NECPSO) approach. In: Convergence of Blockchain and Explainable Artificial Intelligence, 117–144. New York: River Publishers.

21 Sivanantham, K., Vijaykumar, V.R., Krishnakumar, A., and Pradeepa, M. (2023). Cyber-bullying detection for multimodal data convolutional neural network with butterfly optimisation. In: Combatting Cyberbullying in Digital Media with Artificial Intelligence, 98–115. Boca Raton: Chapman and Hall/CRC.

22 Beemkumar, N., Gupta, S., Bhardwaj, S. et al. (2023). Activity recognition and IoT-based analysis using time series and CNN. In: Handbook of Research on Machine Learning-enabled IoT for Smart Applications Across Industries (ed. N. Goel and R.K. Yadav), 350–364. IGI Global https://doi.org/10.4018/978-1-6684-8785-3.ch018.

23 Talukdar, V., Veeraiah, V., Roy, D.R. et al. (2023). An innovative approach: typicality-based collaborative filtering for recommender systems. *Proceedings of the 3rd International Conference on Advancement in Electronics & Communication Engineering (AECE)*, pp. 863–868. https://doi.org/10.1109/AECE59614.2023.10428576.

24 Pandey, J.K., Yadav, A.K., Purushotham, M. et al. (2022). Colorimetric detection of deltamethrin pesticide from biological samples based on the peroxidase-mimic catalytic activity on MnO_2/SnO_2. *Key Engineering Materials* 928: 45–51. https://doi.org/10.4028/p-4ti1i0.

25 Pandey, J.K., Kumar, S., Lamin, M. et al. (2022). A metaheuristic autoencoder deep learning model for an intrusion detector system. *Mathematical Problems in Engineering* 2022: 11, 3859155. https://doi.org/10.1155/2022/3859155.

18

Future Trends and Opportunities in Visualizing Nutrition

Pushkar Singh Rawat[1] *and Shalini Singh*[2]

[1] *Department of Biochemistry, University of Lucknow, Lucknow, Uttar Pradesh, India*
[2] *Institute of Biosciences and Technology, Shri RamSwaroop Memorial University, Barabanki, Uttar Pradesh, India*

18.1 Introduction

18.1.1 Concept and Scope of Nutrition Visualization

Nutrition visualization is the representation of nutritional data in the form of a graph or image in order to allow people to comprehend and evaluate the nutritional value of foods. It encompasses methods ranging from image segmentation and regression models to estimate calories, fats, proteins, and carbohydrates directly based on food images to enhance the precision in nutritional assessment [1]. There are also visualization techniques, including plotting nutrient-specific data, i.e., fiber and protein per calorie, in a two-dimensional graph in order to compare them with the dietary recommendations, to help consumers make more healthful food choices [2]. These visual tools improve usability by making the provision of spatial reference and not mere numeric data easier to understand the nutrient content of the meals in a balanced meal. Visualization of nutrition is also beneficial to weight management, as the focus on the nutrient density allows identifying the foods that can be included in the diet [2]. Altogether, nutrition visualization can enhance the connection between complicated nutritional information and consumer-friendly knowledge through the application of sophisticated and computerized systems and visualizations to foster a healthier dietary lifestyle [1, 2].

18.1.2 Evolution of Visualization Techniques in Nutrition Science

A shift in visualization practices in nutrition science is one that resulted in the use of more sophisticated computational and interactive visualization techniques in place of the traditional ones. Visual images were largely employed as a descriptive capability by early research papers, and this situation changed after the 1960s to a presentation of complicated relationships and logical arrangement of nutritional information to increase the persuasion and scientific communication [3]. In recent years, the concept of computer vision and deep learning has transformed food recognition and nutrition analysis, as it has become possible to automatically classify and estimate the nutrients of food based on images with much greater accuracy and usability [4]. Multispectral imaging combined with neural networks enables the prediction of food properties using standard RGB images, and approximates more complicated multiwavelength practices, at the expense of being less practical [4]. The visualization has also been further developed through augmented reality (AR) and mobile technology, which delivers interactive and real-time nutritional data, enhancing user interaction and comprehension [5]. Also, there is an ingestion of visual and textual data to improve the estimation of nutrition using artificial intelligence (AI) and ingredient-sensitive models, which is expected to become popular among individualized and accurate diet evaluation [6]. This development reflects a shift toward nonmoving, symbolic visual representations to moving, AI-assisted, and immersive visualization methods that would be more useful in nutrition science research and in promoting health among the general population. Table 18.1 highlights the progression from static, descriptive nutrition visuals to dynamic, AI-enabled visualization systems supporting personalized and real-time dietary assessment.

18.1.3 Rationale for Future-Oriented Approaches

The rationale behind future-oriented methods in nutrition visualization is the necessity of overcoming the shortcomings of the old self-reported dietary measures, which can be characterized by the inability to report the food type and portion size objectively and are mostly subjectively biased [7]. The developments of the image-based dietary assessment system based on deep learning and computer vision allow more objective and automated food recognition and volume estimation, which leads to better accuracy and efficiency in nutritional analysis [8]. Adding visual data to nutrition knowledge bases and applying large language models will additionally promote the accessibility and accuracy of food identification and nutritional information, which will support a personalized diet [9]. Moreover, the combination of 3D body modeling and virtual reality technologies also proposes new methods to visualize changes in bodies over time, which may enhance adherence to dietetic interventions through having immersive and motivating feedback [10]. Packaging and educational tools that use visual communication design are also essential to informing more people to

Table 18.1 Evolution of nutrition visualization techniques.

Era/phase	Key visualization methods	Data sources	Primary applications
Early period (pre-1990s)	Static charts, tables, nutrient lists	Dietary surveys, laboratory data	Descriptive nutrition research
Transitional phase (1990–2010)	Graphs, nutrient-density plots	Food composition databases	Comparative dietary analysis
Digital expansion (2010–2020)	Interactive dashboards, mobile apps	Digital food logs, images	Consumer nutrition education
AI-driven era (2020–present)	Image-based recognition, ML models	Food images, sensors, omics data	Precision nutrition, real-time assessment

make healthier food decisions through the prioritized presentation of nutrition-related information [11]. In general, the visualization of future-oriented nutrition technologies will integrate AI, immersive technologies, and effective design to offer more precise, engaging, and personalized nutrition data to address the existing issues and contribute to enhanced health outcomes.

18.2 Technological Advancements Shaping Nutrition Visualization

18.2.1 Artificial Intelligence and Machine Learning–Driven Visualization

Machine learning (ML) and AI are changing the concept of nutrition visualization by making the assessment and advice about the diet more precise, efficient, and personalized. AI-assisted image recognition nutritional assessment systems, which are frequently part of mobile or web apps, identify foods, approximate quantities, and nutrient consumption in real time; a feature of AI-assisted methods that minimizes recall bias associated with traditional methods [12]. Large language models (LLMs) have the potential to lead to personalized nutrition recommendations and visual tools, but the existing scope of issues, such as the lack of recognition of vital nutrients, continuous development is likely to reduce these drawbacks [13]. Precision nutrition is an AI-based product that uses personal biological data, including blood markers and microbiome profiles, to dynamically make dietary recommendations, benefiting the control of various diseases, including obesity, diabetes, and chronic kidney disease [14]. Also, AI models proved to be highly accurate in detecting malnutrition and classifying nutritional status, which is essential in terms of timely intervention in at-risk groups, including cancer patients [15]. No matter how these advances have been made, there are still issues of data privacy, transparency in algorithms, and fair access with regard to the implementation of AI-driven nutrition visualization tools, which highlights the necessity of collaboration between healthcare professionals, technologists, and policymakers [6]. Table 18.2 summarizes emerging technologies that collectively enhance the accuracy, personalization, and usability of nutrition visualization tools.

Table 18.2 Key technologies shaping future nutrition visualization.

Technology	Core function	Visualization output	Major advantage
AI and ML	Food recognition and nutrient estimation	Interactive nutrient dashboards	Reduced self-reporting bias
Big data analytics	Integration of large nutrition datasets	Multilayered visual analytics	Population-level insights
Wearable biosensors	Continuous metabolic monitoring	Real-time nutrient status graphs	Personalized feedback
Augmented/virtual reality	Immersive learning environments	3D food and diet simulations	Enhanced engagement

18.2.2 Big Data Analytics and Interactive Dashboards

Nutritional visualization is being revolutionized by big data analytics and interactive dashboards, which allow the clear presentation and analysis of large and complicated nutrition data. Such platforms as Global Nutrition and Health Atlas (GNHA) can be seen as an example because it is an open-access web-based platform, enabling 500 nutrition and health indicators in over 190 countries to be collected, stored, tracked, and visually presented, and provides users with a chance to interact with data and formulate hypotheses [16]. Reviews of food and nutrition dashboards have identified systematic reviews as significant in terms of metrics of data integrity, completeness, quality visualization, and interactivity that can improve usability and enhance knowledge sharing between researchers, practitioners, and policymakers [16]. Interactive dashboards enable the real-time monitoring and decision-making process by making multidimensional nutrition data accessible (in an easy-to-understand format), and therefore, the gap between the complexity of the data and its practical application is low. Further innovations in the methods of data visualization, such as customizing the dashboard and the ability to estimate multiple gazes, enhance the involvement of the user and the efficiency of interpreting the data [17]. In general, the combination of big data analytics and interactive dashboards is essential in moving forward with precision nutrition, evidence-based policy, and collaboration in the community of nutrition science.

18.2.3 Integration of Wearable and Sensor-Based Data

The data integration of wearable sensors is improving nutritional monitoring through real-time and personalized nutrient intake and metabolism. Recent advancements include wearable electrochemical biosensors, which are noninvasive to study metabolites and nutrients in sweat, including amino acids, vitamins, and riboflavin, and can be used to provide continuous biochemical information during exercise and rest [18]. They can be used in conjunction with microfluidic sampling and wireless communication, where these sensors can be used to dynamically monitor nutritional status and detect early nutritional metabolic abnormalities to enable precision nutrition based on individual metabolic profiles [19]. Chemical sensors are complemented by mobile applications and wearable motion sensors that recognize events of dietary intake and approximate calorie consumption, but their accuracy and integration of data are yet to be developed [20]. Self-powered wearable sensors and IoT environments are also emerging technologies that increase the idea of portability and round-the-clock nutrient monitoring, promising useful tools in individualized dietary recommendations and health control [21]. Regardless of the advancement in technology, there exist sensor sensitivity, data fusion, user compliance, and clinical validation issues to consider in order to fully utilize the potential of the wearable nutrition monitoring system [22].

18.3 Emerging Visualization Modalities

18.3.1 Augmented and Virtual Reality in Nutrition Education

Augmented/virtual reality (AR/VR) technologies are becoming more and more popular in nutrition education to improve engagement, awareness, and behavior change. VR virtual supermarkets can offer these experiences by providing users with immersive and controlled environments with the ability to browse food options based on product placement, pricing, and nudging, and could help them simulate real-world decision-making and enhance nutritional knowledge [23]. AR applications are frequently interactive in nature, particularly among children and teenagers, and can focus on culturally relevant elements such as traditional food to make learning about nutrition more enjoyable and memorable, allowing for improvements in retention and positive changes in dietary behavior [24]. The benefits of these technologies compared to the traditional methods are that they offer a sense of presence and interactivity, a sense of reduction of boredom, and personalized and experiential learning [25]. Nonetheless, the majority of research is committed to short-term behavior change or lab-based testing, and the gaps in the area include long-term effects evaluation, community implementation, and the rigor of the methodology [26]. Further development and studies are required to streamline AR/VR applications into a more efficient way of delivering nutrition education, as well as mitigate the issues of accessibility and compatibility with the current education systems [24].

18.3.2 Immersive and 3D Data Representation

3D and immersive representation of data in nutrition is changing the way that nutritional data is displayed, analyzed, and customized. Developments in 3D and 4D food printing allow the production of custom foods with custom shapes and textures as well as specific nutrient content, and that react dynamically to environmental inputs, such as temperature and humidity, which creates new prospects in individualized nutrition and the minimization of food waste [27]. Knowing that vision-based algorithms with RGB-D fusion networks enhance the capability of automated nutrient estimation by merging depth and color images, they are more accurate in estimating calories and nutrients by the detailed 3D modeling of food [28]. The MetaFood3D dataset of 743 3D scanned food items with nutrition values associated allows better portion size estimation and synthetic data generation, solving food computing problems and allowing dietary analysis with more accuracy [17]. The interactive visualizations created with dimensionality reduction methods such as t-SNE allow users to find detailed nutritional information on a micro and macro level and engage and learn more about nutrition and nutritional content [29]. In spite of these developments, the problem of speed of printing, integration of data, and interaction with users still requires enhancement, and more studies are required to streamline immersive analytic and 3D visualization systems to be used in practice in nutrition [27].

18.3.3 Mobile and Web-Based Visualization Platforms

The development of cloud computing, HTML5, and WebGL technologies has seen significant changes in mobile and web-based visualization platforms and has facilitated high-performance graphics on mobile devices [30]. Such platforms provide a democratization of access to higher-order visualization by offloading heavy calculations to the cloud and relying on mobile devices as interactive interfaces, complex data analysis is now more convenient and mobile [30]. Such tools as ECharts offer declarative visualization construction systems based on the rapid and interactive construction of web-based visualization, and allow nonprofessional users to build rich visualizations with high scalability and efficiency

[31]. Solving difficulties peculiar to mobile devices, MobileVisFixer is a reinforced learning-based algorithm that enhances the visualization design on small screens to make them more user-friendly and user-experienced on a mobile phone [32]. Computational clusters formed by multiple local devices are used to perform visualization computations in an ad hoc fashion; for example, VisHive is used to run its work on smartphones, tablets, and smartwatches without requiring any extra software [33]. New web-based applications also incorporate the use of immersive and augmented reality, allowing situated and 3D visual displays that can be accessed on handheld devices, increasing the range of interactive data exploration of mobile data [34].

18.4 Personalized and Precision Nutrition Visualization

18.4.1 Visualizing Multiomics and Metabolic Data

Precision and personalized nutrition build on the multiomics technologies (genomics, epigenomics, transcriptomics, proteomics, metabolomics, and metagenomics) in order to develop a holistic view of individual metabolic phenotypes and dietary responses. Those methods allow the creation of individualized dietary prescriptions based on the combination of intricate biological data with both environmental and lifestyle elements, the purpose of which is to shift the health outcomes and control such diseases as obesity, metabolic syndrome, and type 2 diabetes [35]. The multiomics and metabolic data visualization assists in identifying biomarkers and metabolic processes that connect genetic and molecular phenotypes with nutrition-related health outcomes, which allows precision intervention [36]. Even though the improvements are promising, there are still data integration issues, high expenses, ethical issues, and the necessity to have strong and reproducible outcomes, which are currently limiting the clinical use [35]. New computational tools and ML algorithms are under development to process high-dimensional omics data, which can be more effectively interpreted and provide an individual approach to nutrition [37]. Altogether, the multiomics data visualization became a key to the development of precision nutrition, yet additional studies must be conducted to convert these results into the form of practice-based and tailored dietary and nutrition recommendations [38]. Table 18.3 illustrates how advanced visualization approaches enable precision nutrition by integrating biological, behavioral, and lifestyle data into individualized visual feedback systems.

18.4.2 Personalized Dietary Feedback and Decision Support

Individual metabolic profile, genetics, and lifestyle data integration into personalized nutrition recommendations is becoming accessible to wearable sensors, mobile apps, and web-based applications, which are progressively providing personalized dietary recommendations through the integration of each of these components. Noninvasive wearable chemical sensors can measure biochemical responses to food intake in real time and modify the diet based on that information to enhance nutritional balance and change dieting [19]. Behavior change-based web-based systems and customized dietary feedback have been demonstrated to be highly acceptable among dietitians and able to increase diet quality

Table 18.3 Visualization approaches for personalized and precision nutrition.

Visualization approach	Data integrated	Visualization format	Key benefit
Multiomics visualization	Genomics, metabolomics, microbiome	Layered network maps	Identification of nutrition–health biomarkers
Metabolic pathway mapping	Enzyme and metabolite data	Interactive pathway diagrams	Mechanistic understanding of diet responses
Personalized nutrient dashboards	Dietary intake, biomarkers, lifestyle	Individualized scorecards	Tailored dietary recommendations
Behavioral feedback visuals	Eating patterns, adherence data	Trend graphs and alerts	Improved compliance and motivation

through customized dietary advice provided to individuals after an individual dietary evaluation [39]. AI innovations are used to examine intricate datasets derived from genetics, metabolomics, and continuous monitoring to create dynamic and personalized dietary recommendations that evolve in response to their needs and environment [14]. Individualized nutrition solutions have proved more effective than the generic recommendations and particularly in combination with dietitian consultations and self-monitoring, resulting in better adherence and health results [40]. However, there are still difficulties in uniting multimodal data, privacy, and scaling these technologies to use them in the clinical and population-wide context [19].

18.4.3 Behavioral Insights Through Visual Analytics

Visual analytics (VA) of behavioral insights in personalized and precision nutrition deals with the incorporation of multiple-layered data of genetics, metabolomics, lifestyle, and environmental data to comprehend and control the specific dietary behavior of individuals. VA can be used to convert multidimensional data into understandable and practical information that promotes individual dietary choices and behavior change to improve nutrition adherence [41]. AI and ML also allow reviewing the diet and behavioral reactions, enhancing the relevance of individual feedback and allowing changes in nutrition plans in real-time [6]. These strategies also prioritize the role of integrating biological information with behavioral and social variables to develop powerful and personalized interventions that take into account the differences in response to diet [41]. There are still difficulties in the domain of data integration, privacy, and fair access to individualized nutrition technologies, which underlie the necessity of cross-disciplinary work and ethics [42]. In general, VA represents a pivotal point between the intricate omics data and operational behavioral interventions, and their development will promote the efficiency of individual nutrition interventions.

18.5 Applications Across Health, Policy, and Industry

18.5.1 Clinical and Public Health Nutrition

Public and clinical nutrition are very important areas that aim at learning the interaction between diet and health outcomes to prevent and control diseases at the individual and population levels. Clinical nutrition studies positively determine the relationship between nutrition and well-being, but have drawbacks, including participant heterogeneity, selection of a control group, blinding, and regulatory overheads that can disrupt the effectiveness of a study and its application to practice [43]. The role of general practitioners and dietitians in the gap between clinical care and population health is also noted in the context of public health nutrition, which focuses on nutrition education to healthcare professionals to promote healthy eating habits and combat malnutrition in the population around the world [44]. The interaction between dieting and other variables, such as sleep, also has relevant clinical and epidemiological consequences that affect the risk of chronic diseases and general well-being [45]. The field of AI is starting to assist in the nutrition of the population, positively impacting the data analysis and intervention method, yet the issue of integration and ethics is still under consideration [46]. Altogether, clinical and public health nutrition necessitates multidisciplinary interventions to help improve dietary habits, prevent non-communicable diseases, and improve health outcomes of diverse populations [47]. Table 18.4 demonstrates how nutrition visualization supports decision-making at individual, institutional, and policy levels.

18.5.2 Food Systems, Sustainability, and Consumer Awareness

Price, taste, convenience, and health are the main factors shaping food preferences among consumers, and their purchasing choices are less focused on the environment and more concerned with food sustainability [48]. A significant number of consumers fail to clearly define what is meant by sustainability and tend to think about it primarily in the context of organic or local food production, and do not have much knowledge about the larger environmental effects of their actions [49]. The acceptance of new sustainable food technology ranges, and urban farming and plant-based alternatives have been well received, whereas technologies such as insect ingredients and cell-cultured meat are not accepted, which depends on cultural and personal differences [50]. Gen Z, in particular, is more

Table 18.4 Applications of nutrition visualization across sectors.

Sector	Visualization use	Impact
Clinical nutrition	Patient-specific nutrient profiles	Improved dietary counseling
Public health	Interactive population dashboards	Evidence-based policy decisions
Food industry	Front of pack labels, nutrient scoring	Health-oriented product reformulation
Sustainability	Diet environment impact visuals	Informed sustainable food choices

environmentally conscious and considers adopting sustainable food behaviors, but the level of understanding the connection between food and the health of the environment has not yet been fully established [51]. Sustainable consumption can be reinforced with the help of such policy tools as information campaigns, nudging, and regulations, but their success will require transparency, consumer education, and attention to sociodemographic differences, including education and gender [52]. To eliminate the barriers and introduce long-term changes toward sustainable food systems, technological advances, corporate responsibility, and collaborative efforts of multiple stakeholders are necessary [53].

18.5.3 Industry Innovation and Product Development

Visualization of nutrition is important to industry innovation and product development as it helps companies to evaluate and enhance the nutritional content of their products systematically. An illustrative case is that Ferrero came up with category-based and progressive nutritional requirements to shape the formulation of its products and refinement of its portfolio, considering the balance between nutrient density and flavor among other factors [54]. Such visual tools as the Nutri-Score have been demonstrated to be efficient in nudging consumers into making healthier food choices as they simplify nutritional information and affect the purchase behavior across demographics [55]. The profitability of firms is also positively influenced by product nutrition innovation in combination with packaging innovation and advertising, which demonstrates the commercial importance of product development with a nutrition concern [56]. Additionally, AI promotes efficiency and innovation in the food sector by positively affecting nutrition visualization, analyzing the data in real-time, quality control, personalized nutrition recommendations, and sustainable food production [14]. Nevertheless, there are still difficulties related to placing nutrition and sustainability information visually high on packaging since the existing designs focus more on branding than informative indicators and provide the consumer with only a limited space to focus on health-related information [57].

18.6 Challenges and Ethical Considerations

18.6.1 Data Privacy, Security, and Ethical Use of Nutritional Data

The issue of data privacy, security, and ethical use of nutrition data is of particular concern, particularly as nutrition research becomes more dependent on digital technologies and AI. Nutrition intervention programs in the workplace raise issues regarding the privacy and confidentiality of the participants, as well as the potential of the participant being pressured into the intervention, lack of data protection, and disclosure [58]. Nutrition apps that can be used publicly gather great volumes of data showing their content users, which food they consumed, and their privacy conditions are not always standardized, have no apparent privacy policies, and transparency regarding the use of personal data, which is a concern in terms of ethics and legality [59]. There are privacy issues associated with an image-based dietary assessment system, in which the sensitivity of food pictures and metadata is a point of trust, and the systems should adhere to data protection laws [60]. The application of

Table 18.5 Challenges and ethical considerations in nutrition visualization.

Challenge	Description	Required mitigation
Data privacy	Collection of personal dietary and biological data	Strong data governance frameworks
Algorithmic bias	Inaccurate predictions across populations	Diverse and validated datasets
Misinterpretation	Oversimplified or misleading visuals	Standardized visualization guidelines
Cultural inclusivity	Lack of representation of traditional diets	User-centered, culturally sensitive design

personalized nutrition services, in particular related to genetic data, is subject to ethical considerations of consumer protection, honesty, and legal assurance because existing laws are not fully developed and are still developing [61]. Also, AI and ML in nutrition research need a cautious modeling approach to prevent bias and ethical traps to guarantee information safety and conscientious use of findings [62]. On the whole, privacy and ethical protection require open-minded consent procedures, the strong protection of data security, and the continuous development and promotion of laws. Table 18.5 outlines key limitations that must be addressed to ensure responsible, equitable, and trustworthy nutrition visualization.

18.6.2 Inclusive and Culturally Sensitive Visualization Design

Diverse populations should be well-reached through inclusive and culturally sensitive styles in the design of nutritional visualization, which enhances health outcomes. Digitization of traditional food knowledge can make personalization more intimate and lessen the degradation of cultural food knowledge, although obstacles to standardization and technological constraints are also problematic: participatory design and algorithms can be culturally informed to alleviate these challenges [63]. Linguistic and cultural customization of nutrition education material, as shown in programs with Pakistani women in Catalonia, increases understanding, interest, and assurance to change to healthy diets without violating the customary diet pattern [64]. Culturally responsive nursing nutrition simulations in health care education also improve awareness of the varied needs of patients to facilitate effective nutritional care [65]. Culturally adapted visualized nutrition education that is delivered in short series and adapted to the language and cultural context of the target audience is more effective [66]. Moreover, it is important to combat algorithmic biases in digital space so that all people can have fair access to culturally relevant nutrition information, particularly underserved populations like African American women with cancers of the reproductive system [67]. In general, user-centered design, development of inclusive and culturally sensitive nutrition visualization, deep socio-cultural adaptations, and co-creation with communities can be considered the primary strategies of the proposed technological solutions [68].

18.6.3 Addressing Bias and Misinterpretation in Visual Analytics

Bias and misinterpretation in nutrition VA are a crucial matter that can enhance the quality and efficiency of dietary determination. Image dietary assessment techniques based on AI have the potential to minimize human error, and the relative error of estimating calories and volume from food images is as low as that of human assessment, but food image databases vary across studies [69]. A combination of segmentation and regression methods will increase the accuracy of nutrient estimates, yet the absence of common nutritional content denominations complicates the validation of the model [1]. Ethical issues promote the possibility of biases in the model of ML, and misleading conclusions may be made unless best practices in AI modeling are strictly adhered to; careful, customized modeling procedures are required to prevent the emergence of these risks [70]. A biased approach to research is not a recent development in nutrition studies, and there is a need to enhance the reliability of studies through the introduction of transparent risk-of-bias evaluation and rigorous research designs [71]. In general, to reduce bias and misinterpretation in nutrition VA, enhancing data quality, harmonizing practices, and implementing ethical principles of AI modeling are needed.

18.7 Future Directions and Research Opportunities

18.7.1 Interdisciplinary Collaboration and Innovation

Future orientations of visualizing nutrition focus on the significance of interdisciplinary cooperation and technological advancement so that the visualization would be more accurate, inclusive, and usable. Recent developments in AI and deep learning, e.g., the incorporation of EfficientNet and Swin Transformer networks, have dramatically enhanced nutrient recognition on the basis of food images with the ability to provide nondestructive dietary analysis within a wide range of food types [72]. The enhancement of nutrition knowledge bases with strong visual food recognition techniques will be used in more intelligent applications that will provide convenient access to detailed food information to the users [9]. To integrate the traditional nutritional wisdom into the digital platform, with culturally adaptive structures, the partnership between nutritionists, anthropologists, and technologists is needed to guarantee the inclusivity and respect of the cultural food heritage [63]. The building of interdisciplinary capabilities in nutrition professionals promotes team science solutions to challenging food security and sustainability issues through the integration of a variety of perspectives and approaches [73]. In general, the further development of nutrition visualization should rely on working with communities, ethical AI construction, and intersectoral cooperation with the aim of developing personalized, equitable, and culturally sensitive nutrition visualization [6]. Table 18.6 highlights priority research areas that will shape the next generation of nutrition visualization tools and their translational impact.

18.7.2 Standard Frameworks and Evaluation Metrics

Typical models and measures of evaluation are critical to further the visualization of nutrition data by promoting consistency, transparency, and comparability of tools and studies.

Table 18.6 Future research directions and opportunities in nutrition visualization.

Research area	Focus	Expected outcome
Standardization frameworks	Harmonized visualization metrics	Improved comparability and validation
Ethical AI design	Transparency and bias mitigation	Increased trust and adoption
Human–computer interaction	Usability and cognitive load	Better interpretation and engagement
Policy-oriented visualization	Translation of evidence for decision-makers	Stronger nutrition and food policies
Sustainability-linked nutrition	Diet–environment data integration	Promotion of sustainable dietary patterns

One systematic review of food and nutrition dashboards determined 48 evaluation metrics in accordance with the principles of evidence, efficiency, emphasis, and ethics, which can help to improve the data integrity, completeness, quality of visualization, and interactivity, thus improving its usability and knowledge sharing among the stakeholders [16]. The evolution of nutrient profiling models of sustainable food evaluation identifies some of the most important points of distinction, including standardization of energy, weighting, and validation of nutritional life cycle assessment, which are vital in standardizing nutritional life cycle assessments and enhancing transparency [74]. Global diet quality scales, such as the Global Diet Quality Score (GDQS), show that there should be standardized, validated measures of both nutrient adequacy and diet-related disease risk in populations of different people [75]. Also, the inclusion of nutritional and sustainability indicators into the food security evaluation requires multifaceted frameworks that integrate ecological, economic, and nutritional aspects to enhance the policy and practice of food security [76]. In general, the visualization of nutrition is expected to evolve in the future by the adoption of standard assessment models, metric harmonization, and the validation of tools to facilitate effective nutrition science monitoring, decision-making, and communication.

18.7.3 Translational Impact on Nutrition Practice and Policy

The translational importance of visualizing nutrition lies in the potential to influence nutrition practice and policy through enhancing communication, decision-making, and behavior change. More efficient visualization tools may be used to convert difficult nutrition science into readily understandable, actionable messages to healthcare providers, policymakers, and the general population and promote improved nutritional health [77]. Nutri-Score and traffic light nutrition labels have been found to affect consumer perception and attention, leading to a healthier food choice, as well as the public policy on the package labeling standards [78]. The conceptual frameworks of agriculture-food systems-nutrition outcomes can help policymakers comprehend the impacts of food policies on the public health of the population so that more specific and effective interventions can be conducted [79]. Highly customized nutrition recommender systems with visual feedback have shown a beneficial impact on dietary habits, and technologies play a crucial role in the translation of nutrition knowledge into practice [80]. Future opportunities are focused on researchers, practitioners, and policymakers working together using evidence-based, user-friendly visualization tools to improve nutrition education, sustainability of the food system, and policy making to improve population health [6].

18.8 Conclusion

The capacity to make complex, multidimensional data straightforward, easy to act upon, and comprehend in meaningful visual forms is becoming the future of nutrition science and practice. As the concept of nutrition visualization has developed since the days of drastic charts and tables, today, it has turned into a dynamic, intelligent, and immersive system that incorporates the principles of AI, big data analytics, wearable technologies, and interactive design. The inventions have greatly increased the field of nutrition visualization, which has positively impacted the scope of dietary evaluation, communication of nutritional data, and user interaction in various settings. As they minimize the use of self-reported information and numeric abstraction, modern visualization tools assist in closing the scientific evidence and realistic daily decision-making gap, allowing people to learn more about their eating habits and control them better.

The combination of AI- and ML-based models is a revolutionary change in the process of gathering, interpreting, and displaying nutritional data. Precision nutrition image-based food recognition, automated nutrient estimation, and individualized visual feedback systems assist in analyzing the unique variation in metabolism, genetics, lifestyle, and environment. Via multiomics and sensor-derived data and visualization platforms, in combination, can give a comprehensive view of the nutritional condition and metabolic health, which opens new opportunities to intervene early and give personal dietary advice. Also, the use of immersive modalities, including the use of augmented and virtual reality, 3D food and body modeling, contributes to experiential learning and increased motivation, especially in nutrition education and behavior change programs. In addition to the applications at the individual level, nutrition visualization has significant implications for clinical practice, the population, policy making, and food industry innovation. Evidence-based policy-making, population-level monitoring, and nutrition intervention evaluation are supported by interactive dashboards and standardized visual measures. Nutritional labeling, through front-of-pack labels and visually accessible indicators to nutrition, has an effect on consumer behavior and promotes healthier reformulation of products, which is the same direction as the population health objectives and industry incentives. Furthermore, visualization systems are essential to enhance sustainability through connecting the dietary norms with the environmental and food system effects, which contribute to responsible and informed consumption. Even with these opportunities, there are a number of issues that need to be looked at in order to be able to guarantee the responsible and effective use of nutrition visualization technologies. Privacy of data, security, transparency of algorithms, and ethical management are much needed as nutrition information is becoming more personal and commercially advantageous. There should also be the necessity of culturally sensitive, inclusive, and user-centered design to prevent misinterpretation, prejudice, and unfair access. Standardization of frameworks, validation, and measures will play a significant role in enhancing reliability, comparability, and trust among the visualization tools and platforms.

To sum it up, nutrition visualization is a field of science, technology, and society. Its future contribution will be tied to the long-term interdisciplinary partnership, ethical novelty, and willingness to transform the complicated nutritional information into visual interpretations accessible to everyone. Nutrition visualization can play an important role in enhancing diet, health results, and sustainability, and develop a more insightful and confident nutritional environment when designed and executed appropriately.

References

1 Zhao, Y., Zhu, P., Jiang, Y., and Xia, K. (2024). Visual nutrition analysis: leveraging segmentation and regression for food nutrient estimation. *Frontiers in Nutrition* 11: 1469878.
2 Pratt, N. and Donovan, S.M. (2015). Improving the usability of nutrition information through graphical visualization. (3710943 PG–144), 144.
3 Magnet, A. (2001). Diachronic analysis of the visuals in the research paper: a corpus-based study of the strategies and semiotics of visual representation in nutrition biochemistry. *LSP and Professional Communication* 1 (1): 55–77.
4 Lee, K.S. (2024). Multi-spectral food classification and caloric estimation using predicted images. *Food* 13 (4): 551.
5 Bayu, M.Z., Arshad, H., and Ali, N.M. (2013). Nutritional information visualization using mobile augmented reality technology. *Procedia Technology* 11: 396–402.
6 Kassem, H., Beevi, A.A., Basheer, S. et al. (2025). Investigation and assessment of AI's role in nutrition—an updated narrative review of the evidence. *Nutrients* 17 (1): 190.
7 Lo, F.P.W., Sun, Y., Qiu, J., and Lo, B. (2020). Image-based food classification and volume estimation for dietary assessment: a review. *IEEE Journal of Biomedical and Health Informatics* 24 (7): 1926–1939.
8 Min, W., Wang, Z., Liu, Y. et al. (2023). Large scale visual food recognition. *IEEE Transactions on Pattern Analysis and Machine Intelligence* 45 (8): 9932–9949.
9 Ming, Z., Xie, Z., Zhang, C. et al. (2025). Robust visual food recognition for enriching nutrition knowledge bases. *IEEE Transactions on Multimedia* 27: 2825–2835.
10 Fuster-Guilló, A., Azorín-López, J., Saval-Calvo, M. et al. (2020). An RGB-D-based framework to acquire, visualize, and measure the human body for dietetic treatments. *Sensors (Switzerland)* 20 (13): 1–20.
11 Jodkiewicz, M., Malinowska, J., and Marek-Woźny, K. (2025). A comparative analysis of graphic models for enhancing nutrition education. *Nutrients* 17 (12): 1947.
12 Phalle, A. and Gokhale, D. (2025). Navigating next-gen nutrition care using artificial intelligence-assisted dietary assessment tools—a scoping review of potential applications. *Frontiers in Nutrition* 12: 1518466.
13 Bergling, K., Wang, L.C., Shivakumar, O. et al. (2025). From bytes to bites: application of large language models to enhance nutritional recommendations. *Clinical Kidney Journal* 18 (4): sfaf082.

14 Agrawal, K., Goktas, P., Kumar, N., and Leung, M.F. (2025). Artificial intelligence in personalized nutrition and food manufacturing: a comprehensive review of methods, applications, and future directions. *Frontiers in Nutrition* 12: 1636980.

15 Alam, M.M., Khan, A.I., Zafar, A. et al. (2025). Advancing nutritional status classification with hybrid artificial intelligence: a novel methodological approach. *Brain and Behavior: A Cognitive Neuroscience Perspective* 15 (5): e70548.

16 Zhou, B., Liang, S., Monahan, K.M. et al. (2022). Food and nutrition systems dashboards: a systematic review. *Advances in Nutrition* 13 (3): 748–757.

17 Chen, Y., He, J., Vinod, G. et al. (2024) MetaFood3D: 3D food dataset with nutrition values.

18 Shi, Z., Deng, P., Zhou, L.A. et al. (2024). Wireless and battery-free wearable biosensing of riboflavin in sweat for precision nutrition. *Biosensors & Bioelectronics* 251: 116136.

19 Sempionatto, J.R., Montiel, V.R.V., Vargas, E. et al. (2021). Wearable and mobile sensors for personalized nutrition. *ACS Sensors* 6 (5): 1745–1760.

20 Mortazavi, B.J. and Gutierrez-Osuna, R. (2023). A review of digital innovations for diet monitoring and precision nutrition. *Journal of Diabetes Science and Technology* 17 (1): 217–223.

21 Ranganathan, C.S., Riazulhameed, A.A.M.A., Swathi, G. et al. (2024). Optimal dietary management via data analytics, ANN-based health insights, and ecosystem for continuous nutrient analysis. *2024 6th International Conference on Energy, Power and Environment (ICEPE)*, pp. 1–6.

22 Shi, Z., Li, X., Shuai, Y. et al. (2022). The development of wearable technologies and their potential for measuring nutrient intake: towards precision nutrition. *Nutrition Bulletin* 47 (4): 388–406.

23 Onita, C.A., Matei, D.V., Onu, I. et al. (2025). Using VR supermarket for nutritional research and education: a scoping review. *Nutrients* 17 (6): 999.

24 Yigitbas, E. and Mazur, J. (2024). Augmented and virtual reality for diet and nutritional education: a systematic literature review. *ACM International Conference Proceeding Series*, pp. 88–97.

25 Paramita, A., Yulia, C., and Nikmawati, E.E. (2021). Augmented reality in nutrition education. *IOP Conference Series: Materials Science and Engineering* (2): 1098, 022108.

26 McGuirt, J.T., Cooke, N.K., Burgermaster, M. et al. (2020). Extended reality technologies in nutrition education and behavior: comprehensive scoping review and future directions. *Nutrients* 12 (9): 1–14.

27 Chauhan, P., Kaushal, M., Vaidya, D. et al. (2025). Personalized nutrition to intelligent foods with 3D and 4D printing. *Applied Food Research* 5 (1): 100863.

28 Shao, W., Min, W., Hou, S. et al. (2023). Vision-based food nutrition estimation via RGB-D fusion network. *Food Chemistry* 424: 136309.

29 Lunterova, A., Spetko, O., and Palamas, G. (2019). Explorative visualization of food data to raise awareness of nutritional value. In: Lecture Notes in Computer Science (LNCS) (ed. C. Stephanidis), 180–191. Cham: Springer International Publishing. Lect. Notes Comput. Sci. (including Subser. Lect. Notes Artif. Intell. Lect. Notes Bioinformatics), 11786 LNCS.

30 Levkowitz, H. and Kelleher, C. (2012). Cloud and mobile web-based graphics and visualization. *2012 25th SIBGRAPI Conference on Graphics, Patterns and Images Tutorials*, pp. 21–35.

31 Li, D., Mei, H., Shen, Y. et al. (2018). ECharts: a declarative framework for rapid construction of web-based visualization. *Visual Informatics* 2 (2): 136–146.

32 Wu, A., Tong, W., Dwyer, T. et al. (2021). MobileVisFixer: tailoring web visualizations for mobile phones leveraging an explainable reinforcement learning framework. *IEEE Transactions on Visualization and Computer Graphics* 27 (2): 464–474.

33 Talukdar, V., Veeraiah, V., Roy, D.R. et al. (2023). An innovative approach: typicality-based collaborative filtering for recommender systems. *Proceedings of the 3rd International Conference on Advancement in Electronics & Communication Engineering (AECE)*, pp. 863–868. https://doi.org/10.1109/AECE59614.2023.10428576.

34 Batch, A., Shin, S., Liu, J. et al. (2023). Evaluating view management for situated visualization in web-based handheld AR. *Computer Graphics Forum* 42 (3): 349–360.

35 Ramos-Lopez, O., Martinez, J.A., and Milagro, F.I. (2022). Holistic integration of omics tools for precision nutrition in health and disease. *Nutrients* 14 (19): 4074.

36 Brennan, L. and de Roos, B. (2023). Role of metabolomics in the delivery of precision nutrition. *Redox Biology* 65: 102808.

37 Keijer, J., Escoté, X., Galmés, S. et al. (2024). Omics biomarkers and an approach for their practical implementation to delineate health status for personalized nutrition strategies. *Critical Reviews in Food Science and Nutrition* 64 (23): 8279–8307.

38 Pandey, J.K., Yadav, A.K., Purushotham, M. et al. (2022). Colorimetric detection of deltamethrin pesticide from biological samples based on the peroxidase-mimic catalytic activity on MnO_2/SnO_2. *Key Engineering Materials* 928: 45–51. https://doi.org/10.4028/p-4ti1i0.

39 Murakami, K., Shinozaki, N., Masayasu, S., and Livingstone, M.B.E. (2021). Web-based personalized nutrition system for delivering dietary feedback based on behavior change techniques: development and pilot study among dietitians. *Nutrients* 13 (10): 3391.

40 Beemkumar, N., Gupta, S., Bhardwaj, S. et al. (2023). Activity recognition and IoT-based analysis using time series and CNN. In: Handbook of Research on Machine Learning-enabled IoT for Smart Applications Across Industries (ed. N. Goel and R.K. Yadav), 350–364. IGI Global https://doi.org/10.4018/978-1-6684-8785-3.ch018.

41 Donovan, S.M., Abrahams, M., Anthony, J.C. et al. (2025). Personalized nutrition: perspectives on challenges, opportunities, and guiding principles for data use and fusion. *Critical Reviews in Food Science and Nutrition* 65 (30): 7151–7169.

42 Verma, M., Hontecillas, R., Tubau-Juni, N. et al. (2018). Challenges in personalized nutrition and health. *Frontiers in Nutrition* 5: 117.

43 Mirmiran, P., Bahadoran, Z., and Gaeini, Z. (2021). Common limitations and challenges of dietary clinical trials for translation into clinical practices. *International Journal of Endocrinology and Metabolism* 19 (3): e108170.

44 Ray, S., Ball, L., and Laur, C. (2016). Nutrition education for public health. *Public Health* 140: 3–6.

45 Frank, S., Gonzalez, K., Lee-Ang, L. et al. (2017). Diet and sleep physiology: public health and clinical implications. *Frontiers in Neurology* 8: 393.

46 Pandey, J.K., Kumar, S., Lamin, M. et al. (2022). A metaheuristic autoencoder deep learning model for an intrusion detector system. *Mathematical Problems in Engineering* 2022: 11, 3859155. https://doi.org/10.1155/2022/3859155.

47 Santos, M. and Assunção, R. (2025). Food choice, nutrition, and public health. *Food* 14 (7): 1243.

48 Sánchez-Bravo, P., Edgar Chambers, V., Noguera-Artiaga, L. et al. (2020). Attitude towards the sustainability of different food categories. *Food* 9 (11): 1608.

49 Pandey, J.K. (2025). Understanding the impact of electromagnetic and optical radiation on physical and biological systems. *Journal of Nano- and Electronic Physics* 17 (5): 05023. https://doi.org/10.21272/jnep.17(5).05023.

50 Srinivas, P., Veeraiah, V., Thejaswini, K.O. et al. (2025). The impact of electromagnetic and optical radiation on physical and biological systems. *Journal Of Nano And Electronic Physics* 17 (4): 04012. https://doi.org/10.21272/jnep.17(4).04012.

51 Halicka, E., Kaczorowska, J., Rejman, K., and Plichta, M. (2025). Investigating the consumer choices of Gen Z: a sustainable food system perspective. *Nutrients* 17 (3): 591.

52 Ammann, J., Arbenz, A., Mack, G. et al. (2023). A review on policy instruments for sustainable food consumption. *Sustainable Production and Consumption* 36: 338–353.

53 Nichifor, B., Zait, L., and Timiras, L. (2025). Drivers, barriers, and innovations in sustainable food consumption: a systematic literature review. *Sustainability* 17 (5): 2233.

54 Canzoneri, F., Charron, M., Drewnowski, A. et al. (2025). Development of Ferrero nutrition criteria: category-specific, progressive guidelines for product innovation. *Frontiers in Nutrition* 12: 1612183.

55 Shingne, S. (2024). Unveiling packaged foods to empower consumer food choices: a comprehensive review of packaged food analysis. *International Journal of Scientific Research in Engineering and Management* 08 (05): 1–5.

56 Cao, Z. and Yan, R. (2021). Product nutrition, innovation, advertising, and a firm's financial gains. *Journal of Business Research* 133: 13–22.

57 Orquin, J.L., Bagger, M.P., Lahm, E.S. et al. (2020). The visual ecology of product packaging and its effects on consumer attention. *Journal of Business Research* 111: 187–195.

58 Pasaribu, A.S. and Basrowi, R.W. (2024). Review of challenges of nutrition intervention research at the workplace. *Journal of Indonesian Specialized Nutrition* 1 (4): 138–152.

59 Maringer, M., van't Veer, P., Klepacz, N. et al. (2018). User-documented food consumption data from publicly available apps: an analysis of opportunities and challenges for nutrition research. *Nutrition Journal* 17 (1): 59.

60 Sharma, A., Czerwinska, K.P., Johansen, D., and Dagenborg, H. (2023). Capturing nutrition data for sports: challenges and ethical issues. In: Lecture Notes in Computer Science (LNCS) (ed. D.-T. Dang-Nguyen, C. Gurrin, M. Larson, et al.), 601–612. Cham: Springer International Publishing Lect. Notes Comput. Sci. (including Subser. Lect. Notes Artif. Intell. Lect. Notes Bioinformatics), 13833 LNCS.

61 Ahlgren, J., Nordgren, A., Perrudin, M. et al. (2013). Consumers on the internet: ethical and legal aspects of commercialization of personalized nutrition. *Genes & Nutrition* 8 (4): 349–355.

62 Pansawira, P. (2025). The role of Artificial Intelligence in nutritional assessment in clinical practice. *World Nutrition Journal* 8 (i2): i–ii.

63 Suarez, C. and Adibi, S. (2025). Integrating traditional nutritional wisdom into digital nutrition platforms: toward culturally adaptive and inclusive health technologies. *Nutrients* 17 (12): 1978.

64 Mohamed-Bibi, S., Vaqué-Crusellas, C., and Alonso-Pedrol, N. (2022). Design of culturally and linguistically tailored nutrition education materials to promote healthy eating habits among Pakistani women participating in the PakCat program in Catalonia. *Nutrients* 14 (24): 5239.

65 Tatum, J.L., Van Hoose, D.E., Esquivel, M.K., and McFall, P.M. (2024). Pilot study of interprofessional learning and engagement in culturally responsive nutrition simulations. *Teaching and Learning in Nursing* 19 (3): 269–274.

66 Li, X., Huang, Y., Yin, R. et al. (2019). Visualized nutrition education and dietary behavioral change: a systematic review and meta-analysis. *Critical Reviews in Food Science and Nutrition* 59 (12): 1976–1985.

67 Abdul-Hameed, Z.L., Ford, G., Burton-Obanla, A., and Rivers, D. (2025). Abstract 4908: exploring the role of algorithms in accessing culturally tailored nutrition information to improve survivorship in women with reproductive cancers. *Cancer Research* 85 (8_Supplement_1): 4908–4908.

68 Livingstone, K.M., Love, P., Mathers, J.C. et al. (2023). Cultural adaptations and tailoring of public health nutrition interventions in Indigenous peoples and ethnic minority groups: opportunities for personalised and precision nutrition. *The Proceedings of the Nutrition Society* 82 (4): 478–486.

69 Cofre, S., Sanchez, C., Quezada-Figueroa, G., and López-Cortés, X.A. (2025). Validity and accuracy of artificial intelligence-based dietary intake assessment methods: a systematic review. *The British Journal of Nutrition* 133 (9): 1241–1253.

70 Pich, O., Bailey, C., Watkins, T.B.K. et al. (2022). The translational challenges of precision oncology. *Cancer Cell* 40 (5): 458–478.

71 Shyam, S. and Salas-Salvadó, J. (2025). The impact of investigator bias in nutrition research. *Frontiers in Nutrition* 12: 1513343.

72 Wang, H., Tian, H., Ju, R. et al. (2024). Nutritional composition analysis in food images: an innovative swin transformer approach. *Frontiers in Nutrition* 11: 1454466.

73 Zheng, L., Garrity, K., Pradhananga, N. et al. (2024). Developing interdisciplinary collaborative skills to thrive in the era of team science. *Journal of Nutrition Education and Behavior* 56 (6): 351–352.

74 Green, A., Nemecek, T., and Mathys, A. (2023). A proposed framework to develop nutrient profiling algorithms for assessments of sustainable food: the metrics and their assumptions matter. *International Journal of Life Cycle Assessment* 28 (10): 1326–1347.

75 Hanley-Cook, G.T., Gie, S.M., Parraguez, J.P. et al. (2024). Cross-context equivalence and agreement of healthy diet metrics for national and global monitoring: a multicountry analysis of cross-sectional quantitative 24-hour dietary intake studies. *The American Journal of Clinical Nutrition* 120 (5): 1093–1104.

76 Siminiuc, R., Ţurcanu, D., Siminiuc, S., and Vîrlan, A. (2025). Integration of nutritional and sustainability metrics in food security assessment: a scoping review. *Sustainability* 17 (7): 2804.

77 Crowley, J.J., Laur, C., Carter, H.D.E. et al. (2018). Perspectives from the third international summit on medical nutrition education and research. *Frontiers in Public Health* 6: 93.

78 Gabor, A.M., Stojnić, B., and Ban Ostić, D. (2020). Effects of different nutrition labels on visual attention and accuracy of nutritional quality perception – results of an experimental eye-tracking study. *Food Quality and Preference* 84: 103948.

79 Kanter, R., Walls, H.L., Tak, M. et al. (2015). A conceptual framework for understanding the impacts of agriculture and food system policies on nutrition and health. *Food Security* 7 (4): 767–777.

80 Hauptmann, H., Leipold, N., Madenach, M. et al. (2022). Effects and challenges of using a nutrition assistance system: results of a long-term mixed-method study. *User Modeling and User-Adapted Interaction* 32 (5): 923–975.

Index

Note: Page numbers in italics denote a figure and those in bold a table relating to the entry on that page. Acronyms are listed at their expanded form unless the abbreviation is the form readers are most likely to seek; see cross-references connect the two.

g

h

i

k

l

m

n

Printed in the USA/Agawam, MA
October 8, 2026

911622.017